THE MORAL SENSE IN THE COMMUNAL SIGNIFICANCE OF LIFE

Investigations in Phenomenological Praxeology:
Psychiatric Therapeutics, Medical Ethics and Social
Praxis Within the Life- and Communal World

Edited by

ANNA-TERESA TYMIENIECKA

The World Phenomenology Institute

Published under the auspices of
The World Institute for Advanced Phenomenological Research and Learning
A-T. Tymieniecka, President

D. REIDEL PUBLISHING COMPANY

A MEMBER OF THE KLUWER   ACADEMIC PUBLISHERS GROUP

DORDRECHT / BOSTON / LANCASTER / TOKYO

Library of Congress Cataloging-in-Publication Data

Main entry under title:

CIP

The Moral sense in the communal significance of life.

. (Analecta Husserliana ; v. 20)
"Published under the auspices of the World Institute for Advanced
Phenomenological Research and Learning."
Contains research work presented at the Third Phenomenology and
Psychiatry Conference held by the International Society for Phenomenology
and the Human Sciences Apr. 25–26, 1984, in Cambridge, Mass.
Includes bibliographies and index.
1. Phenomenology–Congresses. 2. Ethics–Congresses.
3. Social ethics–Congresses. 4. Psychiatry–Congresses. 5. Medical
ethics–Congresses. I. Tymieniecka, Anna-Teresa. II. World Institute
for Advanced Phenomenological Research and Learning. III. International Society for Phenomenology and the Human Sciences. IV. Phenomenology and Psychiatry Conference (3rd : 1984 : Cambridge, Mass.)
V. Series. [DNLM: 1. Ethics, Medical–congresses. 2. Morals–
congresses. 3. Psychiatry–congresses. 4. Sociology–congresses.
WM 62 M828 1984]
B3279.H94A129 vol. 20   142'.7 s   85-23258
[B829.5]   [142'.7]
ISBN 90-277-2085-1

Published by D. Reidel Publishing Company,
P.O. Box 17, 3300 AA Dordrecht, Holland.

Sold and distributed in the U.S.A. and Canada
by Kluwer Academic Publishers
190 Old Derby Street, Hingham, MA 02043, U.S.A.

In all other countries, sold and distributed
by Kluwer Academic Publishers Group,
P.O. Box 322, 3300 AH Dordrecht, Holland.

Printed in The Netherlands

THE MORAL SENSE
IN THE COMMUNAL SIGNIFICANCE OF LIFE

ANALECTA HUSSERLIANA

THE YEARBOOK OF PHENOMENOLOGICAL RESEARCH

VOLUME XX

Editor-in-Chief

ANNA-TERESA TYMIENIECKA

*The World Institute for Advanced Phenomenological Research and Learning
Belmont, Massachusetts*

SEQUEL TO VOLUME XV
*Foundations of Morality, Human Rights,
and the Human Sciences*

TABLE OF CONTENTS

PART III

CIRCUITS OF COMMUNICATION

PART IV

PSYCHIC CIRCUITS OF SENSIBILITY AND MORALLY SIGNIFICANT SPONTANEITIES

PART V

THE LIFE-WORLD AND THE SPECIFICALLY MORAL SIGNIFICANCE OF THE COMMUNAL/SOCIAL WORLD

ANNA-TERESA TYMIENIECKA

The Theme

THE THREAD OF THE MORAL SIGNIFICANCE OF LIFE RUNNING THROUGH THE HUMAN SCIENCES

The present volume contains a collection of research studies marking the second major step in the progress of our work which should be considered as the "phenomenological praxeology of knowledge." In our first major effort, represented by the volume entitled *Foundations of Morality, Human Rights and the Human Sciences (Analecta Husserliana,* volume 15), we surveyed the problems that the human sciences share with phenomenologically — or just philosophically — oriented inquiries. We have encountered a major thread running through the heart of the human sciences, namely, that of moral concern, moral attitude, or moral right which the human individual exercises or is expected to exercise in social interactions as well as in intimately private experiences. This thread leads as a *filum Ariadne* directly to the phenomenology of the human being in a philosophical perspective.

The present collection of studies following this thread enters into the primogenital core shared by philosophy and the sciences of man: the significant differentiation of the individualizing life. With the change of perspective from the nature of human discourse to that of life we gain a unique insight into life's interdependencies and discover that reflection upon life and the enactment of life are intimately interwoven with moral issues.

Concern with ethical codes, rules, and principles are of crucial significance for societal life and have attained particular prominence in the present. The question of "ethics" is the center of attention in the cultural, professional, business, and political life of today. What can the philosophy of today contribute to the enigmatic status of ethical issues within human life? How can philosophy clarify the disarray of points of view, confused issues, and seemingly insoluble paradoxes contained in "medical ethics" and which involve options "for" or "against" life? Can there be an explanation that would remedy the sophisms with reference to which institutions claim to "care" for the rights of citizens while in fact deceiving them as well as themselves?

In the midst of present controversies, we are led to ask whether there is not a basic misunderstanding at the root of the so-called "ethical" attitude. In these ethical conflicts we raise questions concerning the nature and origin of ethical norms, rules, and principles which are explicitly or tacitly assumed to regulate human conduct. Are these regulations established for utilitarian purposes, or do they represent ideals innate to the human mind and meant to raise human life from ruthless egotism? Are they *sui generis*, autonomous with respect to the changing social systems that govern individual existence, or should they be seen, in contrast, as relative to changing circumstances and shifts of power? Whatever the answer, ethical codes appear in these questions as abstract universal prescriptions to which actual human conduct is supposed to measure up but in concrete practice hardly ever does.

This discrepancy between theory and practice — which appears inseparable from the very nature of philosophical perspectives within which ethical theories are formulated — makes us wonder whether ethics *per se* has any but a strictly speculative interest. And yet questions concerning human conduct are of paramount significance. Must philosophy remain absolutely helpless with respect to this discrepancy?

In the present volume we propose to remedy this hitherto persistent sense of helplessness. *In the phenomenological praxeology of knowledge that brings forth the "transactional analysis," the theory and praxis of life will come together.*

We are, indeed, proposing a new approach to their treatment such that the gap between ethical theory, concrete human behavior, and self-experience may be overcome. To accomplish this objective, we must radically alter the direction of our attention. From the ratiocinations about principles, rules, norms, values, experiences, etc., that are pre-scribed, desired, or recommended — assumed to be already given — of "ethical" conduct, we will focus on the origin of *actual human conduct in life affairs.* We will state that every sector of current life and culture is suffused with what we call "moral" language, "moral" appreciations, "moral" judgments, "moral" feelings, etc. We will then naturally raise the question: What is the specific element, quality of feeling, valuation, language, human relations, or self-experience that we call "moral" and that is expressed in infinitely varied nuances? Does this moral element which permeates human affairs, as well as the inward life of human beings, stem from the genius or inspiration of lawgivers who brought it into societal life as directives and constraints necessary to regulate communal life, or did it emerge from the nature of communal ties themselves?

Both of these questions lead us down the wrong tracks. The proper locus for the disentanglement of the links between the human being and his fellowman whence the societal life circuit emerges has to be sought elsewhere, namely, *at the source of the human significance of life.*

In the reseach programs carried on by the *International Society for Phenomenology and Human Sciences* we have, in fact, radically reversed the direction of approaching the significance of "ethical" conflicts in human existence. By going to the heart of things, namely, to the *Human Condition as it reveals itself within the constructive advance of the individualizing life, we find in the Moral Sense the factor of moral valuation, sensibility, inclination, and judgment instilled into foundational human functioning.*

The moral sense, as the virtuality of the Human Condition, allows the individual to establish a link with the Other in a morally significant "transaction". The societal network of the life-world emerges with human transactions, and the individual bringing this network about becomes a fully human person. In human transactions reflection and life-enactment — theory and practice — are just two instruments through which moral benevolence establishes and promotes the essentially societal existence of the human person.

Opening with the focus upon the human person seen in the perspective of the human transaction, our book extends into a vast spectrum of investigations in the fields of psychology, medical ethics, psychiatry, sociology, and philosophy of culture; it pursues the various facets of the moral thread within the personal and societal existence of the human being.

May it open new roads toward the fundamental revision of the status, formulation, and significance of moral and ethical issues that are vital for our time.

A-T. T.

ACKNOWLEDGEMENTS

This volume contains research work presented at the Third Phenomenology and Psychiatry Conference held by our *International Society for Phenomenology and the Human Sciences* on the theme, THE "MORAL SENSE": TOWARD PHENOMENOLOGICAL GUIDELINES FOR PSYCHIATRIC THERAPEUTICS, 25–26 April 1984 in Cambridge, Mass., as well as a selection of studies which were read at our continuing program, the *Boston Forum for the Interdisciplinary Phenomenology of Man*, during the years 1982–83 on the premises of the *Institute* in Cambridge, Mass.

Our precious collaborators deserve our warmest appreciation for their dedication to our common effort.

Parts of Chapter One of Anna-Teresa Tymieniecka's Monographic Study in this volume have been previously published in *Analecta Husserliana*, Vol. XV (1983), and also in the book entitled *Reproductive Technologies and the Human Person* (St. Louis: Pope John 23rd Center, 1984).

A-T.T.

PART I

THE HUMAN PERSON
AND THE HUMAN SCIENCES

ANNA-TERESA TYMIENIECKA

THE MORAL SENSE AND THE HUMAN PERSON WITHIN THE FABRIC OF COMMUNAL LIFE

The Human Condition
at the Intersection of Philosophy, Social Practice, and Psychiatric Therapeutics

A MONOGRAPHIC STUDY

A-T. Tymieniecka (ed.), Analecta Husserliana, Vol. XX, 3—100.
© 1986 *by D. Reidel Publishing Company.*

INTRODUCING THE MORAL SENSE INTO THE CONCEPTION OF
THE HUMAN PERSON: THE CONTEXT OF THE INQUIRY

In the present study, which continues my investigation of the "Moral
Sense in the Foundations of the Social World," I intend to present a
conception of the human person within the unity of everything-there-is-
alive. I hope to justify this new attempt at grasping the beingness of the
human being by bringing together within one naturally intertwined
context all the main lines along which the human person as a living
individual within the system of life, as well as in its unique role in introducing inventively the specifically human significance of life in the form
of socio-communal existence, is originating and unfolding.

From our inquiry it will appear that at the same time as it is inappropriate to fall into the extremes of "anthropocentrisms" prevalent in
our times, by overemphasizing the reach of man's inventive capacities
and his subsequent "powers" over natural forces as well as his ever-
expanding self-consciousness, so it is inaccurate to go to the other
extreme and reduce this uniquely human creative potential by trying to
explain away the significance of its effects with reference to any type of
"conditioning" whether by the laws of life itself or by society.

The complete truth is that *man-the-creator stands out as autonomous
from the chains of any conditioning; nevertheless he is existentially just a
complex link within the unity-of-everything-there-is-alive.*

This amounts to gaining a profound insight into the ways and means
of the life process itself as much as to discovering the particular place
and role of the living human individual within it.

The center of our implicit attention within this context is naturally the
Human Condition as the key to both.

In the first phase of our inquiry we will focus upon the "inner workings" of the functional complex of the living individual, which stands out
from the fabric of the *unity-of-everything-there-is-alive* in the autonomous role of a "person." We will emphasize its main role as the "functional agent" *bringing into the life-system the "specifically human
significance of life."*

In the second phase we will focus upon the arteries of the functional
system of the person spreading into the socio-communal fabric of
human existence. We will show that it is basically through these arteries
that the human individual constructs *his specifically human personal*

*self-interpretative identification. It will come out that the personal ident-
ity of the person proceeds from its essentially communal transaction with
the Other.*

Only the "transactional complex," as I have introduced it in my
conception of the "moral sense," and which will be corroborated further
in the second chapter of this study represents the person in its full-
fledged functional set of inter-linkings and interdependencies. It allows
us thereby to overcome naturally all the otherwise unbridgeable gaps
stemming from the antithesis: nature/spirit. Needless to say that this
antithesis itself will be dissolved in the course of our transactional
analysis.

Our argument will delineate itself through the inquiry into the human
person who finds himself "on the brink" of existence. Such a radical
state between a human existence and animal vegetation is in this perspective identical with the loss of communal life involvement. It will be
envisaged, first, by following its progressive coming about, and then, in
turn, by discovering the postulates of its redressing. The discovery of
principles common to both theory and praxis is unique for the *transactional approach to the nature of the human person.* It reveals, in particular, the essentially *communal circuit of social existence.*

Lastly, "on the brink" situations fall out of the current, regular, or
"normal," life pattern; as such they are considered to belong to the
"abnormal" (that is in this case psychopathological) phenomena and are
objects of psychiatric concern. Leaving the psychiatric perspective of
the investigation to the specialists in the field I propose, however, to
investigate the phenomenon of the genetically transactional complex in
the "on the brink" situation by a strictly philosophical analysis. Nevertheless, considering first the fact that we pursue this transactional
analysis of the person "on the brink" of human existence within the
basically philosophical framework of phenomenological psychiatry
expanded from the inside by my own previously established view, and
second, that we concentrate upon the communal perspective of the
above phenomenon — so far left completely out of sight in psychiatry —
I discover in this investigation natural linkings with psychiatric theory
and practice.

Consequently, the natural conclusion of this inquiry will be a proposal for a new "moral" approach to psychiatry at large in the guise of
some main postulates for a "socio-communal psychiatric theory and
therapeutics."

CHAPTER ONE

THE PERSON AND THE HUMAN SIGNIFICANCE OF LIFE

AN OVERVIEW OF THE ISSUE

At the recent symposium held in France to commemorate the origin of the personalist movement, it was remarked that, among the various concepts with which philosophers have attempted to grasp the specificity of the human being, only one has remained unchallenged and has even gained in importance, namely, the concept of "person." Concepts like "human nature," "ego," "human subject," "consciousness," etc., which have in the past been used to express this specificity of man, have now, in the light of recent insights into the situation of the human being obtained by scientific and scholarly inquiry, proved themselves unsatisfactory.

The scientific inquiry has clearly transformed our view of the living individual and of his circumambient world. From the static conception of the world as a box within which things and beings are "placed," we have moved to view the world as a relatively stable system in process of becoming, whose structure is projected by individual beings. The individual being is no longer seen as growing and developing according to a pre-established pattern of "humanness" from a prefabricated, miniature *homunculus* but as crystallizing from a germinal stage in a self-constructive process.[1] Last, and most significantly, the self-constructive progress of the individual appears to begin with an elementary conjunction of some simple elements within a life-context upon which it constantly draws and in response to which it shapes itself; and its progress is as much an integration into the circumambient life-conditions as a shaping of these conditions themselves. Modern science and philosophy have brought the investigation of nature, man, and the world from abstract speculation down into this concrete flux itself. We have become more attentive to the enormous variety of types of human individuals, to the life-conditions which have shaped them, and to the stages in the history of mankind during which the actually existing types of human beings have evolved. Having become aware of the transforming dynamics of life in which the progress of human beings is caught, we can ask: How far can the transformation of the human individual proceed without man's losing his "humanness"?[2]

The key intuition in the sciences of man consists of an emphasis upon the flexibility, variety, transformability, and vulnerability of the specifically human being who emerges from, and is sustained by, innumerable exterior conditions. This intuition pervades our approach to social and public issues; it underlies our sentiment of "cultural crisis" and "foreboding doom."

The groping for an adequate approach to the human being finds its expression in the search for an adequate conception of the human person. In philosophy, sociology, psychiatry, anthropology, public life or politics, reference to the human person in arguments concerning the crucial issues confronting man, the social world, or culture is identical with the recognition of man's specifically human status.[3] This reference is made to point out those uniquely human features which give man a special right to be considered in a special way by fellowmen and by man's collective institutions. It seems that today we could more than ever approve of Mounier's statement expressed several decades ago, that our civilization affirms the primacy of the human person over both the material necessities of life and the mechanisms of collective life. Last and most importantly, the value of the human person is such that it retains its prerogative, even an obligation, to object to the imposition of manipulating ideologies, tendencies, etc., that could be used to suppress its self-expression and decision-making. This means that to man as a person is attributed a special kind of prerogative precisely on account of the specific humanness which he exhibits. We may reaffirm with Mounier that in this sense contemporary culture — and not only in the Occident — exemplifies a strong personalistic tendency.

However, no single definition of the concept of "person" has yet been accepted. Conceptions of the person abound, and vary so widely that they suggest different, even opposed, attitudes toward man. The human person is, in fact, currently approached in terms of values. Since values are the product of specific cultures, they are themselves conditioned by the tendencies and biases of a particular cultural period. Hence, the concept of person, conceived on the basis of values, fails as a point-of-reference toward mutual understanding; different societies, social groups, and individuals subscribe to different sets of values. At the same time, an overemphasis on the values of the person, or on the status of the human individual, leads to an extreme individualism. Indeed, present-day Occidental culture has produced, in the name of the extraordinary value of the human individual/person, claims which menace the existential equilibrium of humanity itself. This suggests that the approach to the human person is not such a simple matter as it might seem. It is indispensable that we clarify the problem of the person before proceeding to some constructive ideas toward a proper understanding.

To this end, I will raise two sets of questions, which open a twofold perspective on the inquiry into the human person. First, we must look into man's current understanding of his basic existential conditions. Second, we must ask how this view of man's existential interdependencies (within nature as well as within his life-world) accounts for the varieties of notions of the person. The paramount question, however, which I will attempt to answer at the end of my investigation is: How should we conceive of the human person such that we can draw a definite line of demarcation between human beings and the animals, on the one hand, and indicate an adequate way for man to understand himself, on the other? I propose, in fact, that it is "moral sense" as a unique virtuality of the "human condition" within the system of life and the world, which is the decisive factor in man's specificity, the meaning of his life and of his destiny. It is also the sole valid indicator of man's responsibility in private and social matters. Most importantly, however, the moral sense is, in my view, the decisive factor in making man aware of his situation and role within the *existential unity of everything-there-is-alive.*

SECTION I: THE PARADOXICAL SITUATION OF THE HUMAN
INDIVIDUAL AS REVEALED BY CONTEMPORARY SCIENCE

a. Conditioning Versus Autonomy

Contemporary psychology, anthropology, and sociology have made us aware of the extraordinary precariousness of man's life and social situation — a situation precarious to the point that, in investigating the innumerable contingencies of the individual's life-course, the line of pursuit loses track of the crucial factors which direct his differentiation from the rest of life. When viewed within the multitude of existential ramifications impinging on it, human life seems to dissolve into the fabric of both natural life and societal existence.

On the one hand, in the face of the vital forces which constitute the unbroken network of existential conditions, the very reality of man's "autonomy" appears questionable. On the other hand, the precariousness of man's autonomy — whatever preliminary meaning we may attribute to it — would have to be wrung out by him from such an absorbing tissue of essential dependencies and interdependencies within this fabric of life that the traditional ontologies and philosophies of human nature, in the anthropocentric emphasis, offer little more than ideal speculation. Indeed, in view of the innumerable factors shaping the individual life-course, and man's decisions and actions, we cannot assume without examination that any of his social manifestations, his likes and dislikes, aims and ambitions, even his will, are his "own." That is, concurrently with seeking whether he has any say about his course, we must discover what his "own" course would mean.

And yet, the same fabric of life and of the world, which ever seems to be absorbing the human being, is not chaotic. Existence means order. And strangely — but significantly — in the attempts to grasp this order, attempts in which we oscillate between the "infinitely great" and the "infinitely small" as the starting point, the greatest emphasis falls upon the individual as the cornerstone of universal world-order. The human individual is the core of the preoccupations of human science, morality, and the socio-civic laws, regulations, and structures; paradoxically, they are all founded on the assumption of the very thing which they puzzle over in their investigations: man's autonomy with respect to his circumambient conditions.

That is, all human interests and endeavors which are objects of inquiry (practical, philosophical, scientific) — and which ultimately reveal the nature of man — are distributed around the axis of this paradoxical nature of man: his existential selfhood measured over against the power of forces which keep him in existence.

In fact, our renewed phenomenological inquiry into the *nature of man and the human condition* shows that, in spite of the fact that the human individual might be "thrown" into the life-world (as Heidegger points out) without having a choice, he "individualizes" himself nevertheless in various phases which unfold his faculties within the circumambient life-conditions. In contradistinction to all other types of living beings, the human individual appears, in fact, as the only one who takes into his own hands the individualizing course and balances out his tendencies toward selfhood and the conditioning determinants. Moreover, he turns the second to the advantage of the first. Lastly, he deploys his virtualities for existential connectedness into the intersubjective commerce with other human beings from which, in a common effort, he elicits the personal significance of his own existence.

b. The Glorification of the Individual Versus Contempt for His Rights

It is at this juncture that the foundation of morality — morality operative in the intersubjective circuit of man's self-individualization — is to be found. In the present stage of Western culture, in which man's "self-hood" attempts to exercise an inventive control, even over the intersubjective, that is, social conditions, we witness a striking and absurd phenomenon. On the one hand, there is great progress in recognizing human individuality; there is a stress upon and call for the highest possible autonomy and independence of the self (in freedom of choice, direction, respect for individual needs, etc.) in the intersubjective, social life-world. This concern with the individual is expressed by recognition of man's rights and of the need for suitable conditions in which to exercise them. Further, these rights are protected by elaborate, rationally formulated, legal measures: principles, laws, rules, regulations, and precepts support their recognition. These rights form the operative "nervous system" of social life: individual transactions, corporations, societal institutions (educational, practical, political, religious) have pledged to honor them, and pride themselves on their attempts to incarnate their validity. These rationally devised measures designed to protect man's individualized selfhood — which is proclaimed to be an inalienable right — has been increasingly corroborated and developed. Again, paradoxically, as if to challenge the moral validity of this specifically human prerogative, the inventive meaning-bestowing as actually put into practice in modern Occidental culture characterizes itself by a well-founded suspicion on the part of individuals toward those social institutions which, while pretending to implement these laws, are in fact intent on ignoring, abusing, and violating them; that is, there is a widespread temptation to neglect the *moral axis* upon which the very essence of intersubjective sociability is suspended.

c. Overrationalization Resulting in the Loss of Balance in Vitally Significant Estimations

1. Injustice

A few decades ago, Husserl diagnosed the crisis of Occidental culture as involving the estrangement of the Western human being and of his life-world from his existential soil — which he called the "life-world" of natural human deployment. Husserl attributed this estrangement to the excessive, and inadequately interpreted, development of human reason (intellect). His denunciation of reason was, however, restricted to the cognitive experience.[4]

Can it be that Husserl meant to defend an unadumbrated and automatically assumed sovereignty of reason, which is in a specific way responsible for this seemingly insoluble paradox? Is man not much more than a cognizing being? Is the cognitive experience constitutive of the life-world not essentially intertwined with the "moral experience"? The question of paramount importance is then: What is playing the crucial role in the foundation of the life-world in its social, that is, specifically human, phase? What is the relation of the cognitive experience to the moral experience? Certainly, in modern culture — in the objective regulations of the social order, the sphere of the implementation of law and of the distribution of social justice and morality, etc. — it is held to be reason (intellect) which is the sole arbiter of the "objective" understanding, deliberation, decision-making, and implementing of law. Law, which is rooted in morality, is in theory meant to enable the human being to delineate his own life-course in the midst of social "conditioning" upon the unfolding selfhood. In practice, however, the social institutions which implement laws often jeopardize this selfhood and violate man's sense of justice, his rights, and his very conscience by submitting it to the unyielding "conditioning" of social forces.

Is not the intellect, which interprets social laws in their implementation, responsible for this unfortunate development? Does its priority not confound the proper moral postulate of objectivity — to "give everyone his own" — with the false assumption that moral significance is the fruit of a morally neutral "objectivity" of things and of a life-survival system? Indeed, we entrust to reason the differentiation, discrimination, and appreciation of social situations in which moral significance is at stake. The logical conclusion is clear concerning the social appropriateness of this moral significance. Yet, since it is the faculty of the intellect, infinitely dissecting, adumbrating, inductively or speculatively concluding, and projecting new possibilities, which is recognized as the arbiter, then the understanding and appreciation of the social situation and the final decision depend upon the sharpness with which the intellect is exercised. In its infinite possibilities, the abstracting intellect is always capable of turning things to its own advantage. Short of logical contradiction, everything can be plausibly established and justified by reason. That is, there are infinite ways and means for achieving the social "conditioning" of the individual.

2. The Abuse of Life

The superiority of reason over all the other functions of the human being is assumed obviously on the grounds of its capacity to estimate and calculate the elements of life-conditions. Yet, this calculation is performed with reference to values, which are themselves partly the fruit of rational estimation. Values may be situated in relation to each other; regulated, not according to their "natural" role, but by the respective importance attributed to them. Not only does this importance vary from culture and from one historical situation to another but it is itself subject to the rational calculus. With growing faith in the sovereign values of human rational self-consciousness as the specific prerogative of the living human individual, his essential "ingrownness" in the life-system of all living creatures and nature (with the specific life-conditions) has been slowly forgotten. The person, identified with the intellectually suggested conveniences of the rationalized modes of existence, has lost consciousness of his life-community with everything there is alive. To pursue his so-called humanity the individual of today is alienating himself from his innermost grounding in the system of life. Moreover, stressing his unique self-importance, contemporary man has, due to the shortsighted pursuing of his individual interest, broken the equilibrium of his existential balance within the web of the unity of all living beings. The preference given to values uniquely serving his overblown importance endangers his very existence (ecology, arms race, use of natural resources, etc.).

It is apparent that the "rational calculus" of values is blind to the facts, laws, and prospects of life.[5] It is also obvious that reason alone is a misleading guide in matters of such crucial significance. I suggest that all matters of life are ultimately of "moral significance."

d. The "Moral Significance" of Life

To conclude, the crucial issue remains: If it is reason which, overreaching itself, becomes instrumental in the abuses of moral practice, what is it that is being abused? In other words: what is essential to the moral significance of human life? The precariousness of the human sociobiological condition points to the overwhelming role of the "natural" spontaneities (reactions, instincts, feelings, emotions, etc.) and to the limited power of reason in dealing with the opposing forces, competitive situations, conflicts of interest, etc. And yet, not only does the individual delineate his life-course, but in addition he manages to cooperate with others in ways which respect their various interests. The modalities of this "transactional" understanding establish the social world, which brings a new "meaningfulness" of life into the natural orbit. In the light of what we have just observed, it could not be in virtue of the rational faculty alone.

What is significant in fact for this inquiry is the search for the origin of moral meaningfulness in human feelings and actions. The fact that communication within human life and the social world is suffused with a "moral language" (concepts, judgments, values, etc.) is obvious. It is equally obvious that this language is not the "language of objects." Currently in philosophy it is assumed to be the "language of values." In my own contributions I am challenging this priority of values. I submit that this priority is the real culprit in the current moral disarray: the discrepancy between the striving for freedom of the human conscience (against social conditioning) as the guarantee of the highest accomplishment of human beingness (guaranteed by laws), and the cunningly subversive coercion of this very individual into submission to the practical ineffectiveness of the laws and social conditions. This submission deprives man even of the "freedom of conscience." Its effects impinge even upon the self-interpretative meaningfulness of his life, his very selfhood!

Indeed, the moral significance of our intersubjective relations with other beings neither stems from, nor remains protected by, the jurisdiction of reason alone. We have to seek the origin of the moral significance of man's self-interpretation in existence in the autonomous faculty of the *Moral Sense*![6]

SECTION II: THE NOTION OF THE "HUMAN PERSON" AT THE
CROSSROADS OF THE UNDERSTANDING OF MAN WITHIN THE
LIFE-WORLD PROCESS

a. *The Notion of "Person" as the Point of Reference for the Understanding of Man Within his Life-Conditions*

As mentioned above, in contemporary thought the notion of the "human person" plays the role of a point of reference for understanding the human being. The human being is in our times viewed in concrete terms, that is, not as an abstract model of an entity, but as a living individual struggling for survival with the organic life-conditions on the one hand, and the world-conditions on the other. Concreteness and flexibility in the notion of "human person" appear to be most appropriate to account for various features of the human individual, which are approached from different perspectives. Fundamentally, this notion is meant: first, to grasp and indicate the distinctiveness of the human being with respect to other living individuals and things, and the modalities of organic and social life; second, in appreciation of man's conduct, aims, rights with respect to the perspectives of his innermost nature. I would venture to say that, in general, too much stress is placed upon the unique accomplishments of the human being and not enough upon his role among other living beings which this uniqueness compels him to play. That is, in articulating the notion of the "person," we seek to establish a new *meaningfulness* (understanding) *of the specificity of the human being with respect to his organic conditions as well as to conditions which the world within which be delineates his life-course sets upon it.* In our times, in which little is taken for granted, we seek for an ever more adequate understanding of the world and of our place in it. Significantly, we have come to discover that not only the "brute" organic/cosmic/vital facts have no "meaning" unless we ourselves as sentient beings turn them into the conditions of our existence, but also that we might even transform these conditions by our own inventiveness. Therein lies the greatness and the peril of our age.

To establish the significance of the human being within his *vital*[7] conditions, within nature and his lived world, the notion of person is instrumental from various perspectives. First, it appears at the center of the investigations conducted by human science (psychiatry, psychology, the social science, etc.).[8] Second, it serves as a center of gravitation in the public debates on cultural, social, and political matters. Third, it remains a crucial notion in personal and religious practices. Although it is conceived in a great number of ways, and in terms of various approaches, we may distinguish in all of them one of the following three functions attributed to it.

First, the person always appears as a system of organization (or articulation) of the functioning of the living individual within his life-conditions. Second, it is taken as a pattern centralizing the fundamental faculties and virtualities operative in the individual's life-progress. Third, the notion of the "person" expresses through its structure and virtualities a specific phase of the individual's developmental achievement. The epitome of this third model, which includes the other two, is, or culminates in, man's self-conscious functioning. It pinpoints the specifically sociopolitical[9] significance of life.

b. The First Two Basic Models for the Conception of the Person

We may see the *first function* of the notion of the person as basic to psychiatry. Introduced into psychiatry by Freud and Jung[10] the person plays an increasingly central role in diagnosis and therapy. It is intuited as a specific functional pattern by means of which the human being organizes his vital operations at the level of the life-world. Starting with organic processes, the individual unfolds a network of processes relating him to his circumambient world, by means of which, beyond strictly organic growth and subsistence, he projects around him a spatiotemporal dimension. Within this network he himself acquires a meaning as a living being and his circumambient conditions acquire the meaning of a "life-world." This projection by the living individual of interworldly relations with other living beings, things, events, and processes, endows them with a significance that reaches beyond that of the brute organic survival that is attributed to a *specific functional system:* the "person" (Binswanger).[11] Through his interrelations with other living beings, persons, events, and processes, the individual and his life-world are simultaneously sustained in existence, grow, and expand. Mental illness is here viewed as the dissipation of this functional system: the person — the central functional pattern of the interworldly relations — is disturbed; its functional ties disintegrate (Henri Ey).[12] With any degree of disintegration of the person some corresponding dimension of the life-world loses its significance. The mentally ill person becomes "confused" or "disturbed" but does not leave this world; the physical and social world "is there" for the others as it was before. Yet for the mentally ill it is reduced to its bare physicality. The significance of interworldly relationships which previously sustained the person within this world now vanishes.

The specific role of the person in giving meaning to the world within which the human individual pursues his existence is equally obvious in the *second type* of role attributed to the person. Indeed, from a sociocultural perspective, we attibute to the person a set of faculties, which accounts not only for the organic existence of the human individual but also for his sociocultural forms. These are organized in a coherent pattern comprising constant as well as variable features. Intelligence, imagination, will are the faculties which all human individuals are assumed to possess. They are the constants. yet the industry with which individuals use them, capacities to apply them to different circumstances, adaptability to life-conditions, etc., seems to account for the vast variety of cultural and social differentiations which distinguish humanity as such. Moreover, the various "gifts," "talents," "virtual propensities," etc., which belong to this pattern, are distributed unequally and in their respective development account for the uniquely different "personalities" of individual human beings. The meaningfulness of life, which as a result of human creativity, inventiveness, etc., takes different cultural forms — as well as different forms of interpersonal and social relations — is the result of the person so understood.

We see, then, that in the process-like views of the world (and of the step-wise unfolding of life) to the notion of the person is attributed those functions that allow the individual to establish and pursue a coherent, meaningful existence within the flux of changing conditions.

c. The Third Model of the Person as a Subject/Agent Within the Social World

The radical shift from the assumption concerning the stable situation of man in the cosmos maintained in antiquity and the Middle Ages, to that of a fluctuating role which man develops for himself within the social world (an approach that began with modern philosophy and finds its culmination in present-day thought), motivates the third model of the conception of person. In fact, when it comes to the issue of public life, we find that philosophy, social science, political thinking, etc., almost unanimously refer to the person as the *relatively stable system of self-conscious manifestation:* an "agent" from whom the initiatives and their realization within the social world stem; as the "subject," who is the direct or indirect recipient, victim, beneficiary, etc., of these actions. Seen simultaneously as agent and subject, the person is the cornerstone of public life: the bearer of responsibility toward others as well as of individual rights. Whether it be responsibility or rights in the private, legislative, judicial, political, or religious sectors, in all of them it is assumed that these are responsibilities or rights of the human person.

Both as the agent and as the subject, the person is assumed to be a concrete, fully developed, and self-conscious being.[13] "Self-consciousness" means, in the first place, *the capacity to relate the significance of circumambient conditions to one's own vital needs,* Second, it means to endow one's vital course with specific meaningfulness of existence. Third, and foremost, it means *the capacity to rise above the concrete acts of achieving one's vital development toward the principles, evaluation, and planning of those acts, and to invent new means and ways to advance that development.* In this sense we talk about the person as "transcending" man's biological, social, and political conditions: as a self-conscious agent the person may encompass their singular, concrete significance, and accept or reject it; or, invent and propose a new one. It is the conception of the person as the actor within the social world that gives rise to the enigmatic question: To what degree is man sharing his life-course and his life-world, and to what degree is he shaped by them? The stand on this matter inspires different formulations of the notion of the "person."[14]

The three abstract models of the person as distinguished above are operative in the conception of the person as an agent/subject. We cannot fail to see that all three of them fulfill this special task in man's functioning as a living being. Contemporary philosophy unanimously agrees that the specifically human feature is to be able to establish the web of meaningfulness accounting for the self-conscious entity of man, as well as for the meaningfulness of others and of the common life-world. However, in the appreciation of the faculties of man which enter into his meaning-bestowing, priority has so far been given to the intellect. To the work of the intellect alone is attributed not only the orchestration for all other faculties and the establishment through an intentional network of consciousness of the objectivity of life and world existence (Husserl, Max Scheler), but to intellect is also attributed the highest adjudicating role.

Although it is also universally accepted among contemporary philosophers of various persuasions that it is the *ethical significance* of actions and reactions, feelings and reactions, feelings and decisions which marks the unique threshold between the vital meaningfulness of life and the specifically human, cultural significance of life, yet this ethical turn in man's self-interpretation is also attributed, in the final analysis, to his rational faculties.[15] In the introductory remarks I have denounced the abuses of reason; this denunciation makes the understanding of the human being an open question. In light of the foregoing analysis of the notion of the person as used in contemporary thought, the question about the specifically human feature of the living individual boils down to this: What is the origin of the significance which marks a turning point in human development? This passage is indicated by the passage from the *vital meaningfulness* of circumambient conditions to the *sociocultural* one. But what the specific sense-factor that brings it about is, has to be clarified. We have also to ask how does this "sense" originate.[16] In my attempt to answer this question. I will challenge the sovereignty of reason in three respects. First, I will propose that the decisive factor in the specifically human significance of life (in the *vital, social* and *cultural* world) is not the intellect, but the *Moral Sense.* Second, the essential feature of the human individual — of his humanness — does not reside in his highest rational self-consciousness, but in his *consciousness of the universal life-conditions.* Third, in view of man's awareness of them, his individual rights have to be balanced against the interests of *all other living beings.* The human person *indeed* crystallizes the works of the moral sense and thereby becomes the *custodian of life.*

SECTION III: THE PHENOMENOLOGY OF THE HUMAN PERSON
IN ITS ESSENTIAL MANIFESTATION

In the preceding discussion I have emphasized, first, the crucial role attributed in contemporary thought to the notion of the "person" and clarified the reason for this; second, I have emphasized that this role culminates in its "meaning-establishing" function; and third, I have proposed that it is the "moral meaningfulness," which the human person alone unfolds, that leads a living being to become truly human. In brief, to be "human" is to see life in moral terms.

Last, I have claimed that this moral significance stems from a unique factor. That factor — the moral sense — is, in my view, a "virtual factor" of the "Human Condition", which is decisive for human "Nature." It is not a ready-made code of moral conduct to be applied in action. On the contrary, it unfolds together with the vital, psychological, intellectual, and spiritual development of the individual: an unfolding, which culminates in the emergence of the person. It is within the person that the moral sense functions. It imbues the actions of the person with its quality. Through the person it speads into the social world and life. It is my claim that the life of the spirit which lifts the human being above the strictly human confines and Nature, surges and develops as an inner stream of the moral life. Lastly, I submit that it is by means of the moral exercise that the soul weaves the thread for the "radical leap," to use Kierkegaard's expression, toward the encounter with the Divine.

It is now time to give a succinct phenomenological view of the person as it, sustaining the forces of life, invents the social world and, turning its back upon Nature, weaves the thread of the "transnatural destiny of man," aspiring thereby to enter directly into the great game of creation and redemption.

a. The Phenomenology of the Human Person in a Fourfold Perspective

When we want to give a succinct phenomenological account of the human person, we have to distinguish three main perspectives. In the first place, the human being appears in its concrete "manifestation," first as an organized, stable core, marking by its substantial persistence a "place" in space and time, as the *sense-giver* and as the *moral agent*. It is manifested, first, in the "substantial persistence" of its "presence" within the world of life and human interaction: as the "body."[17] Second, the person manifests within the life-world the human being in his "self-identity." This self-identity is partly manifested in the role which the person assumes, namely, in maintaining an identical center from within which man's interaction with other living beings in the external life-world is consistently organized and from within which they spring forth. The person as the identical center reveals itself also through the "forces," "powers," and strivings which lurk behind the interactions and signal the existence of an invisible realm of the person, which reposes in itself. Indeed, although caught in the incessant turmoil of *actio et passio* within the circumambient world, on one hand, and within the irreversible course of an inner transformability of its own capacities, on the other, the person still remains the "same." This self-identity reveals itself indirectly through the persisting pattern of sameness in the external interactions within the world of life. Through these, however, appears an equally "substantial persistence" of the person's "invisible," "inner" life of passions, emotions, feelings, drives, nostalgias, etc. In this perspective, the person appears as the *psyche*/the *soul.*

The third perspective upon man opens when, focusing upon this identical pattern of the person ascertaining itself most powerfully, although in an indirectly "visible" way, through life-participation, we witness it in the role of an "ordering factor." The human being through his cognitive and inventive powers assumes in fact the role of an architect of the life-world and of the social world. Through cognitive means he projects a system of articulations into the otherwise indissociable, opaque maze of forces. He discerns and measures their intergenerative powers and calculates their effects; he plans and projects. He basically projects the meaningfulness of life.

In fact, in this perspective the person is conceived as a cognitive and inventive apparatus: mind or reason. With the faculty of the intellect at its center, a vertiginous living system of rational ordering, applied to man's individual life-course as well as to that of his circumambient milieu, springs forth. This meaningful system reposes in the scheme of consciousness which spreads over the person's entire realm and penetrates all through the rational articulations of intentional inter-connectedness. Thus the person "embodies" the system of the conscious mechanism which generates the *rational meaningfulness of the life-subservient sense.*

But the question arises: Is the person a "sense-giver" of only one — the rational — sense?

From the above-described self-identical center of the human being made visible through the substantial persistence within the dynamics of life through which the body and the soul are present within the world, there opens up the fourth perspective in which the person — or his humanness — asserts himself within the interactions of the life-world. Indeed, the person asserts himself by the self-enactment of his life-course. Not only are all the vital operations, by means of which his physical, organic, and psychic faculties are unveiled, the very expression of the person, but their modalities, directions and aims are the person's "choice." The person acts: the person is an *agent.* Although most of the vital choices are situated within the play of conditioning forces, yet in the midst of this conditioning the *personal agent* not only deciphers the possible choices from the life-situations, but he also introduces *his own distinctive sense into the evaluation of alternatives:* the *moral sense.* It is as the *moral agent* that the person stands out as "human" within the business of life.

Let us now envisage how the body, the soul, the conscious mechanism and the moral agent manifest together the nature of the human person.

b. The Manifest Person

1. The Body-Complex

In approaching the human being from the standpoint of his process-like nature we may appear to go against common sense. Do we not experience the human being, whether as another man or ourselves, as a "being" that is a consistent entity, reposing in itself and "occupying" a position in space as well as centralizing the passage of temporal phases of the past and future in a presence? We experience man, indeed, as the cornerstone of life and as continuously "present" in life's flux — not only participating in it, but, as it were, challenging it by his own life-directions, devices, etc. Hence we experience ourselves and others in what has always been considered a "substantial" persistence. The person representing the human being is then accountable for the ways in which he is experienced and manifests himself in the progress of life. He is credited then with accounting for *stability of self-enaction, and "substantiality" in manifestation.* These two attributes of the person manifest themselves through the body.

In principle, we distinguish in our experience of the body between (1) the body as an object (whether it be someone else's or our own body); (2) our body as experienced by ourselves: the organs in their functioning that we experience as our own, e.g., sight, hearing, etc.;[18] (3) our body as "ourselves," that is, our originary (basic) feeling of ourselves as extending through our organs (e.g., movements which we command by *our* will, and which make us an integral segment of life in acting and "suffering"). Considered as such an experienced complex of functions, our body (or the body of another man that we experience through the impact of his bodily manifestation) as an automatized highly complex functional system "carries" the human person; the body is the "ground" in which the person lives and through which he manifests himself. How is this intimate interweaving of the "dumb" life-mechanism of Nature with the sentient expression of the psyche to be accounted for? In fact, the body as an organism is interwoven with the vital, psychic, "substantial" system of the soul.

2. *Mute Performance Versus Sentient Interiorizing: The "Voice" of the Body*

Not only do scientific observation and experimentation of the way in which our body — the human body as such — is carried on by innumerable operational circuits show how the so-called "inorganic elements" take part in the organic life-carrying mechanisms and operations, but we experience it in direct observation (e.g., medicinal treatment of our vital organs by inorganic substances, etc.). This cooperative interplay of both occurs under the aegis of the individual's life-process.

The organic processes which carry out bodily stability and sustain us as an entity (in contrast to a process which consists merely of a series of transformations), are themselves so automatized — as Bergson already emphasized — that experiencing ourselves as "our body" we remain completely oblivious to them. Only when their automatic circuits break down (e.g., illness or bodily injury) do we become aware of their role. And yet they seem to "carry" this being of ours and to establish and maintain in existence the outward appearance of ourselves, which we call "our body." They carry also the movements of our organs which we experience as ours and under our command; which organs themselves are established and carried on by these mute processes. Each and every one of these operations is "ours." In this sense, our body is a result of each and every operation, which constitutes an integral link in the circuits of a person's life and manifestation.

In fact, the functional circuits organize the vital operations and lead them to unfold organs; these latter play the role of establishing constructive centers. They all enter fully into the enactment of the life-process of the individual.

Contrary to misleading appearances, nothing just happens to us in an "anonymous" way; each functional segment participates fully in our progress and we, as a self-individualizing living beingness, stretch in it and through it. Here it suffices to note how some of the operations we remain totally unaware of, breaking down in efficiency, disrupt the entire functional balance; and we are thrown off our usual unawareness of our organs to feel a pain so acutely "localized" in one single area of our vital operations, that we feel our entire being concentrated in this one segment, hitherto ignored (e.g., the toothache). We may distinguish, however, within a vast spectrum of their differentiation, "organically significant operations" and "vitally significant acts." The first ones are "mute." Their emergence and mechanical performance is so automatized and repetitive that they do not "stand out" to make themselves "see," "hear," etc. They raise a "voice" only when the regularity breaks down and upsets the entire system. Their coming together occurs on the basis of a constructive need that "need" not affirm itself, i.e., make itself known. But in contrast with this type of operations, are the *vitally significant* acts which supplement them.

The life process in its spreading calls for operations which release *vitally significant* reaction/responses to circumambient conditions. The release of the responses is not rigidly repetitive and uniformly established; they surge with respect to the ever-varying elements of the flux of life from which the human being differentiates his own course from the one in which he himself progresses.

The operations surging "in response," in reaction, to the elements of circumambient conditions, emerge from the already established organs "on behalf of which" they "respond" by signals of alarm, signs of satisfaction, calls of need, etc. Thus, the nature of these operations is more complex; we call them in general "acts." Whether we talk about the most elementary "acts" (e.g., recoiling from a life-threatening contact, as in the lowest, pluricellular organisms), or pulsations within our more complex being of joy, forcefulness, like or dislike, etc., we mean operations endowed with expressiveness, standing out, attracting our attention; that is, uttering a "voice." With the development of its expressiveness, the voice of the vital acts intensifies into a coherent unity. The field upon which this unity of expression manifests itself — its ground — is what we experience as the "body." Indeed, what we experience as the body is the unity of a life-sustaining complex which spreads in space and time. As such it is the primary manifestation of the basic identity of man. It also maintains man's self-sameness. In its operation, as well as in its manifestation, the body establishes and sustains the spatio-temporal continuity of the "presence" of the human being within the world.

From the substantial but mute manifestation of the body in space, we have, with the vocal presence of the body, proceeded to its temporal spread. However, in moving from the mute organic operations to the "vocal" physiologico-psychic acts, we have almost imperceptibly penetrated into the middle-ground territory of sensing, feeling, desiring, etc., which the body shares with the psychic, or the empirical realm of the soul.[19]

c. The Body/Soul Manifestation of the Person

The body is indeed neither experienced nor externally manifested as a "neutral" or inanimate "thing." Unles we see it lifeless as a corpse (which does not maintain its form in space and time), the living body is not only "animated" in the sense of reacting, moving, but above all, it is "animated" as expressing a surplus over what it merely appears to be. "Hidden" behind its frame appears an "invisible" concentrated "agency" which feels, desires, strives, decides, etc.[20] This hidden, invisible, and yet "substantial" complex of powers and forces constitutes a forceful "inward" presence. It is manifested by the bodily acts and motions as an equally, although differently "substantial," driving force. In fact, the human person is *experienced* most prominently in its "inward" presence. Our superficial experience of the overt activeness of the individual shows us already that it is organized and oriented from a "center." We become aware of the peson through the experience of some or other strikingly individual act of a living being; this overt, bodily act enables us to glimpse the inward agency from which it stems; the act in its quality manifests the inwardness of the person. The multiplicity of acts sketches the field of this inwardness. With it we are moving upon the common territory of body and psyche.

To psyche, however, belongs also the "intimate" dominion of the soul, which constitutes the nature of this inwardness. We have now to describe the soul itself in its essential nature.

d. The Essential Nature of the Soul

Let us now consider the essential nature of the soul as it manifests itself.[21] Edmund Husserl, the founder of phenomenology, has explored in unparalleled depth the pre-eminent significance of the spirit in human life. He has also emphasized the crucial role of the soul in the mediation between the body-complex and the spirit. There are in his thought three different functional realms that are interwoven but distinctive: the body, the soul and the spirit. At the borderline of the bodily functions emerge those of the soul, while at the borderline of the functions of the soul emerge those of the spirit. This diffusion of all three of them as if along one continuous axis occurs because Husserl (and later phenomenologists, e.g., Scheler) place themselves on one, single plane: that of the ordering function of the intellect. Although I fully recognize the indispensable role of the cognitive/constitutive apparatus (consciousnes with its faculties), still I approach these functional complexes (including the rational apparatus) from a more fundamental point of view than that of rational ordering; namely, from the point of view of their role with respect to *man's unfolding from within the Human Condition.*[22] Only in its perspective may the nature of the soul appear in its fullness.

In agreement with Husserl, I see the soul, first, as the passional ground of forces nourishing the bodily mechanisms; second, as the center of the self-identity of the individual. I agree with Husserl, that the crucial role of the soul lies in being the middle ground between the body in its vital and passional (passions and strivings) resources and the spirit. Yet — and here I part radically with him — although it is this natural, empirical wealth of the soul that makes it the middle ground of the human make-up, its nature, its resources and its role have to be interpreted differently from Husserl: although Husserl is right in seeing in the soul the ground of the spirit, yet in contrast to his view, it is not at the *borderline* of the soul that the spirit originates; rather, it surges from its *center.*

(a) The soul gathers into itself, like into an experiential receptacle, all the life-operations of the living body, the organism; from the soul, as from a center, spring the prompting forces and powers that galvanize the entire living psyche. In this fashion, the soul is the "substantial ground" of powers. Concretized in these powers, the soul reposes in itself; however, the soul is not — as Leibniz, Husserl, Ingarden, and others after them, thought — self-enclosed by its substantial content, like the Leibnizian "monad," which had "no windows" or "doors."

(b) On the contrary, while the body is open to the influx of externally conditioned energies and substances, so the soul is opened to stirrings, nostalgias, strivings, longings, revelations, which do not belong to its natural ground; they stem from the abysmal realms of prelife-conditions. They do not remain encapsulated within the soul or merely pass through it. In fact, they galvanize and stir the most essential resources of the soul; through them they ignite the entire apparatus of man's functioning oriented rationally for the sole sake of survival — for the propagation of the designs of the animal nature — is by their influence prompted to enter the workings of nature itself and to invent new avenues of life. This inventive work leads to the specifically human meaningfulness of life.

Indeed, it is through the crevasses of the otherwise opaque passional ground of the soul that there enters the "initial spontaneity" that has originated life as such, with all its resources.[23]

(c) It is with the Initial Spontaneity that there enter into the code of the natural life of the individual — its *entelechial* code (to be rationally "deciphered") of natural unfolding — the "virtualities of the Human Condition," of which the most significant for making life "human" — for endowing it with a "human significance" — is the *Moral Sense*.[24]

(d) The soul is the battlefield upon which, in the turmoil of life-energies and influxes of the Initial Spontaneity, the Human Condition concretizes itself within an individual, concrete, living human being. Beyond that, the soul provides the ground and the field for an extraordinary, "extranatural" turn within the unfolding of the human condition: the turn toward the birth of the personal spirit.[25]

To summarize: (1) The soul appears as an empirical life-promoting and sustaining factor. (2) The soul, which is orchestrated through the intellectual apparatus of the intentional consciousness — with the self as its axis — appears as the factor of the self-identity of the human being. In this sense, we as human beings identify ourselves with the totality of our experience. (3) The soul appears with respect to the Human Condition as the "middle-ground" into which the decisive virtualities of man flow and within which their individualizing unfolding generates and develops. That is, the soul appears as the ground for the origin of *all the types of meaning* by which man endows neutral and anonymous nature with *his own* meaningfulness, with *his own* sense. (4) Finally, although it seems that the soul extends and remains in an intimate interplay with the body, on the one hand, and with the life of the mind or intellect, on the other — articulating and animating the one, and being informed and processing its dynamisms through the filters of the other — thus encompassing the entire human person, it is far from enclosing it within itself. Although the human person might be self-enclosed like a "monad," the soul — contrary to the views of some phenomenologists — is not. It is, at one extreme, recipient of the ungraspable, inexplicable Initial Spontaneity, and at the other, processor of the existential thread breaking through all its natural frontiers toward the Transnatural. In this crucial role it proceeds by the intermediary of the moral agent.[26]

Now, it remains for us to bring together our presentation of the person in its substantial manifestation. It appears that the person is manifested: first, in the self-sustaining, "animated" bodily complex; second, through the substantial self-identity of the soul; and, third, as an agent presenting the person as we experience it in the "real" self; all three of these complexes are informed by the mind. This analysis of its modes of manifestations shows "what" and "how" the person is. Nevertheless, if we want both to understand what makes the individual specifically human and to account for humanity, we must approach the person from the point of view of the various types of *functioning through which he becomes and unfolds as a living being and accomplishes his human telos.* Advancing in degrees along the line nature-spirit he reaches his full dimension with respect to his interplay with the Other.[27]

The reason for this priority is clear: the human being becomes human through the introduction of *his* type of meaningfulness of life into an otherwise anonymous, "pre-human" nature. He does it as the "Creator of his own Interpretation of Existence."[28] The sense of this interpretation and of its unfolding emerge from and through his functional system. The modes of the manifestation of the person in reality, in life, and in the world are but the result of its meaning-unfolding functions.

Among them, the one which constitutes the instrument through which all lines of his constructive meaning-bestowing upon brute facts proceed, is the moral sense.

It has to be emphasized at this point that the crux of the present conception of the human person lies in its being the moral agent. *Yet it is a moral agent only insofar as its life-enactment is, throughout all the "vocal" circuits of its functioning, informed by the moral sense.* We have now to attempt to trace the origin of the moral sense and its role in constituting the human person.

SECTION IV: THE MORAL SENSE OF LIFE AS CONSTITUTIVE OF
THE HUMAN PERSON

a. *The Person as the Subject/Agent Within the Life-world*

We have so far emphasized the role of the person in the living individual's organizing, articulating and acting; that is, in his functioning through which he unfolds by delineating his individual life-course. When it comes, however, to asserting the point at which this life-course takes a turn of a specifically human sort, it seems most difficult to single out from among the factors entering into human functioning an element that would account for the specificity of this turn which both differentiates man from other living beings, and maintains the line of continuity with other functional circuits. When we ask, What accounts for the specificity of the human being? we cannot consider the human being in an abstract set of features by means of which he "presents" himself; we have to seek this specificity in the network of functioning by means of which his manifestations occur. That is, we have to seek it within the life-world which he establishes as the system of meaningfulness of his existence. We have to seek it in the various types of interrelations, meanings and corresponding "languages" (e.g., the language of art, the moral language, the religious language, etc.) which serve as means of communication within the human world. Furthermore, as is obvious from the first two models of the person presented above, the person draws upon and participates in the entire system of life and nature. Unlike the notions of the "subject," "consciousness," or "ego," which stress the separation from concrete nature, the abstraction of human thought, the person emphasizes the *unity of all living factors within man.*

Although we could say that contemporary philosophy in general agrees that it is the ethical factor or the spirit that accounts for the specificity of the human manifestation (Husserl, Scheler, etc.), the problem is far from being solved in a satisfactory way. It depends on, first, how we conceive of the origin and nature of morality, and second, how much validity we attribute to it. The question of the specifically human factor within the life-world remains an open question. We have prepared the ground for taking it up afresh. First, we will pursue it as the question concerning the *origin and nature of the uniquely human meaningfulness of the human existence and of the world.* Second, we will approach it as the question concerning the *specific meaning-bestowing function of the person as the subject/agent within the social world.* It will appear from our analysis that it is up to the human person to introduce the moral sense into the understanding of the *life-world* as the *social world.* Man's self-consciousness, thereby established, entails *consciousness of the conditions of its progress,* i.e., man's *responsibility for life's survival.*

b. *Man's Self-Interpretative Individualization*[29]

In fact, we may seek for the source of morality by retracing the phases of *man's self-interpretative individualizing life-course.* In my previous work on the self-individualizing (interpretative) progress of the real individual I have distinguished the following phases: (1) the "pre-life" virtualities coming together in the life-individualizing process; (2) the *entelechial-oriented organic/vital phase;* (3) the *vital sentient phase:* (4) the *sentient/psychic phase;* and (5) the *psychic/conscious phase,* initiated by the "source experience," in which *all* of man's "virtualities" unfold.[30] In each of these phases of the dynamic constructive progress of the individual, that is, in the unfolding complexity of the functional mechanisms and systems, the following, crucial issues arise: first, the various types of ways and means of coordinating the elements entering into the operative and generative systems; second, the principles of these coordinations; third, the potentialities of the elements (and of the operational segments) to unfold their functioning and to assume their respective roles in the constructive advance of the self-interpretative process. I have maintained that it is by these various types of articulations of processes, by which the individual differentiates himself from the circumambient conditions — while benefitting from the otherwise neutral elements, but which he may turn into essential resources of his own progress — that he establishes the meaningfulness of this progress, and creates the meaningfulness of the circumambient conditions with respect to their relevance to his needs. It is the element of constructive differentiation from the life-conditions while transforming them into *his conditions* of the "life-world." In the first phase of the pre-life conditions, we may consider this coordination of needs and means as an automatic response of virtually loaded pre-life elements coming together in trial and error or seemingly haphazardly.[31] There is no valuation present there, not even in a germinal form.

We can, however, talk about a principle of "fitness" according to which the coordinates will occur. It begins, as it seems, with the origin of the individualizing process of beingness at its *organic vital* phase. There we are dealing with a solicitation response situation, in which the "need" of the emerging complex of living individualizing elements — under the aegis of the entelechial principle intrinsic to it — seeks and "solicits" other elements for its "satisfaction" toward the further progress of life in its unfolding.

With the phase of the *vital sentient* self-individualizing complex of processes there enters the acquiescence/rejection principle of the constructive discrimination of vitally significant elements — a far more complex significance. Here life's need for further life-prompting elements is not automatically and mechanically satisfied: it is qualified by the sentient discrimination on the side of the individual, who qualifies the elements of his circumambient world by distinguishing those which may satisfy, or are congenial with, his needs, and those which are not. It is, however, only upon reaching the complexity of the *sentient/psychic* phase of man's self-differentiation in the constructing process that we witness a specific significance brought in by the *acquiescence/rejection* principle of articulation. Indeed, beyond the mechanical functionality present in the individual's sentient/vital seeking for, and "recognition of," the elements needed for his organic functions up to the point where satisfaction occurs — observed in the second and third phase of the constructive differentiation of life — we find in the sentient/psychic functionality brought in by now a more complex existential interaction. It involves an evaluative complex of *recognition/estimation/appreciation* on the one hand, and a responsive acceptance or qualified refusal, on the other.

The discrimination/fitness system proceeds in a pluri-directional "sensitivity," and establishes "significance" consisting in "psychic" relations to elements of the virtual fulfillment of the individual's existential needs. This need/satisfaction system crystallizes in the network of existential gregariousness of the higher living beings. Its existential significance lies in communicating by protective reflexes, signals, single and chain-acts of care (belonging to the instinctual/psychic life-protective set) the same existential "life-interests" shared by individuals. It is rooted, however, exclusively in the *self-interest* of each member of the group, with the addition of an existential-affective reliance upon the affective presence of other individuals.

The above-mentioned types of coordination of life-promoting elements, operational segments and functions establish the distributing order of the individualizing progress. At each of the phases they establish the *meaningfulness* of the elements which enter into the individualizing process. Each type functions by establishing sense-giving. Yet, its sense comes from, first, the *vital*, and second, the *gregarious* life-significance of the life-serving process. In its coalescent/fusional/organic way it functions as sense-giver; as viral sense-giver in the vital/psychic selecting mechanisms; as *vital/gregarious* sense-giver in the sentient/psychic appreciative and interest-sharing selectiveness. At each of these phases there emerges an appropriate significant *novum* that has been released from the progressing complexity in functioning, which stimulates the virtualities intrinsic to its components.

The previously enumerated coordination principles carry on the life-progress in all types of selectiveness which they serve, whether by response, acquiescence, or even by an individual initiative. For their being they merely need to put into operation an *"exciting" reason*. But even the touching "devotion" to the care of little ones shown in animal behavior has its *reasons* in instinct and affectivity which *"excite"* the functional system and prompt its operations and direct the "actions" of the animal towards these goals. Exciting reason is applied in its full extent in the use of affectivity and instinct as specific life-prompting functional complexes.

With the emergence of the full-fledged conscious functioning of the individual, the exciting reason, which prompts his selective mechanism toward acquiescence or rejection, does not suffice by itself. Full consciousness means not only the instinctive sharing of self-interest with other individuals, but also the propensity to *expand one's own individual meaningfulness into transactions with other individuals.* The dominant limitation by the *universal scheme of life* — identical for each species — is broken down and recedes before the *inventive function* by which the individual devises his own way of existential self-expansion. This expansion may be accomplished only in transaction with others.

In *transactions* among individuals we deal with multiple and partly conflicting interests; each of them demands his own; each of them is prompted by *individual life-interests*; each of them seeks to promote the new significance of *his* devices for his own self-interpretation in existence; each of them is, by his own spontaneous impetus in this *existential expansiveness* — and even while encroaching upon those of others — going in directions that are naturally prompted to intrepret the transactional components according to his own life-interest "carried" by his *expanding spontaneities*. Thus, he is prompted to interpret his own significance upon the transactional network: the transindividual social world, which is nevertheless common to all of them. Were we left with the coordination principles of the exciting reason, hitherto valid, in which the decisive factor is the drive toward one's own life-interest — even already significantly expanded into that of sharing in the *preserva-*

tion and *propagation-of-life-significance* with other individuals — the expansiveness would have, in the first place, remained limited to the functional circumference of vital sensibility. The individual would share with the other beings in the "law of the jungle," as penetratingly analyzed by Kipling. In his analysis the gregarious order appears partly as a "law" based upon the instinctive/vital/sentient/psychic/operational circuit, in which the sharing of common vital interest, survival and propagation instincts, the affective needs, etc., establishes a vital-interest circuit which harmonizes with the overall system of life. With the advent of full-fledged conscious experience within which emerges the *intellec-tual sense*, marking a new individualizing phase of the individual life-progress, an *objective order* of the life-progress is released. The *inventive function* of consciousness — and cooperation with it — being added to it, a *communication* among individuals is instigated and spontaneously unfolds.[32] The emergence into operation of the inventive function of the human being not only explodes the life-subservient directional scheme for the coordination of functional operations, but it gives them a new focus, an imaginatively *self-enlarging inventory of possible ways* to unfold and stretch *one's own meaningful existential script* over the inter-subjective life network. The release of these factors would certainly prompt attempts at transactional undertakings by individuals in concert. Yet, would the available coordination principles be adequate to such a common effort?

The operative-coordinating principles which give significance to the life-promoting operations — *organic, vital, gregarious* — are geared to the self-interest of each of them alone. They establish in the individual's self-interpretation its *vital sense.* The objectifying reason (intellect) releases a new sense — *the objective Sense.* This latter is altogether neutral to individual survival interests. The rational deliberation which it allows for the sake of estimating purposes, means, circumstances for action and undertakings in common appreciates the individuals appro-val toward an "agreement" or an individual decision to commit oneself to its implementation. In such an agreement, the life-interest of the individual would be, necessarily, as much satisfied as curbed or renounced. The "exciting reasons" which serve individual striving and express its needs recorded by instinct and affectivity toward life-preser-vation would fall short of the mark. In the striving of individual interests could a transactional agreement ever take place? The "law" of the strong or of the cunning would prevail.

c. The Moral Sense in the Intersubjective Interpretation of Life Affairs

Seeking for a new factor which, in the face of the neutrality of the intellectual sense versus the individual aggressivity of a pretransactional situation, appears indispensable for entering upon a neutral deliberative analysis and for inspiring an interpretative turn toward mutual agreement, consensus, and commitment to implement its terms, we discover the *Moral Sense*. In fact, the surging of the *Benevolent Sentiment* of the Moral Sense endows the interpretation of the transactional component variations with a *justifying reason*.[33] Justifying reason, as Lord Shaftesbury so penetratingly saw, demands the "sense of right and wrong." This sense is presupposed by the cognitive function of deliberative operations; it is also independent of other extraneous sources (e.g., religious). It is by the working of the Moral Sense that the benevolent sentiment applies itself to the interpretation of conflictual situations. It surges from, and differentiates qualitatively in, the self-interpretative progress of the individual himself. Its effect manifests itself primordially on the significance of the transaction. The transactional self-interpretation goes together with the "neutral" informative and cognitively objectified set of elements for deliberation. The benevolent sentiment being brought in, the *valuability* of these elements for the significance of the purposive end of the transaction has to be established, not strictly individually but in common; not for the sake of any one of the partners alone, but transgressing their strictly self-centered interests. This *valuability* resides in the threefold relevance of the transactional interests of the involved individuals. It resides, first, in the relevance of the given transaction to each unfolded individual interpretative script and in the prospect (with an implied necessity) of promoting in full or in part the life-significance of each individual. Secondly, it resides in the valuability of the elements of this expanding striving/adjusting/surrendering "negotiating" complex with reference to the given circumambient life-world situation (ecology, social system, etc.) of each partner in the negotiation. Lastly, it resides in the valuability of the elements for selection to the *universal life-system*, which the selection might serve, simply accommodate, or jeopardize in some respect.

However, the switch from the existentially significant coordination category of mere "fitness" in the automatic or "exciting" phases of the self-individualizing complexity, to that of valuative significance in the selective process of coordination is a further indication of the radical transformation within this process. Here we are hitting the threshold of the passage through which — as a discrete and progressively extended phase in the spontaneous self-interpretative progress of the living individual — from the merely life-promoting meaningfulness of self-individualizing life, we cross to the *human significance* of life. The sharpness of this threshold is marked by the question: On what basis does an individual make a deliberate selection of alternatives which are against, in conflict with, or simply a surrender of, his own *life-interests* for the sake of those of others? In other words, what gives "valuability" to the alternatives that oppose self-interest and in terms of what may we justify our selection? If the threshold to the human significance of life is marked by the new relevance of life-promoting deliberations to the significance-axis of "right" and "wrong," how does this axis originate in the Moral Sense? As the basic significant factor in the deliberation and valuation context, the right/wrong axis elevates this significance from the level of the strictly "exciting" mechanisms — serving the self-interest drive of the self-enclosed individual — to that of the intersubjective "justification."

It is the Benevolent Sentiment at work, introducing the ultimately *moral* axis of *right/wrong*, that establishes the intersubjective life-sharing. It allows the balancing out of the conflicting self-interests.

The justifying reason which directs the decision of the transactive significance cannot indeed be founded on the automatized relevancies; it is rather the result of, and a conclusive step in, a deliberative process. Although deliberation involves all of the conscious faculties — which have to be released in the source-experiences — none of them is capable of bringing in this *novum*. Where does it make its original appearance? I suggest that we discover its presence first in the *valuative process*. The principles of selectiveness along the line *valuable/unvaluable*, operating in the valuation process, with respect to the components of transactional deliberation — that is, concerning basically our relation to the Other — are conduits of the Moral Sentiment. The selective decision is not a mere calculus of convenience but conveys the moral sentiment by means of conscious moral acts of *approbation* or *disapprobation*.

Approbation/disapprobation, conscious acts, as the manifestations and carriers of this *significant novum* we are concerned with, are neither based upon, nor consist in, an intuitive instance of the cognition of values. They are judgments which manifest the new up-lifting sense-giving factor: the Moral Sense. It is the vehicle of man's significance, of his self-interpretation in existence: of the *social world*.

d. Valuation and Moral Sense

In fact, the moral sense enters the valuative progress in man's self-interpretation in existence as a specific and irreducible *novum* of the spontaneous *Benevolent Sentiment* that surges from the full-grown complexity of the evolutive significance in the individualizing (and through the individualizing) progress of life. Nevertheless, we cannot assume either that it is due to this progress that the moral sense, as a meaning-giving factor, in its crucially significant sentiment of benevolence toward the Other, operating in intersubjective interactions within the life-world, is released from the functional vital complexity, nor can it be reduced to the latter at any stage of its unfolding. It must have been "lying there in waiting" as the virtual element of the *Human Condition* — a virtual element for the *specifically human meaningfulness of the life-progress* and of the life-world.

With the recognition of the moral sense as the new meaning-bestowing factor we can at last "exfoliate" properly the nature of valuation, understood as a morally significant, selective experience. In contrast to the ethical emotivists who see moral valuation as the expression of feelings, it appears that valuation emerges together with its own form, the *moral context*, in which the "factual" components appear virtually in a moral perspective. Although imbedded in a network of sentient, instinctive, intellectual and other life-promoting factors, these elements, with the concrete emergence in this complex of the valuative process of the Moral Sense, appear fraught with *virtual* moral aspects.

In opposition to the intellectualism in ethical theory which assumes that all evaluation in the field of ethical action goes back ultimately to the intellectual/estimative (deliberation and choice) rationalism represented so forcefully by Leibniz, we see from our analysis that the intellect could never account for the specific effects of the intrusion of benevolence. Would a moral act of approbation/disapprobation be possible without the benevolent sentiment introducing a basis for the differentiation, along the axis right/wrong? Leibniz' fallacy lies in his erroneous stress on the strictly personal/subjective significance of "morality." We have, on the contrary, seen that morality surges into the interpretation's subjective realm as an *interpersonal* affair.

As for the "affective intuitionism" represented by Max Scheler and his followers, the controversy with the moral sense position is more nuanced. Yet the distinction of the "states of mind" and the "intuition of values" does not free the intuitionistic position from the essential fallacy of attributing an undue role to cognition in the morally valid estimation. The undue role attributed to cognition enters via the direct relating of moral experience to values. In contrast, the foregoing analysis has transferred the crux of the nature of moral experience from the role of value to valuation and shown that the moral elements reside in the valuative process. It has thus moved it from the direct reference to values to that of the direct emergence into valuation of the Moral Sense.

By these differentiations within the nature of the basic moral experience identified with the valuative experience, we have also found a middle way between conceiving moral experience as either an intellectual or an affective perception. The intuitive factor which gives to valuative experience its specific moral significance has been identified with the spontaneous sentiment of benevolence brought in by the Moral Sense. It is not *related* to this intuition; it is this intuition itself!

e. *The Meaning-Bestowing Proficiency of the Moral Sense*

We have attributed to the moral sense a specific sense-giving and promoting function which is responsible for the meaningfulness of "moral life," "moral conduct," and "moral language." We must now try to analyze it directly as it manifests itself within the self-interpretative *system* of the individual. These analyses lead us to contest the identification of the moral sense with a psychic faculty or with any faculty for that matter. On the contrary, the moral sense using *all* the conscious faculties is a *unique* type of spontaneous function, which is virtually present in the *Human Condition*, in which the psychic faculties in their mature form also crystallize. It actualizes in its functional proficiency in the *source-experience*. It works out its modes and its way with and through the psychic faculties; yet in itself it is and remains a unique *operative* spontaneity, a "subliminal spontaneity" insofar as it belongs virtually in the Human Condition.

The moral sense is, indeed, not an innate or ready-made *functional factor*, but a virtual *sense-giver*. It is released within the individual's evolutive progress marking the threshold from the gregarious to the specifically human life; it unfolds its meaning-bestowing role when the faculties to promote its proficiency are ready for it. Once unfolded, it does not develop or remain at one and the same stage automatically.

Due to the potential for deviation in the valuative processes and their interpretative application to the individual's conduct, it may weaken to the point of losing its strength and interpretative proficiency.

f. Moral Sense and "Human Nature"

With traditional ethicists, we may raise the question whether the moral sense, as the source of morality, belongs to "human nature." This question leads to some instructive distinctions. It all depends on what we understand by "human nature." In the first place, it certainly does not belong to human nature if identified with the basic phase of man's self-interpretation in existence. As I have suggested elsewhere, in its vital (organic/psychic) phase individualization is controlled by an intrinsic *entelechial* principle. Thus the moral sense does not belong to "human nature" if we understand by it the entire individualizing complex, which the entelechial principle unfolds through its operational ramifications. That is, the moral sense does not manifest itself through and in the "animal nature" of the human being. It belongs, however, to the essential factors through which the human being unfolds his *specific beingness.* It is the key factor in founding the individual's *social significance of life* in a specific quality of his commerce with other human subjects; furthermore, it belongs to these factors insofar as this commerce concerns the essentially human interpretation of life-significant interaction: distribution of goods, opportunities, services for further self-individualization of human beings. Nevertheless, we could envisage the moral sense as belonging to "human nature" if we understand the latter from the perspective of our inquiry, namely, as *the set of virtualities to be unfolded in the conscious development of the living individual into guideposts of moral significance for his self-interpretative progress in intersubjective interaction with others.*

   To conclude this argument: insofar as we conceive of "human nature" as a conundrum of virtualities necessary to initiate and promote the dynamic intersubjective progress of individualized existence, human nature contains the moral sense as its essential and decisive factor.

g. Perception of "Good" and "Evil"

As pointed out above, in its qualitative complex the Benevolent Sentiment contains, on the one hand, a "directive" from its "center," a "center of qualitative orientation" which resides in its *sense*. On the other hand, this *sense* is not an instance of a neutral awakening comparable to the sensuous "senses" (e.g., "instinct," stirrings, feelings, etc., which are being consumed in their spontaneity). On the contrary, this type of sense has a *prompting* impulse and bears in its nature a "prompting for . . ." moment, a germinally meaning-bestowing significance. This significance resides qualitatively in "benevolence" (for the sake of the Other). Although the benevolent sentiment lacks an explicit indication of that in which its "benevolence for the sake of the Other" principally consists, yet it bears it germinally as a proficiency to be crystallized in the exper-iential exercise. In this exercise it also reveals a universal principle. In fact, in the ever-recurring centers of reference within the infinite var-iations in which benevolence is subjectively experienced and applied to actions concerning intersubjective reciprocity (as well as in the transac-tive interpretation of the life-world), we find an ever-repeated residual sense: the "good" of the Other. Hence, with the use of our aesthetic contemplation and intellectual objectification functions, we arrive at the intellectual fixation of the *good* as the ultimate directedness of the moral sense. In extreme opposition, the deviation from the positive "for . . ." of the *benevolent*, to the negative "against" of the *malefic* valuation, leads us in its infinitely extended and varying spectrum of instances in moral praxis to posit cognitively moral *Evil* as the opposed final sense-direction.[34]

With these elucidations of the moral sense we have paved the way to an adequate approach toward the explication of the origin, nature, and role of moral values. However, before we come to this central point of our investigation, we must first consider how the exercise of the moral sense on behalf of the Other "flashes back" upon the interacting subject.

h. The Origin of Moral Conscience

The "fulfillment" of the moral sense within the subject involves more than just attentiveness to the promptings of the moral sense. Its ground-laying moment lies in moral valuation; its central agency is the moral *conscience;* its proper implementation lies in the authenticity of the moral script: the existential script of the moral subject.

In the first place, the moral sense reaches the self-interpretative process in its surge only if it is adequately acknowledged in the valua-tion, that is, if, in the transactional complex of conflicting interests, the

Benevolent Sentiment prevails over the life-tendencies of self-interest. That it may gain this upper hand, however, a deliberation of its role is needed. The deliberation comprising the moral sense as one of its valuative principles unfolds a specific *valuative-preference mode:* to attribute to each of the conflicting elements its *due.* Thus its exercise prompts the unfolding within the interpreting subject of the corresponding *valuative-attributing* modes, the basic one being *righteousness.* The next mode of valuative-attributing concerns the appropriate "recognition" of the respective "merits of the case," which prompted by the moral sense for the good of the Other, in spite of all misgivings which one might otherwise incur, and in spite of all emotional resistance on the side of the appreciating subject — yet taking them all into account — has to express *his own* stand, *his own* perspective: it calls for the *mode of appreciative sincerity.* Moreover, the stirrings of self-interest which hinder "impartiality" in decision-making, bending the tendencies of decision-making functions, stay under the promptings of the moral sense, which moves the pendulum toward a golden middle in the conclusive act toward the *probity of judgment.*

Thus complete moral valuation brings together a full-fledged appreciative and deliberating apparatus: *Moral conscience.* The above enumerated modalities of (valuative) moral appreciation constitute the axis of conscience essential to the proper implementation of the moral sense in our moral self-interpretative system.

Finally, we cannot forget the significance of the *transactional modality* with which the moral judgment prompts its implementation in the interworldly situation. In its prompting spontaneity, the moral sense indicates its own criterion to be applied to the application of the moral judgment in action. This criterion, which we recognize in current life practice under the name of *fairness,* prescribes that the active implementor should follow all the appreciative aspects of the moral judgment with detachment from all circumstantial pressures. This model axis of appreciation/valuation/judgment (decision-making), which constitutes the basis for the deliberating moral conscience, is at the same time the vehicle of our social interaction as the guarantee for our self-fulfillment. Sincerity of recognition, righteousness of appreciative attribution, and probity in meting our judgment pave the way for the self-devised, self-interpretative *authenticity* of human existence in intersubjective relatedness. Regarding the genesis of conscience as a morally deliberating and judging agency, a further point is to be observed.

In the actual exercise of moral sense in transactional deliberations, we do not reassemble our valuative and deliberative system of principles, etc., each time anew, bringing in the moral sense directly and spontaneously into significance-molding operations. It is left to the initially unfolded conscious agency, with its indispensable moral orientation — if not an "ideal" model of moral "probity" — to repeat the modal moral "attitudes" in repetitive kinds of situations. In current usage we understand by "moral conscience" a conscious agency that represents a previously acquired set of attitudes.

i. Some Fallacies Concerning the Origin, Cognition, and Role of Moral Values in Moral Valuation and Experience

A paradoxical question emerges at the start: How can we explain that, in a cultural period like our own, which distinguishes itself from previous periods of Western culture by a highly refined conception of moral values and by sophisticated institutional (legal, etc.) methods to implement them in current interpretations of human transactions, we still witness in social practice the violation, abuse, neglect, ignorance, or outright contempt of them? This issue, which I stated at the outset of my inquiry, will now be explained by way of a contrast between the origin and role of values in moral practice and the moral sense.

First, we must draw a succinct conclusion from the previous analysis to the effect that, apparently, values do not function as *a priori* points of reference for moral valuation, but conversely, they emerge from the valuative process. In fact, values as objectively graspable "entities" or "objects" are — and in this we disagree with most phenomenological ethicists — the fruit of the constitutive function of cognitive consciousness under the aegis of the intellect. Yet, it is not as such that they functionally contribute to promote or to direct the implementation of the moral sense in action. It is true that we refer to values in our cognitive deliberations of transactional situations as points of reference. Yet had values been the decisive moral factor, how could we explain the current phenomenon of moral life, that in spite of our awareness of all the moral aspects which they offer us for evaluative and elective judgment, we may still choose to ignore or to circumvent their practical moral implications and decide not for the right but for the wrong; not for a benevolent stance but for self-interest. Having made this decision how could we have found a satisfactorily justifying "moral" reason for it?

Although we might "know" perfectly well all the moral aspects, implications, and subtle nuances of the values involved, yet we might remain
as if "blind" — Max Scheler would say — to the moral significance of the
transaction, in the interpretation of which we would be applying the said
values.

We may return here to the previously mentioned role of reason. If we
conceive of moral practice as related essentially to the "intuition,"
emotive or intellectual, of values — that is, to their cognition — we
surrender ourselves to the power of either reason or emotions or both
conjoined. There is, in fact, no end to the objective strategies which the
intellect may invent in the rational estimation of the transactional situation. They may be such "devious" strategic interpretations that the
"objective content" of the values involved may be one or the other side
used to serve self-interest. By applying values as points of reference to
interpret our self-interest, we confer upon our judgments, decisions, and
actions a seeming "moral" justification. Yet doing so, we may totally
abuse the "authentic" *moral significance* of the transactional situation.
Indeed, when we refer to the cognition of the previously intellectually
constituted values as points of reference of moral action, we naturally
may presume that by our cognitive attention we may revive within our
evaluative process the moral sense which they are meant to represent.
That is, it is not just a revival of emotions which this moral content of
ideal pure subjectivity may entail. However, such an expectation is
fallacious, a *non sequitur*. It is not in the power of reason to conjure up
the moral sense, which, being intimately "subjective," nevertheless
possesses a strictly universal validity of its own: its own "*sense,*" a sense
most universally valid because valid for man as such. To conclude, the
cognition of values is morally proficient *only insofar* as the "moral actor"
is in a position to *revive* through them the corresponding modal form of
the moral sense. The awakening is not necessarily in the power of values
themselves.

The moral experience, as well as the cognition of values, possesses an
ambivalent significance due to the *aesthetic sense aesthetic taste.* The
subject's enjoyment accompanies not only the experiential crystallization of the moral sentiment on the side of spontaneity, but also the
cognition of values on the side of constitutive structuralization. No
doubt we enjoy our good, virtuous, benevolent feelings, attitudes,
actions. We have recognized in this enjoyment one of the spontaneities
prompting the exercise of the moral sense: its major force. The cognition

of values, because they express the benevolent/malefic sentiments in *objective* forms of the moral sense, proportionately manifests the *aesthetic* load of "enjoyment virtualities."

We find indeed a gratifying and uplifting aesthetic enjoyment of *positive* moral values in contemplating them rationally — as much as we find morally "destructive" the enjoyment of *negative* (malefic) moral values. And yet this contemplation does not entail — or necessitate — the *lived spontaneity* of the benevolent/malevolent sentiments, nor of the moral sense in general. Hence the well-known fallacy of moral practice: we take our uplifting enjoyment in the contemplation of moral values for the moral state of our conscience and our subjective beingness. This contemplation should "naturally" animate the moral sense's spontaneities by stimulating the multiple aesthetic, intellectual, and sentient chords of the psychic functional system, in which these spontaneities find their resources and which carry it as a psychic phenomenon. Yet, this natural direction might be, and too often is, diverted by self-reflective concentration (narcissism) on the benefit of the enjoyment. Thus, to modulate our functioning "morally" remains the exclusive task of the practical exercise of benevolence and of the moral sense in its fullness.

Hence: Let us stop talking about values and revive the moral sense!

CONCLUSION: THE MORAL PERSON AS THE CUSTODIAN OF THE EXISTENTIAL BALANCE WITHIN THE "UNITY OF EVERYTHING-THERE-IS-ALIVE"

In the above inquiry I have not attempted to establish in the first place the "essence" or "nature" of man, as has been customary in traditional philosophy. I do not deny that the "essentialistic" approach possesses considerable merit; undoubtedly the human mind seeks to "understand"; it delights in universal principles and concepts which give intellectual enjoyment and satisfaction. Moreover, such principles and concepts leave the impression that man's understanding encompasses the entire universe of his concerns and that he can easily locate his own place in it. Lastly, universal forms, fixated in the fluid and elusive progress of the life-experience, and principles offer an — illusory — support to fall back on. However, man's life scheme, which clings to certitudes that are unwarranted by concrete events, is invariably shattered by the unfolding course of affairs; and man's expectations are then disappointed. In short, called to the task of bridging the gap between the abstract explanations of the speculative intellect and the concreteness of the ever-changing life-progress, traditional philosophy, as well as classic phenomenology, becomes hopelessly entangled in controversial generalities. It fails to shed light upon concrete facts through which human life and man's destiny advances and delineates its course. A philosophy which offers universal explanations that fail to enlighten man in concrete life issues and to give him direction toward solving them is a mere game of the intellect.

Instead of aiming at an abstract theory of the essential nature of man, I have approached man from the perspective of the *Human Condition.* In this perspective the gap between the more or less concrete phases of unfolding life (brute nature-bios-mind) is overcome by pursuing its becoming in terms of *conditions* from within which the living individual and the human person unfold. The human person emerges as the source of the life-significance of "brute facts," in enacting this significance on the one hand, and on the other hand, inventing it as self-conscious agent. Although we have differentiated the specifically human person by his crucial meaning-bestowing role in introducing the moral significance of life, this specific role does not distance itself from, but merely expands, deepens and renders more flexible the life-significance brought about by the "natural person" proper to all higher types of living individuals. In this respect two crucial points must be emphasized. First, in delineating his meaningful life-course the person manifests himself in his inventive role as an outstanding type of beingness from among the entire chain of living beings. The human person in its role of self-conscious giver — receiving, giving and promoting agent — on the one hand manifests a *universal consciousness of life-conditions*, and on the other hand, its self-consciousness culminates in the capacity to appreciate, calculate, and plan them.

It is not only in virtue of reason that this self-consciousness emerges, but equally — and most importantly — in virtue of the moral sense. While reason and imagination open up possibilities for unfolding and growth, the moral sense opens the perspective of the moral accomplishment and of the spirit. This accounts for the unique autonomy of the human being and his "transcending" the narrow limits of natural life-significance.

Second, we must emphasize that it is in view of the universal self-consciousness of the human person that its ties with the "great chain of living beings" acquire a new significance.[35] From a mute and matter-of-fact participation it is transformed into a conscious and moral *unity-of-everything-there-is-alive*. Indeed, the moral sense brings into the natural unity of life a benevolent sentiment toward all living things. The human being does not rise from the natural anonymity of life for the sake of aiming at an ever greater autonomy and self-awareness. The moral sense reminds him that he is an integral part of living nature. Moreover, in his universal self-consciousness of the life-conditions he is intimately united to *everything alive*.

Lastly, the moral awareness of the universal life-conditions prompts the person toward responsibility for the progress and well-being of all living things. Our approach steers the middle course between the extremes of naturalism and spiritualism, but remains open to both. Our conception of the person emphasizes, indeed, man's integral participation in and his unique role with respect to the unity of life.

The specificity of this uniquely human role consists in three points: (1) the human person is self-conscious of the universal life-conditions and capable of appreciating, inventing and planning the routes of life; (2) in virtue of the moral sense, the human person is capable of making judgments and decisions for conduct and action; (3) the human person manifests benevolence and moral responsibility for the well-being of all living beings.

It is on account of our self-consciousness (developed in *these* respects) that the human person originates the consciousness of its unique "human dignity." It is also in its name that we claim a singular respect for human dignity.[36]

In the perspective of the human condition the human person emerges in its highest significance as being THE CUSTODIAN OF THE EXISTENTIAL BALANCE OF EVERYTHING-THERE-IS-ALIVE!

CHAPTER TWO

THE SOCIO-COMMUNAL IDENTIFICATION OF THE HUMAN PERSON – THE INTRODUCTION OF THE MORAL SENSE INTO PSYCHIATRY

Introducing "Transactional" Analysis as the Key to the Exfoliation of the Human Person in the Full Expansion of His Identity Within the Communal Ties

Drawing conclusions from our previous analysis we may state that the human individual as a person cannot be adequately understood unless in his very core he is envisaged concurrently with respect to his essentially significant attitude toward the Other, to his self-devised and enacted spread through ties within the life-world, and to the virtualities of the Human Condition unfolded and concretized within his self-interpreting functional agency. This means that the human person in his unfolding, growth, development, etc., cannot be adequately understood unless he is approached basically through the disentangling of this entire personal complex through which the basic functional system works and, with respect to each of its circuits, shapes its workings. This amounts to submitting that neither anthropology, sociology in any form, psychology, nor philosophy of culture may on its own or in conjunction with each other understand the human person within his life without essential biases and distortions.

The human person can be approached adequately only within this complex, that is, within the complex of what I have in the previous analysis called the human "transaction."[37] The complex of the transaction between the person and the Other brings together in inventively projected ties and at the most basic level, namely, that of the very origination of the human significance of life, the workings of the virtualities of the Human Condition. These workings are, however, in principle already operating with response to the forces, trends, and interdependencies of the vitally as well as the socially significant life-world within whose progress they enact an integral part.

It comes thus clearly to light from our preceding inquiry that the so-called "autonomy" of the person with respect to the circumambient life-and-social world cannot be conceived of in terms of an ontological, self-reposing structure comparable to the traditional metaphysical notion of "substance." Yet we have emphasized over and over that

45

the person is autonomous in his self-interpretative, self-directing function with respect to the circumambient forces upon which it draws and
which carry its progress. I submit that the person's "autonomy" is of a
unique "transactional" nature. It is through his capacity of spreading
specific meaningful segments of life *from his own inventive powers into
the transactional network with the Other* that the person acquires and
exercises his autonomy with respect to the anonymous forces and strivings of the life-system. It is only within the transactional complex that
the person comes into his own. Lastly, it is in his full-fledged communal
nature and role that the person "transcends" the "conditioning" of
Nature as well as of society.

With these conclusions drawn from our analytic work we enter into
our present argument, which centers upon the exfoliation of the uniquely
"communal" nature of the human person.

This argument will have a twofold scholarly significance. In the investigation of the human person in his "transactional genesis," lies the point at
which the functional apparatus of his progress breaks down and the
already accomplished task falls apart (which appears most revealing);
we would consider such a breaking-down phenomenon as a limit situation of human existence. The human person finds himself at this point,
indeed, "on the brink of existence." The expression "to find oneself on
the brink of existence" means simultaneously a radical social, as well as
psychosomatic, crisis.

The sociological interest of our approach meets here the psychiatric
one.

Yet it is the psychiatric approach which establishes a framework of
inquiry allowing a fuller reach into the heart of the matter. Consequently
we will set up our philosophico-phenomenological analysis with a
direct relation to phenomenologically oriented psychiatry. The "on-the-
brink-of-existence" situation of the human being, which culminates in
the human person's "falling out" of the socio-communal life-world
circuits, is to be seen, indeed, in a sociological perspective (e.g., the situation of the so-called social "misfits" or "homeless" etc.). Yet, at its
critical phase it is brought back directly to the inner psychic malfunctioning of the individual being. Identified as the "disintegration" of
consciousness it is considered by some leading figures of contemporary
phenomenological psychiatry as the pathological deformation of the
inner functioning. Moreover, it is considered to be the reason for the

majority of psychic disturbances and diseases. With this inner condition of malfunctioning we really reach the most radical and revealing existential situation of the human being — his roots in the virtualities and laws of the inventive becoming which I used to call the "Human Condition."

As we know, psychiatry in all its trends refers basically to the philosophical conceptions of the human being, his life-world situation, and the nature of life itself. This interrelationship between the two is particularly clear in the case of phenomenological psychiatry. The philosophico/psychiatric conception of phenomenological analysis of human existence (*Dasein*-analysis) of Ludwig Binswanger establishes a field of intimate interplay between philosophical and psychiatric investigation. It is corroborated further by his followers, and finds new perspective in the recent above-mentioned psychiatric investigation by Henri Ey and his school.

Drawing upon the main points of their work I propose for the present endeavor the following framework or context: (a) the spatio/temporal framework of the life-world, which Binswanger has established as the field of *Daseins*-analysis. It will be expanded from the inside (b) by my conception of the *inventive/creative self-interpretation* in which the human being constructs his existence; as well as (c) by my basic conception of the human person as the *meaning-bestowing functional system with an inventive/creative agent as its center.* Leaving the discussion of relevant trends and ideas in psychiatry itself to the specialists in this field, I will concentrate in the following upon a philosophico-phenomenological description of the "on-the-brink-if-existence" situation of the human being. In the first section I will show how the vital socio-communal significance of the life-world, together with that of human existence, is progressively distorted/dissolved with the advance of psychic disease. This advance itself is seen as the dissolution/distortion of significant communicative links between the person and the Other. It will be argued, first, that these links — which carry the self-interpretative identification of the human person — are of a socio-communal significance. Second, corroborating my conception of the person, it will be shown how the Moral Sense promotes the *communal significance of life as the intimately human life-sphere.* Third, the analysis of the "disruption of communication" between the person and the Other within the communal life-world reveals in its genetic advance that the *identity of the human person may be accomplished only within and with reference to a communal network.*

The transactional postulates toward the psychico-communal re-construction of personal existence, derived conjecturally, will be proposed in the second section. It will be argued that the *re-construction of the communal life-world consists in the communal re-identification of the human individual.*

The relevance to sociology and to psychiatry is obvious. The thesis submitted in the second section, that the aim of psychiatric practice is the re-identification/re-integration of the person "on the brink of existence" within the communal network, will be, in conclusion, complemented by a philosophical proposal of a "socio-communal psychiatric therapy."

SECTION I: THE ANTITHESIS SPIRIT/NATURE AND
ITS RESOLUTION IN THE COMMUNAL SIGNIFICANCE OF LIFE

a. *The Antithesis Spirit/Nature in the Perspective of Existential Analysis*

The phenomenology of Husserl together with the Heideggerian conception of man's being-in-the world gave the foundational insights to the existential "*Daseins*-analysis" which became a landmark in the development of psychiatry.[38] In a conversation between Ludwig Binswanger, who initiated it, and Sigmund Freud, recorded by Binswanger in "Magna Charta of Clinical Psychiatry," Freud called it a "re-introduction of the spirit into psychiatry." Freud said:

> Yes, spirit is everything. Man has always known he possessed spirit; I had to show him that there is such a thing as instinct. But of course I do not believe a word of what you say. I have always lived in the parterre and basement of the building. You claim that with a change of viewpoint one is able to see an upper story which houses such distinguished guests as religion, art, etc. . . . If I had another life-time of work before me, I have no doubt that I could find room for these noble guests in my little subterranean house.[39]

Yes, the spirit. But it is of paramount importance to see *how* it is introduced by Binswanger into psychiatry and whether the understanding of its role in human existence in Binswanger's conception is complete. In the present study I intend, first, to sketch the main points through which the manifestation of the spirit in human existence is laid down by Binswanger's existential analysis. Second, with the intention to complete Binswanger's conception, we have to examine the structuration/dissolution (disintegration) mechanism of its life-world functioning. Third, I propose to link it with my investigation of the origin of sense and to introduce into it the perspective of *human communicative transaction within the communal world*, as described in the first chapter of this study. We will begin by investigating the main arteries of the spirit revealed by existential analysis.

1. The Universal Patterns of Human Existence as Interwoven with That of the Life-World

This "re-introduction of the spirit into psychiatry", so-called by Freud, does not consist, as he interpreted it, in considering its highest manifestation as shown by art, religion, etc. On the contrary, when we view Binswanger's conception of the human being properly interpreted and in addition expanded by our views, we see that its essential significance consists in showing that the framework as well as the main arteries of human functioning *bring spirit into nature and generate it further as the prerequisite of human life or as the expression of the Human Condition.*

In the first place, what I mean is that into the Freudian conception of man as *homo natura* reducible in all his developmental expansion and in their forms to basic instincts of Nature, Binswanger brings in a philosophical conception of human existence — *Dasein* — which makes it explode from the inside.[40] The "natural man" remains as the "basement" of human existence, but this existence itself is brought forth as a nucleus of "humanness" which reposes in itself and is the central part of the edifice; it establishes a uniquely "human" modality of life, such that it distinguishes the human being from all the other animals who remain in the "basement," without constructing anything above it. Human existence — *Dasein* — is then analyzed in its main functions and in their interrelations. Existence-analysis, or existential analysis, reveals interrelations between and amongst the elements of the human functional system. In Binswanger's existential analysis, human existence spreads through the entire functioning of the living being. In this way it: (1) accounts for the entire spread of this functioning without introducing any prejudice as to the predominant role of any one of them over the others (on the contrary, the complexity of human manifestations merges in a harmonious co-operation the instinctual, sentient, conscious, intellectual, ethical, aesthetic, spiritual and religious significances in the unfolding of human existence); (2) the universal pattern presenting the ways in which human existence unfolds may then serve as the basis for differentiating between its singularized forms assumed by individuals to promote life, which are therefore considered as "normal," and their distortions which hinder the regular life progress, distortions which are considered "pathological". (That is, it offers the basis for psychiatric diagnosis and eventual therapeutics).

This philosophical conception of the human being as human existence or *Dasein* centers in the structural conundrum of the network of interrelations between the living existence of the individual and other living beings within the life-world. It conceives of human existence as being in its pattern interwoven with existences of other human beings within the fabric of the life-world. The living human being originates, unfolds, and progresses upon the spatio-temporal axis which is simultaneously the axis of the life-world; this latter is, in turn, in its progress shaped by the interweaving and interlacing processes, developments, and transactions of living beings, of the human being in particular.

Temporal and spatial forms of human experience form the life-world. These forms express in turn normal or distorted patterns of individual existence. The spatio-temporal axis of experiences situates them within the total design of the human individual and personal life. Yet this design delineates itself in its concrete features, while the functioning of the living individual/person interweaves with the fabric of the life-world. Therefore experiential forms are the measure of the individual person's proficiency in enacting his life. Yet the crucial point of this measure is the reference to the *patterns of the life-world as the universal network of all individualized life.* The living body, the organism, the soul, and the person unfold — progressing or regressing — in the interrelatedness with this network.

Since the time of Binswanger great progress has been made in discovering how the functioning of the individual human being/person is interdependent with the basic pattern and forms of the life-world.

2. *"Destructuration-Structuration" Mechanisms of the Person's Life-World-Texture*

The most significant progress in this direction has been made by the French psychiatrist Henri Ey and his school, who focused upon the differences between the state of the living being/human person in the regular "*coordination*" of experiential and functional forms within the life-world situation and its various stages of "destructuration" in pathological situations.[41] Readers should refer to Ey and his collaborators' extensive body of investigations;[42] here it is enough to point out that the studies of Ey's school reveal that personality, the fortress of the spirit, is essentially affected in the process of the pathological disintegration of the consciousness/life-world pattern of functioning.[43] This seems to indicate a prevailing continuity — even if distinct — of the process of life at all its levels. This striking insight is further corroborated by recently developed pharmacological findings. It appears that organic and inorganic elements may have a direct effect upon the functioning of the human being such that it results in stimulating or stifling the processes which conduct or generate the work of the spirit in its aesthetic, ethical, spiritual, or religious modes. The use of pharmacology in the pathological disintegration of consciousness/life-world patterns corroborates this insight. Yet its effects do not solve the issue of the aforestated antithesis: instinct—spirit; we cannot conclude hastily that all levels of significance may be reduced to the vital one. On the contrary, in view of these landmarks of inquiry in philosophico-psychopathological anthropology we see opening an intermediary field into which the investigation of the nature of this seeming antithesis and its significance is moving.

3. The Soul-Body Territory

In our preceding analyses we have introduced a factor crucially significant for the understanding and interpretation of the continuity of functioning in the constructive individualization of life. To avoid distorting the entirety of this functioning, we may, as I have proposed above, only approach it in its *basic life-significance, that is, in its constructive advance as a meaning-bestowing ordering system.* In this perspective, as we have seen above, this life-promoting continuity breaks down into a series of significant circuits. Its essential core I have called above, in phenomenologico-philosophical terms, the "soul-body territory;" a "territory" wherein lies the linkage between the main functioning along which the establishment of the spatio/temporal bodiliness of the human individual on the one hand, and of his experiential-creative consciousness, on the other hand, takes place. The first represents the "natural life-world expanse of the individualized life"; the second manifests the specifically human *creative* expanse of the new, man-invented avenues of the life-world, the latter being a special province of human culture and of human historical "meaning of life." The soul-body territory in which the ties of this linkage are projected is thus one of crucial significance. It comprises the network of life-functions within which a transformation from the vitally significant life-world — merely gregarious and remaining in submission to Natural designs of life — is transformed into the *specifically human intersubjective communal significance of life.*

What we intend to show within this context of inquiry is that the *life of the spirit is initiated into life and carried on throughout its entire expanse by means of the entire functioning pattern of the living individual, which itself is essentially correlated — and concretely interwoven — with the pattern of the individualized life-world.* Hence, it is my contention that clinical psychiatry, which in principle deals with the vitally significant circuits of functioning, is closely intertwined with psychotherapy, which emphasizes the life-world interlinkage of psychic significance. Both of them should, however, look for their point of reference in what I will provisionally call a "socio-communal psychiatric/or anthropological dimension." With this in mind we will now sketch a *philosophical* analysis of the human functioning/world pattern envisaged in its *full extent as the pattern of the arteries through which the life-constructing meaningfulness of existence is distributed,* together with the communicative links among living beings.

b. The Fourfold Meaningful Life-World Pattern as the Constructive Pattern of Human Existential Functioning

Moral Sense, Intersubjectivity, Human Creativity

Hönigswald, the founder of medical anthropology, considered psychopathological symptoms as being "primarily disturbances of communication." Although this statement appears to affirm an obvious state of affairs, there yet remains the question of how "communication" should be understood. At first, and it is the most striking manifestation of these disturbances, we take it at the level of discourse or dialogue; "disturbances" refer to the "regular" system of meaningfulness of the life-world at the most rationally developed level of which the human language is capable. However, upon more careful consideration, I propose that the disturbances so manifested are to be brought back to the distortions or dissolutions of the functional system of the living individual as manifestations (or consequences) *following distortions or dissolutions of the basic, vitally significant interconnections in which the isomorphic struc-*

tures of the life-world are meaningfully articulated. With the dissolution or distortions of these interconnections, the life-world of the individual shrinks in its functional expansion and becomes topsy-turvy with respect to the order of the world he is supposed to be functionally expanding, so that he may progress and unfold his existence. Within this established context we propose now to switch from the antithesis — instinct/spirit or nature/culture — formulated at the level where the life-world of the individual and the individual himself are approached in their already objectified accomplishment, to the *purely functional perspective* which shows us how the meaningfulness of this world culminates in its objective form.

Going back to my analysis in the first chapter of this study, there are three main meaningful circuits of the functional life-world pattern to be considered. First comes the meaningfulness of the "world of vital exist-ence" being established by the organic life-functions. Although its work is concerned mainly with the "natural" forces, yet it projects arteries of significance through which instincts, drives, pulsations, sensibilities, etc. will be brought into intimate coordination with the meaningfulness of functional circuits in which the circumambient world of the living indi-vidual is organized into group-interests: gregariousness; thus the mean-ingful circuit of the "life-world of gregarious existence" is established. These arteries of functional meaningfulness expand then into a further and, for our purpose, crucial circuit of meaningfulness, namely, that in which the *intersubjectively shared and communicable life-world circuit, that of the "communal" life-world, emerges.* I call the "socio-communal existence." Let us emphasize that this meaningful circuit exhibits the world of the new and, in the development of life, unprecedented phase in which the surging of the Human Condition with its creative function introduces the three novel factors of sense: *aesthetic, moral,* and *intel-lectual.*

Although it is only from the constructive co-ordination of these three factors within the source-experience, marking the nature of the specif-ically human significance of life, that the full-fledged functional pattern of the human life-world emerges and is expanded through the construc-tive course of human life, yet it is in particular the moral sense which accounts for the most significant linkage in this pattern, the linkage between the various and infinitely varied functional systems of living individuals at the level of their direct communication. This linkage resides in a specific communicative system which we call "intersubjectivity." As proposed above, the moral sense is the means by which this system may spontaneously come together and unfold to its full measure.

It is the meaningful circuit of the socio-communal existence — with the intersubjective linkage system at its center — that lays down the main functional arteries of the life of the spirit within the individual life. The creative function of man which orchestrates all the life-significant functions is the source par excellence of the life of the spirit. It also forms the intermediary ground upon which all the lines along which the forces of nature — organic, psychic, gregarious — enter into the *constructive and intergenerative co-ordination with the new forces which surge with the creative/inventive/imaginative function.* This means that the socio-vital world circuit is the source and foundation of the human individual/person's meaningful functioning within the intersubjective communal life or life-world. Its full-fledged exfoliation, expansion, and objective/subjective manifestation take place within the fourth meaningful circuit, that of the cultural world of man.

Thus the world pattern of the living human being projectively enacted in his meaningful life-functioning gives the measure of his functional life-proficiency. The socio-cultural world pattern of the given epoch, period, and social group in its meaningfulness presents the concretizing schema of integration for the individual functioning.

SECTION II: THE CONSTRUCTION VERSUS THE DISSOLUTION OF THE INDIVIDUAL'S LIFE-WORLD FABRIC

We may consider that Martin Heidegger, in his conception of the human individual life-progress, explicated the ways in which this integration either is distorted and thus handicaps life's natural unfolding, or is in the state of progressive *dissolution* (disintegration). In our present concern — to reveal the roots of human communication — it is most instructive to follow the progressive phases of this dissolution of the world-structure within the human life expansion which indicates to us the circuits of meaningfulness of which it is woven. Leaning upon the above guidelines, I will attempt in what follows a philosophico-phenomenological investigation of the course of dissolution as it appears in various types of mental disturbances and diseases and as it may be intuitively grasped.

a. The Dissolution of the Culturally Significant Meaningfulness of the Life-World

In the psychopathological dissolution/distortion of the consciousness/ life-world structure, the interlinkings of the meaningful elements which make the circuit of the cultural life pattern become twisted in their linkage or dissociate. This phenomenon is precisely what manifests itself in the distortion of communication in the linguistic sphere. (a) Meanings used by the psychotic patient have partly the same and partly different denotations and connotations with the cultural phenomena of everyday life; they differ from the usual, current ones. Hence, hindrance in inter-subjective communication results. The divergence in significances goes so far as to overturn the law of contradiction: the "yes" may mean "no," the word "black" may mean "white," "large," may mean "small," etc. (b) The dissolved conventional relatedness of words (meanings) with their objects reveals itself, however, to be *the result of a deeper phase of distortion or dissolution, namely, that of culturally significant reactions, feelings, appreciations,* and, lastly, of *culturally significant emotions,* which ultimately may become extinct. This is partly the effect and partly the cause, or is at least simultaneous with (c) the weakening or extinction of the culturally significant interests (e.g., in reading, study, music, etc.), and, ultimately, of any type of interest related to the socio-cultural world.

This phenomenon is furthermore certainly to be linked with the diminished span of attention and distortion of the memory function. In fact, the extinction of the culturally significant interests is the result of the weakening — if not of a complete disorder — of the *memory functions* and is simultaneously observed. As a consequence, the reservoir of the past experiences retained is either cut off from the present field of consciousness or, at least, does not reach it through its main pipelines but only intermittently or as a random backflash. Yet it is not only the extremely diminished span of attention that does not allow the retrieval of the elements of the past in the proper order, but we first observe the weakening or dissolution of the SYNTHESIZING function of memory, the rememorization (retaining) as well as the retrieving.

In consequence of these three circuits of the dissolution/distortion phenomenon, the entire cultural life-pattern of which the human person is an experiencing center as it stretches throughout his personality, is either twisted and becomes topsy-turvy, or, in a far advanced stage of

the disease, altogether dissolved. Before such a stage of complete dissolution is reached the personality of the individual is already being essentially affected.

b. The Socio-Vital Interconnectedness

(i) The previously enumerated phases of distortion/dissolution have
already considerably impoverished the personality by stripping it of its
essential culturally determined features (e.g., a "poetizing" aesthetic
attitude; a "reflective" or a "thoughtful" approach with respect to life,
events, other people, one's own destiny, etc.). Beyond it, however, the
distortion of communicability with the Other affects the personal life-
world integration at an ever deeper level. When the renewed efforts of
communication to reach the Other seem with the distortion of the
communicative levels to fail, "autism" follows, together with seclusion
from, and ensuing loss of capacity to "trans-act" with, others. Thus the
basic socio-vital interconnectedness is dissolved.

(ii) This falling out of interworldly linkage causes feelings of acute
frustration; it pushes the human being to the verge of his limits and
causes him to experience "beyond" them a "void" or an "abyss" (a "hole"
opens up in front of him). This radical state of frustration transforms the
natural as well as the acquired dispositions of the person. Their change
may reach as far as the originary natural disposition to confidence which
underlies the seeking of companionship; this disposition — of basic socio-
psychic significance — turns into defiance, suspicion, hostility, fear,
jealousy and envy.

(iii) With the hopelessness which creeps in with the disruption of
communicative links indispensable to achieving a significant contact
with the social world, when the ever-renewed attempts irremediably fail,
then the entire conundrum of functioning, which subtends the expansion
in the "all-encompassing," in the "transnatural," that is, of religious
experience, vanishes; it is in fact subtended by the socio-communal
involvement which had vanished from under one's feet. The individual
feels "lost."

(iv) These changes in basic attitudes, which cause not only the disintegration of the integrative socio-vital co-existence pattern but affect
also the basic dispositions of character, begin to hinder, weaken, and
inhibit the very basic *emotive striving* toward interaction and communication instead of promoting it.

Our discussion reveals how the universal pattern of living human beings functions as a crucial system of reference in the consciousness/life-world system of life's progress; conformity of the individual's personal/life-world progress with universally human life-world schema appears in the foregoing analysis as decisive for its advance or for its retreat. The meaningfulness of this pattern stems from the life-promoting moral sense, the exercise of which fosters the communal life of human beings, thus preparing the arteries for the unfolding of specifically human life/existence. It is through these arteries that living beings are enmeshed in communal existence and are dependent upon it for the normal constructive progress of human life.

c. The Descent Into the Elementary Vital "Sub-basement" of Human Life

When the socio-communal transaction, which marks the functioning of what I have called the "soul-body territory," disintegrates, the consequence is, as pointed out above, a progressive impoverishment of personal existence. When one has lost the significance of cultural as well as socio-communal ties, that is, when one has fallen out of the cultural as well as of the socio-communal self-interpretative schema with reference to which one seeks to establish one's own position or place within the ties of community, the human being's synthesized "understanding" or grasp of himself/herself is dissolved; by the same stroke one loses one's bearings within the human community.

Two crucially significant consequences follow.

(1) First, having lost the connective links and avenues necessary to re-establish these ties with his fellowmen and with the transactional system of life, the individual is reduced in his manifestation to elementary vital drives. What remains are, basically, appetite, sleep, and sex. There, however, dispossessed of his communally significant ties, the individual finds himself in a pit. The possibility of satisfying the needs of sleep, appetite, etc., is in the main provided by societal networks; however, the sex drive, which is naturally oriented toward another human being, remains essentially dependent upon the communal life-ties within the significance of which the Other may be met. But the communicative network being broken, the significant encounter fails. Its natural communal outlets are either completely out of sight or remain distorted, deviated in their communal significance (reliability; stability of relationship; e.g., a family frame, etc.). The frustrations that result are of

the most acute intensity. The sexual urge, deprived of its natural arteries for satisfaction in the communal network and finding no proper means to be linked to an adequate channel, is either prone to a violent outburst of one's natural spontaneities, or, deviated into the flux of spontaneities of nostalgias, imaginatively colored yearnings, etc., permeating the entire psychic horizon. Too much has been said and written on the topic of sexual drives to make it worthwhile to continue this theme here. In this situation it seems that the life-world existence is reduced to a disconnected bundle of vital strivings manifesting that the dissolution of personal life pattern is complete.

(2) With the dissolution of the arteries of the individual life-world through which the individual projects his tendencies, wishes, ambitions, desires, etc., none of his naturally spontaneous strivings "travel" in such a way that the tentacles he throws out through his wishes and strivings to "expand" toward other people reach the target or are answered. They are either aborted before entering a regular course or they are so distorted that the other does not receive the "message"; he does not "understand" what these seemingly "absurd" — or "out of the way" — gestures or expressions mean. It is pathetic to witness how the patient "struggles for life," trying again and again to invent a way to reach the proper attention and response of others, and invariably fails. Experiencing his failures as a "rejection," he is, on the one hand, shrinking in his own functioning; on the other hand, having lost faith in his inventiveness, prospects, and capacities he tries less and less frequently. As a result he feels progressively "cut off from," isolated, and ultimately "insulary." His circumambient world is experienced as totally shrunk, isolated, and disconnected from the rest of the human world: an "Alcatraz Island"; the prison of a total isolation. We would consider him to be ultimately bound to succumb altogether did we not have reason to question whether, in spite of this profound shrinkage with its annihilating processes, there did not still remain within his frame some latent and retrievable pattern of humanity. By this I mean not merely the universal virtualities of the Human Condition; I have more specifically in mind *the core of the pattern of the once-personally-established person/life-world* that the particular human being has been unfolding throughout his/her course of existence. The question is how far, how, and what of this core may be retrieved.

d. Here Is Where the Issue of Therapeutics Naturally Comes In

Although our analysis is, as stated at the outset, restricted to a philosophico-phenomenological description of the destructuration phenomenon, we have grasped this phenomenon at its heart. We have been
describing its phases as they may be intuitively grasped, not focusing on
human consciousness which, through the workings of its constitutive
powers, structures the circumambient conditions into a life-world;
neither did we single out the empirical life-forces as a counterpart to
consciousness; lastly, we did not focus upon human action as the means
by which the two would be coordinated. On the contrary, we have
focused on the *crucial functional system in which the human being and
the Other open themselves together to a specific significance of the life-
world within which they bring about a unique complex of interworldly
operations: the transaction. Within this complex I and Thou emerge in
an intersubjective communication of the moral sense that bestows a
unique significance upon the circumambient life-interdependencies and
constructs a communal mode of life.*
  Within the complex of transaction the complete constructive network
situating the human individual within the life-world and his communal
existential ties is contained. Thus, with the analysis of the destructurizing
progress of the communal life-world circuit, we are *simultaneously
laying bare the principles and the constructive schema of the communal
life interpretation which the individual projects.* Description of the one
conjecturally indicates the other; the rules grasped in a theory yield in
reverse postulational indications toward practice. We are thereby from
our philosophical analysis of a pathological dissolution brought naturally toward a philosophical analysis of a *postulational situation for
therapeutics.*

e. The Motoric Resources and the "Psycho-Neurotic Potential" of Constructive Development

While we discuss the circuits of the communal life-world disintegration/
dissolution in terms of transactional forms or functional complexes
which the constructive development of the individual has unfolded, we
seem to assume implicitly the "motoric" aspect of this constructive
advance. In fact, the progressive stages of the transactional formation
of the person/communal life-world pattern indicate that within human

functioning there must be an intrinsic prompting toward this constructive advance; I mean by that the *conjectural presence of a special type of
functional resource*, namely, that which prompts the constructive developmental process toward proceeding in a special way, namely, by
successive phases of construction and deconstruction. It would be
erroneous to consider the process of deconstruction as such to be essentially negative; on the contrary, while the human being advances by
unfolding his being through the projection of ever more complex life
patterns, he dissolves the one already outlived to project another, which
will be more appropriate. It becomes pathological when the phase of
dissolution does not prepare an automatically new reconstructive pattern.

A considerable emphasis has been put upon the role of this construc-
tion/disintegration mechanism operative in human development in the
work of the psychiatrist Kazimierz Dąbrowski. His theory is relevant to
our philosophical inquiry and merits attention here.[44]

In the first place, Kazimierz Dąbrowski proposes that such a mechanism of deconstruction/reconstruction of the personal psychic pattern
is the vehicle of human growth. Second, he attributes the motoric
prompting of this mechanism to the "psychoneurotic potential of development." Third, Dąbrowski points out that it is *owing to the mechanism
of the psycho-neurotic potential toward development that the creative
function to which we attribute a crucial role may itself develop toward its
life-constructive role.* Furthermore, he attempts to show that it depends
upon the strength and intensity of the psychoneurotic potential toward
development as to how the human individual will unfold with respect to
the circumambient interdependencies (physiological as well as social); in
other terms, as to whether he will show enough inner strength to deal
with them, overcoming the hindrances in a "positive" way by reconstructing his life-world pattern at a "higher" level than the preceding one
— which would mean he grows — or whether he will succumb to the
degeneration of normal functioning, in which case the potential toward
development turns out to be "negative."

The main feature of Dąbrowski's inquiry is, first, his stress upon the
fact of the construction/disintegration mechanism as the means of the
developmental progress. Second, there is also stress upon the type of
condition which either promulgates or hinders the positive potential
toward reconstruction. He emphasizes, in fact (and this is of great
support to our philosophical inquiry), that it is the circumambient world

and its attitude toward the individual which may either support or essentially hinder the exercise of the psycho-neurotic potential at the "outside"! But it is also clear from his considerations that it is the strength and intensity of the *creative urge* within the individual's functional endowment upon which the constructive/disintegrative line of growth depends from the "inside."[45]

This psychiatric theory draws our attention to a very important point. We must carefully weigh the thesis that in principle psycho-neurosis, *far from being an illness, is, on the contrary, a vehicle for individual growth.* Furthermore, we must consider that the phases of disintegration of the disturbed individual *may with appropriate societal conditions and creative stimulation lead to stages of great spiritual development within a life which would simultaneously come to the "normal."*

These two points of support which our philosophical investigation thus receives from the perspective of psychiatry toward the postulational therapeutic guideline are of considerable value for *its* validity.

f. Is There an Indestructible Functional Pattern at Whose Revival We Could Aim in the Therapeutic Postulation?

One might, indeed, as pointed out above, wonder whether there could have remained any subjacent network of functions that would represent the original personal pattern, albeit in an ineffective, merely latent way. Those who are familiar with the dissolution of personality which makes the human being, so to speak, "unrecognizable" — because we cannot relate to him in any of the ways in which we used to communicate with him in matters of taste, affectivity, reactions, etc., and because we can hardly identify the individual as the "same" since his demeanor and physiognomy, down to the style and nature of movements, are drastically changed — must ask themselves this question. What we will attempt to show in our subsequent analysis indicates that it is of paramount importance to raise this question at this point. It is here and now that we meet the borderline of our concern. Should we assume that, with the complete disintegration of the individual world-fabric and the descent into the instinctive-elementary level of animality, no *personal core* remains and therefore we are no longer dealing with a human person, would there by any reason to continue measuring the individual by "normal" human standards? In what sense would we consider him as "malfunctioning"? Would it not be enough and proper to leave him at this stage just complying with his needs: material comfort and affection. As a matter of fact, we frequently hear people repeat with conviction that "all we can do for mentally affected people is to show them some affection." In what follows I will argue to the contrary that, first, it is completely false to assume that in principle the work of a life-long — whether shorter or longer — development could be insignificant and that we would, in case a therapy were applied, have to deal merely with some basic rudiments. Second, I will suggest that, even to retrieve functionally the rudiments of skills lost in the process of disintegration, it is of great importance to take into consideration that they have to be retrieved not from a *tabula rasa* but from *within the pattern of the developmental unfolding of each individual person:* it has to proceed from the outset with reference to a LATENT SCHEMA OF THE PERSONAL LIFE-WORLD which has once been established. In short, since the dissolution of his person/communal and life-world pattern *means the loss of his personal identity, it is obvious that in the reverse process of reconstruction we have to aim at the re-identification of the individual.*

SECTION III: PHILOSOPHICAL GUIDELINES TOWARD THE RE-IDENTIFICATION OF THE HUMAN INDIVIDUAL BY THE PROGRESSIVE RETRIEVAL OF HIS BEARINGS WITHIN THE COMMUNAL LIFE-SIGNIFICANT SYSTEM

a. *The Threshold: The Soul/Body Territory*

In our foregoing discussion we have reached the point of the most intimate interplay between the somatic and psychic functioning of the living being. In fact, in our perspective it appears as the threshold between animal and specifically human functioning. As we have mentioned above, it is already within this complex network of interplay that inorganic means working upon the somatic functions may have, in turn, a direct effect upon the specifically human dispositions of personality and its functioning. It is at this point that the psychiatrist/pharmacologist meets the psycho-analyst, the latter being often helpless without the work of the first. But leaving aside the obvious fact of accomplishing the slowing down of intensified reactivity, emotive intensity, violently expressive fears and anguishes, etc. — which belong to the stationary aims, namely, the facilitation of the minimum of vital contact with the patient indispensable for his survival — the work of the pharmacologist maintains a very low ceiling. It could be said that, on the constructive or "re-constructive" side, observation seems to indicate that pharmacology may in fact reinstate the capacity for some basic types of emotive reactions as well as facilitate the reinstatement of the interconnectednesses of some emotive/affective functions — predispositions for the regular type of experience. Observation seems to indicate also that as crucial as these pre-dispositive somato-psychic effects may be, they prepare merely an indispensable condition for introducing other means and ways through which the human individual would retrieve his human existence from a disorderly bundle of drives.

It is from this threshold that we have to envisage such a reconstructive process. I propose that it will consist in principle in RE-IDENTIFICATION of the individual as a complete human being; second, that this re-identification may be accomplished by no other means than through the retrieving of the communal ties by sharing the life-world in a "human" way with others. For a human being there is no substitute for weaving together with others a communal life-significance, a fabric which gives sense, shape, quality to life and, giving it, transforms the elementary vegetation into LIFE.

Third, and foremost, in drawing from the description of the dissolution corresponding guidelines for the re-identification process — guidelines which are supposed to indicate further therapeutic ways and means to be applied to this end — we have to advance the great and enigmatic issue of the patient's will to engage in the process of reconstruction and to choose to perform each of its steps. A human being, in contrast to the merely animal being, which, as I have emphasized earlier elsewhere, is with a quasi-automatic spontaneity enacting the steps of its life-course predelineated by its entelechial "code," becomes "human" in a very special way *sua sponte*: he himself/herself *invents and creatively projects each of his humanly significant steps.* He may project it only "actively": that is, he is moved toward the enactment by his own will. The question of crucial importance then concerns the awakening and/or stimulation of the will.

With this issue we face a twofold dilemma: the will "to life," the will to "get back into the world," the will "to find oneself again," the "will to accomplish" each of these innumerable, tiny, psycho-societal moves in order to achieve these main goals. This means that the will must (1) find

an orientation free of pathological distortions, bends etc., and simultaneously (2) find a responsive "outside" communal attitude in order to be
exercised.

    With these two points in mind I will now proceed to sketch briefly
the SIGNIFICANT CIRCUITS OF THE RE-ESTABLISHMENT OF
THE INDIVIDUAL COMMUNAL WORLD — or, as we are accustomed to calling it, of the patient's "rehabilitation" — as they are
indicated through the personal/communal life-world pattern laid down
in our preceding analysis.

   b.   The Set of the Decisive Psychic Moves (Moves of the "Spirit")
Emergent Within the Vitally Significant (Somatic) Circuit: Attention, Will,
                        Thrust Toward the Other

Parallel to the emphasis placed by Edmund Husserl upon the role of
attention in the perceptual/cognitive process, is Eugene Minkowski's
stress upon attention in the psychopathological condition of the human
being.[46] I would like to emphasize here again that the act of attention
focusing upon one single point being thus discerned from among the
entire surrounding maze in the perceptual process in which the objective world-order is being established, is certainly the essential and
decisive step in establishing the objectifying interplay of functional
operations.

    (i) This objective order at which perceptual attention aims is "precon-
ditioned" in its objectively significant rationale by the vitally significant
operations which are already established by the valuative processes
relevant to survival needs. (This means that attention must be already at
work at the sentient/psychic level to focus upon the need to be satisfied
and the means which correspond to it.) It is already at this pre-objective,
pre-intellectual or prerational phase of life's enactment that the valuative processes work with the elementary types of attentiveness to life's
business. These valuative processes pre-condition the functioning of
objectifying attention; they also pre-condition the "regular" or "irregu-
lar" implementation of forms with which conscious objectifying perception results. We observe that with the distortion/dissolution of
consciousness the objectifying perception is distorted. Obviously it is at
the level of the basement and "sub-basement" that reside the reasons for

this. We have to seek these reasons and attempt to restore the proper configurations of the forces, processes, and functions interplaying in the adequate ordering of our functional life-enactment in such a way that the individual may perform it autonomously, "on his own." In the case of pathological deformations of perceptual forms and of the disappearance of attentiveness to them we may consequently, on the one hand, improve the conditions for a proper perception by applying pharmaco-logical means directly to somato-organic deficiencies. On the other hand, we need to find other means to restore them.

(ii) As Husserl and Merleau-Ponty have established, we have at the same time to consider that the perceptual constitution of forms in experience is performed also by our *kinesthetic operations in space and time*. Furthermore, as I have pointed out above, this constitution refers also to valuative operations. It is, however, the functional orchestration of the innumerable operational elements which enter into the interplay distributed mainly along the three lines mentioned above that must take *such a form* that the "concentrating" moment of attention and the "focusing" may be triggered off. That amounts to saying that in order to articulate the somatic level of functioning, stirrings, drives, pulsations, passional forces, etc., which otherwise would stream forth "blindly" because without direction, and "wildly" because without being chan-neled into constructive arteries of the life-processes, the pre-condition for a such a constructive coming together — constructive in the sense that it will enter into the process of life — is twofold; the release within the patient of the capacity to focus attention is not enough, if it does not find, as concomitant condition: (a) *a positively valuating reaction*, such that this attention will become relevant to a desirable end (or it will be maintained finding no articulating schema in which it would acquire a life-significance by entering into life's play; the natural impetus would be extinguished); (b) in order to maintain itself the moment of triggered attentiveness necessitates an *articulation-outlet*; it needs to enter into a kinesthetic segment of operations; it needs to acquire a rudimentary spatio/temporal co-orchestration with the life-world system. Without entering into such an articulated schema, it cannot acquire a relevance to the vitally significant constructivism of life's progress. Missing the life-world spatio/temporal network, it will be repeatedly exercised at ran-dom until, its impetus consumed, this exercise will vanish.

Here we encounter the core element for the re-constructive or re-constitutive interplay of operations and functions. Before analyzing it, we have, however, to return to our initial point and emphasize that the phases of the *self-individualization-in-existence* at which there occurs a radical switch from the merely vital/psychic significance of life to that of specifically human significance were established *by the surging of the creative function.* Let us recall that this establishment proceeds with reference to the creative designs devised by the creative imagination and emergent from the Human Condition. Only with the onset of the creative function are the vitally significant circuits of operations expanded from within by new circuits bringing in the aesthetic, moral, and intellectual senses. Let us recall also that while the vitally significant factors of sense were correlated exclusively with the relevances to the direct vital needs of the individual's life-progress, the moral sense brings in a special type of relevance: *the thrust toward the Other human being,* which carries its own intrinsic relevance to a constructive accomplishment of a *communal existence with others in all its rays of expansion.*

In order that the pre-established individual conditions for a constructive co-ordination of the natural forces may lead toward a reconstitution of the main arteries of the life-world, a third, and decisive, condition has still to be fulfilled. This condition, the thrust toward the Other — which is expected to lend significance to the evaluative attitude that will in turn spur the desire, the impulse, and the will to use the available forces so that they may enter into the proper co-ordination of the interplay and lead to re-establishing the life's arteries — *has to find a relevant response.* This response has to come from the Other, *from the communal world "outside."* (For example, the instinctual sexual drives informed by moral sense are "tamed" into an affective force and call for an affective response.) If this thrust does not find the relevant response, the creative impetus working itself through the emergent activization of the moral sense cannot take off. The thrust is stifled in germ. No re-establishing of life's networks toward re-construction/re-constitution of human personality within a communal setting may take place. We leave the living individual at the sub-basement of the aborted state — to a vegetation which does not allow even a fullness of animal life.

To wind up the argument underlying our analysis, let us again stress that the moment of attention and the capacity to maintain it over a sequence of functioning — the "attentiveness" which is here at stake, and which is usually identified with the attention complex allowing the perceptual ordering of the circumambient world into a common objective and intersubjectively shared pattern — is itself rooted in a most complex functional orchestration in which the organic functioning, the kinesthetic articulations and, lastly but most significantly, the vital/psychic/moral evaluative processes are of paramount importance. This interplay takes place in the body-soul territory — a functional conundrum where the vitally significant factors, the organic, sentient, and psychic functions, enter into an intergenerative interplay with the virtualities of the specifically Human Condition. In the course of this interplay an intermediary functional phase emerges where the creative function brings in attentiveness in its complete form.

In other terms, the capacity to stimulate all the forces at play toward the release of attention — understood in its full sense as the functional instance which is decisive about specifically human existence — depends upon a complex of functions in which the *will to communicate*, to "relate," to "share in common" (a piece of food, a walk, a game, an idea, an impression, etc.), is already "informed" with the benevolent sentiment, or "inclined toward" the Other in the form of a "confident thrust."

It is upon the capacity to focus and to maintain the attention (a span of "concentration") that the appropriate articulation of our bodily functioning and kinesthetic movement, through which our otherwise impassive physical organs learn to operate in the spatio-temporal circuit of the life-world (e.g., hands learning to grasp objects, legs to move along objectively established trajectories, etc.), depends; it is indispensable in order that our body may become an instrument performing a life-course within the vital life-world pattern. And, let us emphasize again that it is through restoring this interplay within the system of the life-world pattern, through sequences of an articulated behavior intrinsic to the coherent life-world schema, that the human individual will re-establish his bearings within the life-world.

Retrieving his bearings allows the human individual to acquire a vital "equipoise" and a foothold within the unity of the system of life. However, as it has been emphasized here over and over again, the crucial interplay of forces which takes place in the "basement" involves the workings of the "spirit" and can be conducive to a positive, constructive advance only if it finds a responsive continuation in its communal life direction.

We will now leave the body-soul territory in its physiologico-pharmacological circuit of pre-conditions and will follow the lead of our analysis in drawing from it some postulational guidelines for the "socio-communal therapeutics," as it should be appropriately called, which would be conducive to the self re-identification of the patient who is retrieving his personal bearings within the communal network of existence.

c. The "Freedom of the Will" in the Re-Identifying Process

The following conclusions of our analysis reached so far have to be kept in mind: (1) the work of the spirit emerges in "the basement"; (2) at this threshold between the animal and the human significance of life, the work of the spirit picks up the threads virtually present in the Human Condition for the restoration of the communal life-world network which the organico/natural means working upon a person's physiology, etc., prepare; (3) consequently, therapeutics, which is directed toward the restoration of the physiological pre-conditions, cannot be restrictive in its aims by cutting itself off from these very constructive threads which it prepares to being forth (on the contrary, its aims have to be ESSENTIALLY CO-ORDINATED with the re-constructive trajectories that the work of the spirit has to follow); (4) these trajectories of re-individualization or re-identification of the human individual lead through the network of arteries by means of which the elementary elements of everyday life acquire a communal significance. Challenging the Freudian identification of the works of the spirit with its highest points, culture, fine arts, religion, etc., it *becomes clear that the essential and fundamental modality of the spirit is the communal existence of the human individual: the spirit, continuing the work of natural forces, emerges, unfolds, and flourishes through the human person within the communal fabric of life.*

Psychiatry in all its branches has, thus, one aim: *to help the patient toward his reconstitution of his communal world, toward his re-identification within the communal existence, which means his return to the human community.* Clinical psychiatry — pharmacology as well as psycho-therapy, all types of psycho-analysis as well as the so-called "behavioral" and "cognitive" therapies — have all one common aim: the re-individualization of the patient within the communal world and his retrieving of a constructive role and "place" within it.

Leaving aside the specific methods of psychiatric therapeutics, I will in the following part of my reflection outline what in the philosophical perspective should be the common schema of their procedure in which they all should naturally communicate with each other.

SECTION IV: THE SOCIO-COMMUNAL THERAPY AND THE
QUESTION OF "INNER FREEDOM"; THE "FREEDOM OF
THE WILL" IN PSYCHIC DISORDERS

a. The Psychiatric Patient to Be Defined in Socio-Communal Terms

There is still one more conclusion of paramount significance to be
drawn from our foregoing investigation. In our analysis we have established the framework within which we may define the object of our
concern, namely, the psychiatric patient. To begin with, we should
distinguish between the patient with "recurrent phases of psychic disorder" and the patient with a "chronic condition." The first distinction
seems obvious; it concerns the human being who off-and-on needs psychiatric help to function within the community. The second case is more
complex and less easily generalized. We deal here with either a physiological condition which having once surged does not retreat to a latent
position that could be kept in abeyance but remains to seriously hinder
a complete retrieval of the individual's life functioning; or, the individual
has retained his virtual capacity to reconstitute his regular order of functioning but this capacity may be maintained only under specific circumstances that control the functional equilibrium.

Needless to say, these circumstances concern not only the physiological circuit of therapeutic help but foremostly the communal situation.
The question emerges: Is it not the type and kind of communal life, the
already established outside attitudes of the community as well as the
particular situation of the person hindering him/her from re-establishing
and maintaining his regular functioning or maybe even forcing him into
malfunctioning? We cannot enter here into a detailed discussion of
these extremely important but difficult issues. It is enough to bring to
light the double edge of our query: the question of finding ways toward
the redressing and retrieval by the individual of the inner communal
ties, who, having lost them, might *seek ways toward the redressing and
retrieval of some threads lost within the communal network itself.*

Within the framework and on the strength of our analysis we may,
in fact, identify the mentally ill as individuals either isolated from
communal life or/and rejected by the communal world. It is unimportant for the present argument to decide whether the isolation or the
rejection began with an inner condition of the individual who "could not
cope" with demands of the "regular life," or whether his pathological

predisposition, being in a latent state, has developed *because* of the rejection. The crucial fact remains: the isolating and rejecting attitudes of the objective communal world have to be transformed simultaneously with the progress of the individual therapeutic; their transformation is as much a condition for this progress as any specific therapeutic means which might be applied. In the long run, all such means, as well as any progress achieved, will fail unless the communal circuit of human trans-action is ready and willing to respond.

We will substantiate these views, approaching them first from the point of view of freedom as it appears in a double perspective.

b. *"Freedom from", "Freedom for" and the Double Perspective of the Socio-Communal Therapeutic Axis*

Hegel coupled freedom with the will. The "inner" freedom of the patient suffering from a psychic disorder that would allow him to take — or to "choose" — the right step without distorting it or misdirecting it, is as much of the nature of will as is the surging "from within" of a prompting to take a step at all. Such a prompting to become efficient has to acquire a "directional" tendency; sustaining this latter becomes an exercise of "will." Yet this volitional prompting assumes the freedom, first, to give it a proper direction, second, to be exercised and to promote thereby a course of action; both are, it is obvious, most intimately fused. Address-ing most recently the issue of the "inner" freedom of a human being within a psychiatric perspective, Wolfgang Blankenburg proposes it as the foundational question of psychiatry; he states that "the inner free-dom of the human being is the secret, although unavoidable reference point for every psychopathology and psychotherapy."[47] Furthermore, he points out that freedom "is not an isolated individual fact, but an inter-actional phenomenon."[48] These important statements find their fullest significance within our analysis: *the individual functioning for which freedom is a point of reference — if not a measure of adequacy — is in its significance linked with the arteries of transactions within a human inter-worldly network.*

Lastly, while distinguishing between freedom from compulsions, external freedom with respect to the outer reality, on the one hand, and internal freedom with respect to our own self, on the other, Dr. Blankenburg proposes a most important distinction between "freedom from" and "freedom for."[49] We have to give it some attention because it may offer us the axis of the differentiation among the basic categories of the concrete therapeutic guidelines which follow from our socio-communal determination of the chronic psychic deficiency and from the definition of the corresponding aims of psychiatry.

With reference to this basic distinction we will also analyze anew some basic issues concerning the so-called "rights of the patient," especially the "right to life." These rights, guaranteed by the laws of every Occidental society, call for a new exfoliation of their real significance.

(i) *"Free Will" versus "Compulsions"*

Coming back to our description of the psychic dissolution of the individual life-world simultaneous with that of personality, and the view that the therapeutic procedure is meant to help the individual to re-constitute it, we have already hinted at some important points. First, this dissolution is connected with, or results from, the disappearance or distortion of the ordinary forms of experiences. We have also intimated that both of them are at least partly the effects of "compulsions" received by the individual from the "external world," to be seen ultimately in a truncated or distorted rapport with the Other, and are partly due to his own previously established dispositions. These dispositions play in this respect the role of "inner compulsions" (e.g., the patient is not capable of receiving the message of friendly overture as it is intended because he has already established a disposition toward receiving everything as a rejection or failure; he is not capable of seeing the food or material which he is using in its right proportion because he has lost the ordinary perceptual forms of things, etc.). Both of these seemingly opposed sources — the external and the internal — result in the same phenomenon: the patient is innerly hindered (not "free") to redress "at will" the forms of experience.

Furthermore, when these distortions as well as the dissolution come together, as we have hinted, the inner disposition to bring forth the prompting for experience at large is inhibited (the patient feels innerly "paralyzed"). That means that he is not free to perform an act of will. As Dr. Blankenburg says, "psychoanalysis essentially aims at the uncovering of the unconscious determinants of our behavior" (e.g., in repetitive compulsions); but in our present perspective we will substitute here for the conception of the "unconscious" that of the "functional pattern" as the platform from which the entire animal as well as the specifically human life-world's unfolding takes off and with reference to which it constructs its universally intersubjective forms. Thus the "freedom from" compulsions — whether coming from the "outer world" or stemming from "within" the inner functioning of the individual — is to be measured by the capacity of the patient either to receive experience in the ordinary objective forms or to redress the distorted forms according to the universal functional pattern. However, as the ultimate measure of freedom we have to consider the capacity of the patient *not only to perform automatically the basic vital functions of the life-preserving elementary devices but beyond that to bring forth acts toward transaction with others:* to share, to plan, to project, and to undertake. That means the most intimate transactive correlation between freedom *from* the compulsive functional deficiencies paralyzing the prompting or distorting its direction, which are deficiencies with respect to the platform to which they refer for life's equipoise, and the simultaneous opportunity toward functional fulfillment. The redressing of a functional deficiency may occur, as innumerable instances in the biographies of patients demonstrate, only if together with an "impulse" to perform an act, or a "drive" to undertake a course, goes a "responsive occasion"; that is, when "outward" or "inward" occasioned means are available for this fulfillment. If they are failing, the impulsive intent is not reached, and the surging moment of "will" missing its directional aim aborts and vanishes. In short, in order to redress the dcfcctivc forms of experience, or to *forestall the proficiency* for the springing forth of the experiential impulse to assume a regular form, outward conditions have to be fulfilled: the patient has to encounter an opportune situation within the communal life-world accounting for his "freedom for" in the redressing of his functional pattern.

(ii) *The Functional Pattern of Experience-Formation (Sensory, Moral, Intellectual) as the Latent Residuum for Re-constitutive Reference*

While we stress so strongly the issue of freedom as the constant point of reference for psychiatric therapeutics, there is still more to be said about the already proposed universal functional pattern into which the natural energies and forces entering into interplay with the moral/aesthetic/intellectual lines of constitution fall together to assume the role of a formal/material model which works as the directional reference for all the individual functional schemas that the human being may have previously elaborated, and established, and then, during his life-development in reverse, distorted and dissolved. As Dr. Blankenburg points out, in order that we may exercise the "freedom for," ". . . there has to be something there, which can be freed and toward which liberation can move. This, and with it a healthy core, is constantly presupposed by psychoanalytic procedures."[50]

This "something there" which remains would be in our perspective this functional pattern. In fact, as we have suggested above, when we envisage the developmental circuits of life-significance through which the unfolding growth of life in the human individual proceeds, this functional pattern cannot be envisaged only in the form of a universal and uniformly shared system. Although such an abstract functional design has to be conjectured in order to account for the universality of the objective forms of the life-order, as well as of the human experience in which this life-order is being over and over retrieved in an intersubjective consensus, yet the growth of the human individual into a communal person indicates that the circuits of life-significance and corresponding forms of experience concretize within an individual life-course the general outlines of the universal functional pattern in a uniquely individual fashion. I have here in mind the numerous circuits of significance which establish communo-cultural experiential schemas of the human individual through which and in terms of which he differentiates himself from others within his circumambient interdependencies. To the cultural forms within which his experiences are formed belong the way of life, the form of adaptation to climatic conditions, the level of civilization of the community within which he/she has developed, the educational/cultural level of his milieu with respect to which he has molded his own emotivity and affectivity, the ideological — e.g., ethical, religious, etc — modalities of valuation of everything pertaining to the everyday life struggle as well as the aims of life, etc. In short, this universal functional pattern is in the concrete human individual differentiated into a specific *culturally molded skeleton* according to which the entire experiential reactivity of the human being has been modulated. Although distorted and even dissolved, it still remains in the background, as it stands for the meaningfulness of its course representing a particular person as he had been inventively constructed in the main stream of his existential effort. As I have pointed out before, we may observe with fascination over and over again that after a seemingly complete dissolution of personal unity, the re-individualizing process brings back its main characteristics. It indicates that this *culturally significant skeleton of personality remains in its essential lines a latent point of reference for the re-identifying progress upon which the therapeutic efforts may and should try to fall back, and which they certainly cannot ignore.*

(iii)   The "Right to Life" of the Human Individual

There is no precedent in the history of humanity known to us of the stress upon "human rights" as extensive and vigorous as the one pressed today. We claim the rights for political freedom, social equality, rights of children, rights of the unborn, and, of course, the rights of those who do not fall into any standard category, being at the brink of the regular: psychiatric patients. The question is, however, how these rights should be properly understood. Among the numerous particular points upon which the rights of the "mental patient" are claimed are those falling under the following categories: first, there is the patient's claim upon the society in which he lives to *vital subsistence and medical care*; second, there are numerous points of his "rights" concerning the fulfillment of his individual desires. The former take a two-fold societal significance: individualized and institutional. Since the situation of the mentally handicapped individual within communal life might be extremely res- trictive, these latter rights boil down ultimately to the fulfillment of instinctual drives. This shrinkage of the communal potential to satisfy is one-sidedly acknowledged already in the currently practiced institu- tional policy: in the name of the patient's rights the usual moral code breaks down (e.g., sexual morality in its cultural determination vanishes before the purely naturalistic requirements of the "mutual consent between adults," the culturally established evaluation of sexual practices is considered as equally invalid and left to the choice of the individual). According to this policy the patient is also supposed to know what sexual instincts should be or not be stimulated by the freedom to indulge in excitements stemming from pornographic sources; this free- dom is seen to belong also to this right. Yet at the basis of the principles governing the current interpretation of rights seems to lie an intrinsic contradiction. The patient is supposed to know best whether he should accept or refuse medication; he is expected to know when he should seek the help of a psychiatrist or therapist and is expected to turn to it himself (unless he asks for it himself there is no way to provide it for him). As long as he is a part of the communal world, it seems to be,

in fact, assumed that he is capable of a straightforward, regular judgment concerning his condition; that he is, in fact, endowed with the
ordinary perception of life's business and capable of estimating his own
situation and state, and, beyond that, is in a position to formulate a clear
judgment about it. These assumptions are radically at variance with the
state of the individual estimated as having a mental disorder. How can
the patient in a disorderly state of mind know what is good for him?

Here we encounter the striking contrast. On the one hand, the societal network guarantees the disturbed individual a number of individual
rights without having clearly investigated whether their satisfaction leads
the patient toward the basic goal — a goal at which society should, if
fulfilling its obligations, aim. On the other hand, the rights of the
individual to *any* transaction within the socio-communal are drastically
withdrawn from him when he is judged medically to become a public
danger to his fellow citizens. That is, until this stage, no preliminary
steps are initiated within the communal network for a disturbed person
who in his condition is not capable of seeking help of his own accord,
but is at the breaking point of his control over the situation and is by
definition not responsible for his actions. The communal security forces
step in, equating the person with an individual who has committed a
crime against the community. In point of fact he has to deteriorate to the
point at which the community judges him as not fit to be a part of it
before he will be admitted for the hospitalization which he needs for his
health. At this point the community isolates him in an institution. The
necessary first step toward admission to a mental institution means
deprivation of all human "decency" and "dignity" which the nature of
the person commands and which is consequently due to the citizen of a
civilized community. It is performed by the same means (arrest by the
police, handcuffs, etc.) by which a criminal who breaks a contract with
society is treated. What should mean a *natural turn for help* takes on the
form of a criminal *rejection and isolation* from the community on the
grounds of an "offense" toward it already perpetrated or expected. It is
only at the moment of this drastic breaking down of the societal proficiency of the human individual that society takes on its responsibility
for its own member. Yet this responsibility amounts to a withdrawal, a
rejection, of the individual from the regular natural network of the
communal ties and forcibly places him into the confines of the artificially shrunk and restricted framework of "institutional" existence;
the mode of "life" of the institutional framework is truncated in the

natural spatio/temporal axis of expedience and consists in a range of interests reduced to the present moment in a miminal space. Only at the distant horizon of the real world outside of it, and from which the patient is cut off, lurk the vitally significant aims of human reality.

Let us recall that one of the categories under which this responsibility recognized by law falls is to secure the continuation of life. But the "right to life," as currently interpreted, is restricted to the provision of rudimentary medical care and the minimal means of subsistence, if need there be.

Let us now turn to the proper interpretation of these rights. The basic and inalienable right which the society in which he was born and in which he developed guarantees to the human individual is the *right to life*. Thus the basic and inalienable right which this society should simultaneously grant him is the fully interpreted freedom to unfold properly and to conduct his life.

Within the fabric of our foregoing investigation we may now in a synthesizing way approach properly these two claims so basic and so elementary that no other may take precedence over them without perverting the view of the fundamental communal contract between the human being and his group.

Bringing together all the elements of our inquiry shows that the "*right to life" means a right to a fully human life within a communal network*; as was shown above there is no possibility for the human individual to be "human" outside the unit of communal life; the smallest unit of communal life, the family, will try its best and yet left to itself cannot offer a communal network of transaction. It lacks the main communal arteries through which the communal interaction has to spread. No "half-way" institutional circuit may on its own offer it; both of these groupings may be indispensable stepping stones but yet remain inneffectual on their own to conduct the spread and sustain the radiating effort of re-individualization. If it be not spread, it will be extinguished. Taken out of the communal network and put aside for a long period of time into an isolated institutional framework, the individual is deprived of the possibility of retrieving or developing the communal world and is condemned to the extinction of the life of spirit which can flourish only within the *interhuman* networks of the life-world; he is doomed to a vegetative existence which is not only restricted to that of the lowest circuit of the basement-functioning but lacks even the natural freedom and fulfillment of animal life.

If society means to fulfill its basic obligation toward its members who are mentally disturbed, the subsistence as well as the medical care support concepts should be *re-interpreted, re-directed and re-organized in order to accomplish this obligation properly.*

Here emerges the decisive question: What does the notion of "society" mean for us? Are we not too easily prompted to forget that it is a communal network of people of which we ourselves are part and are thus co-responsible personally for the well-being of others, and, instead, to entrust institutional representatives and governments with all responsibility? How can a depersonalized, neutral body enter into the network of personal ties? In view of our communal nature, is not, on the contrary, the responsibility for the communal life of every individual member of our communal network of life our innermost personal task? How can anyone be rightfully excluded from it? What is at stake is obviously not only the material means for subsistence and the safety of human life but the means by which a rightful member of the communal life may sustain himself as human; it is the full human quality of communal existence which is the "right to life" of its individual members.

The individual members of the community have to become aware of the "right to life" of which *all of them are naturally holders as well as custodians;* this awareness itself if put into practice will improve the quality of their own existence.

By opening themselves to the communal needs of their fellowmen — communal needs which go beyond that of animal subsistence — they open themselves to share in the spirit. No greater advance may be accomplished in the life of a community at large than by making the opening for the spirit through sharing in personal modes of life. Allowing a being, condemned otherwise to vegetation or to an existence of continuing despair and dejection — a life wasted — to unfold and flourish in the communal life by sustaining in our own benevolent attitude the efforts of the Other who offers us his innermost self to share, the human being establishes an inward platform of the spirit of incomparable wealth and finds the most fulfilling meaning of his/her existence.

SECTION V: THE MAIN POSTULATION PRINCIPLES FOR
THERAPEUTIC STRATEGIES TOWARD THE SOCIO-COMMUNAL
RE-IDENTIFICATION OF THE DISTURBED PERSONALITY FROM
A PHILOSOPHICAL PERSPECTIVE

*a. The Continuity of Aims and the Need for the
Coordination of Methods*

It is clear from the above that there cannot be one single set of therapeutic means (e.g., pharmacology, psychotherapy, socio-therapy, etc.),
which in one concentrated effort could lead the patient's progressive
life- and communal re-identifying process. This process, which has to
move from one circuit of significant restorations of communicative life-
involvement to the next, calls first for a complex strategy unified in aim
and continuity. Like life's thrust itself it has to be multidirectional at one
and the same time. From the first diagnostic observation of clinical
psychiatry through all the concomitant psycho- and socio-therapeutic
appreciations, follow-up methods and means applied, there must be
one continuing thread of intent: the *retrieval of the significant human
network of life-involvement.* This postulate denounces the inefficiency, if
not the outright harmfulness, of therapeutic methods which would be
discoordinated and thus possibly result in staying at odds with each
other. Furthermore, it indicates that in a random approach that is made
without proper biographical investigation of the case — and this not
merely concerning the record of previous treatments but most importantly the socio-cultural biography in terms of his interrelations with
others in the communal life (school, family, work, church, associations,
etc.) — no plan for a serious therapy can be outlined. The common aim
of the entire therapeutical strategy, which is meant ultimately to help the
patient to "find himself" (even if this would have to occur in a more or
less societally diminished fashion within his communal world), must, in
the use of means and ways for prompting the thrust into the world and
reknitting the interworldly ties, keep in mind the *entire picture.* It has to
aim at the redressing of the person. There is a continuity within the line
leading from the physiological, co-organically motivated functional dissolution or distortion of experience, through the somatico-psychic restitution or retrieving of the regular forms of kinesthetic operations — as
well as of those of sensory experience oriented toward the restoring of

the affective links — *to be simultaneously* rekindled in a transaction with the Other, a transaction which will bring him into the socio-communal network of life's significance. Only through a continuous and concurrent checking of the patient's reactions over against each of these therapeutic lines and reactions to them may a therapeutic strategy in which innumerable factors simultaneously enter and act, be devised.

It may be objected that it is an "ideal" far, far removed from the actual practice. This does not change the validity of the claim toward its realization. Within our present-day civilization we have developed the most complex and accurate care for our organico-physiological functioning. Why should we ignore and neglect the psychic realm of our beingness which is decisive for the very meaning of life as well as for its enjoyment?

The therapeutic strategy, seen within its entire socio-communal context, is in its overall schema naturally *two-directional.* First, it is and continually remains to be directed toward the "inner functioning" of the individual. Yet, after the initial therapeutic efforts had resulted in a sufficient "opening" of the individual toward the "outer reality" and his interest in reknitting ties with it, as well as the minimal re-ordering of his experiential forms having been revived to the degree that it allows for the unfolding of interpersonal relations (for which the capacity has been thus rekindled), there sets in a need for a skillful and progressive second phase, that of the *"inner-outer"* oriented therapeutical work. It belongs to the phase which is currently misleadingly called "rehabilitation." This name is, as a matter of fact, misinterpreting the situation of an institutionalized individual as being compromised with respect to the community, in need of improving his image and of being reconciled with it, as if it were not — on the contrary — *the responsibility of the community* to make up for the need to isolate him for a while from its life by helping him/her to resume a proper role and place within it. Yet this help coming from the "outer" world has to concern the inner self of the patient.

Third, when the inner-outer orientation of functioning has already taken a foothold within the functional promptings themselves, the stress in the therapeutic methods has to go toward the regaining or acquiring of skills, capacities, work and play motivations along which the transactional involvement into the communal networks may spread and be received and prompted by this communal life itself. That is, we are at this phase of re-individualization of the human individual within the communal significance of life dealing with the actual "re-entry" *into the regular intersubjective community* by the patient, hitherto isolated within himself.

The point which comes to the fore is the twofold nature of these therapeutic guidelines. When we plan the strategy of means to revive and promote the inner predisposition of the person, we have conversely to ask whether the community into which the patient is going to strive to re-enter is ready to receive him? When we review again the reasons of his initial falling out of it into the self and the consequent shrinking of this self, we must say that the community, as we know it in present-day Occidental societies, is to a considerable degree responsible for his falling away from it. The community did not respond sympathetically to his sensibilities and difficulties in coping with his current life and, at the breaking point, rejected him/her outright. This is to say *that to open the communal network and help the fallen-out link restore its transactional ties again necessitates reviving the moral sense not only on the side of the individual but on the side of the communal life as well.* Considering the moral foundation of intersubjectivity, they have to go over the naturalistic prejudices in order to meet ultimately in the *moral significance of life.*

b. Some Philosophical Ideas for the "Socio-Communal Therapeutic" Project in its Main Guidelines: Toward the Pilot Project of a Communal Integration of the Psychiatric Patient

The above-discussed philosophical postulate for a full-fledged therapeutic project that would do justice, first, to the nature of the pathological situation of the human being and, second, to the nature of the complex process in which the retrieving of the person's life-promoting faculties simultaneously with his common-social re-integration has to proceed, indicates three basic principles for the therapeutic approach at large.

We are dealing here, as discussed above, with a twofold aspect of the therapeutic procedure: first, the work upon the "inner faculties" of the patient enabling him to transact with the Other; second, his relatedness to the Other within the life-world. We are in both essentially concerned with the most complex process of the *transition between the insulary stage and full-fledged "re-entry" into regular life-commerce.*

The first aspect of the therapeutic project, which is oriented toward the second, concerns the radical situation in which the patient, incapacitated to function "normally" within the world, has remained isolated from the regular traffic of life; in most cases, this isolation means his seclusion in a mental institution.

Simultaneously, the inner isolation of the patient from the Other manifests the inner malfunctioning which goes right down to the phys-iologico-biological circuit. It involves then the basic pharmacological therapy. This latter is to be left to specialists in the field. The complete socio-communal therapeutics, in contrast, delineates itself naturally within our philosophical context of inquiry with reference to the previ-ously established postulates. It is from its inception and throughout the last phase twofold: it will be directed simultaneously at the inner workings of the patient's functioning as well as taking into account the demands for the prolonged re-identification of the person with respect to communal existence. This second line of emphasis will necessitate at several stages a *socio-communal supportiveness.* It will call for various steps of "institutional" modalities as well as *stipulate various forms of adjustments on the part of the community into which the individual will aim to re-enter.* I shall now proceed to outline the philosophical point of the project. It does not claim to be in any way "revolutionary." As we all know, in clinical as well as in the institutional realm much experimenta-tion has been done in the last fifteen years. Much insight has been gained and many excellent methods applied. At the same time, however, these experiments have shown basic shortcomings owing to what I believe is the neglect of the *fundamental nature of psychic disturbance.* Its basis certainly lies in a latent inner pre-disposition, but at the same time this inner disposition is prompted to unfold within the communal world with its basic lack of communal spirit.[51]

In the present project I am addressing myself to this specific point which, as the philosophico-phenomenological analysis conjecturally indicates, lies at the bottom of the experimental failures.

The process of reconstituting the natural life-world of the patient, from its "basement" phase onward, aims at the redressing of experiential forms, because these forms belong to the fundamental universal order-ing of the objective life-world. A question emerges: Since we have attempted to show that the socio-communal life-world cannot be reduced to the natural circuit of the life-world, but is in contrast the manifestation of the work of the human spirit, what else could account for its ordering but the workings of the Moral Sense through human transaction?

In fact, human societies and socio-communal groups form and order their transactional co-existence by rules, laws, taboos and ideals, all of which express the transactional development of the Moral Sense. With respect to the specific modalities of their particular moral code, individuals unfold their transactional responses in the form of especially tuned sensibilities and lasting moral habits. They also unfold a type of "communal conscience" which is geared to this moral ordering of community life. In this way the inner workings of the Moral Sense within the individual functioning of the human being prompting the moral sentiment, benevolent or malevolent, *fall in their transactional shaping and its development into the referential framework established already by the community of which he is "by birth" and becomes with his growth an active member.*

Could then the therapeutic approach in the re-identification of the patient follow strictly naturalistic lines, ignoring the moral axis of each and every communal order? Reconstituting his functioning with respect to the Other on strictly *naturalistic* guidelines, "beyond good and evil, right and wrong", could the human individual develop the responses and sensitivities toward human transactions allowing him the adaptation to the specific "way of life" of any normal socio-communal group?

With these questions the problem of morality is introduced into psychiatry at large and into the psychiatric therapeutics dealing with the disintegrating personality in particular.

At this point, however, reappears the question of freedom. In point of fact, together with the problem of ethics, there always arises the question whether any stress upon ethical rules, prescriptions, and sensibilities is not in their discriminative character hindering the "natural" personal freedom to which the human person is entitled. To resolve this question within our hitherto exfoliated perspective it is enough to point out the radical difference between an exercise of "natural freedom" by making decisions on the spur of the moment and without being able to deliberate their constructive or negative effects, and the freedom constructively oriented by the will. The prompting force of the first type of decision-making is not will but mere whim. A whim does not carry responsibility with respect to the constructive outline of life's progress.

Whereas the "inner freedom of the will," which implies the sustaining directional system of the will-exercise as such, demands deliberation in which the general lines of personal project-directed conduct and their aims have to be taken into account. In other words, the *act of will in which personal freedom is exercised is in the service of life,* within our perspective it is the service of the individual's reestablishment of the socio-communal life-network.

By introducing a moral axis within the directives of the therapeutical approach we are, in fact, introducing the essential directive for the stimulation and sustaining of the patient's "right to a human life."

The specifics of the moral axis to be maintained within the socio-communal psychiatric therapeutics, as well as the ways to introduce them, which should refer to the patient's original life-world pattern remaining latent in his functional system, require an extensive inquiry which lies beyond the limited scope of this study.

The Two-line Concrete Therapeutic Goals and Means: Inner-Directed Therapeutics and "Socio-Communal Re-identification" Therapeutics

In general, traditional psychiatry has always aimed at the retrieval of the communicative links of the patient with Others. This meant the retrieval of his capacity to re-integrate into the life-world; the community used naturally to integrate and absorb him. Our contemporary communities refuse to do it. It is the introduction of the pharmacological means which has brought about the undue division of tasks: here the emphasis upon the inner functioning of the individual, there upon his capacity of communication. Within our renewed philosophical framework we are addressing the need for the unifying of the two at the deepest level, namely, that of the very Human Condition in its communal nature where both of the two tendencies find their common root.

Within this perspective we will, concerning concrete methods, still distinguish between the "inner-directed" therapy, in which the basic physiologico/organic functioning taken into consideration is in focus since it addresses the person at the "breaking point" from whence the communication capacity will have to be rekindled altogether, and the "socio-communal re-identification" therapy, which will aim not only at (a) stimulating the re-constructive process in which the life-world involvement will unfold, but (b) will give it direction offering appropriate outlets for successive steps of the communal involvement.

In has to be, however, strongly reemphasized that both of these therapeutic lines, the "inner" as well as the "outer" life-world-directed therapeutic lines, are to a considerable degree interchangeable, which means that promoting the activation of the patient in one way serves simultaneously to further his progress in the other way, and conversely. Furthermore, the goals set by the first line of therapeutic approach serve simultaneously the other. The one postulates four main points or steps; the other, four phases of development. Such is the profound truth about the nature of the psychic disturbance of functioning and its communal core. In the discussion of the particular postulates for the two distinct therapeutic lines, their mutual interconditioning and interplay will clearly appear. We will limit our discussion to bringing out the specific points of this interplay, leaving all the less striking therapeutic methods in current practice aside.

The Four Steps of "Inner-Directed" Therapeutics

"Inner-directed" therapeutics, as it is now practiced, is meant to prepare the functional means for the person's "socio-communal" expansion. However, what has to be emphasized in the present perspective is that to accomplish that effect, not only the functional means have to be animated, awakened from their atrophied state, but a proper ground has to be prepared for the solid and lasting reconstruction of the personal pattern, now twisted or lost, of communal expansion. Bearing this in mind, four goals should be proposed toward which the "inner-directed" therapeutics should aim. All of them will take into account this crucial *period of transition between the stimulation of the functional means* toward the retrieval of the "will to live," the capacity to "respond," as well as to "thrust" one's interest toward the Other, and the ways of its implementation and means of its exercise. These goals may be briefly stated in the following fashion: (1) the retrieving of the basic spatio/ temporal life-world ordering, which is instrumental for any elementary intersubjective life routine; (2) the awakening of basic "everyday" personal care and the enactment of the elementary procedures of personal life-maintenance; (3) the stimulation of and the engaging into a "trans-actional" linking with the Other; (4) foremost, however, is the stimulation of a "creative" potential. These goals have to be related to the phases of the communal re-identification progress.

We may then consider as the *first therapeutic phase* that dealing with the patient's *physiologico/biological malfunctioning.* Yet as has been emphasized above, the individual isolated in himself must, with the application of pharmacological means, simultaneously be attentively

observed as to his progressive retrieving of communicative responsiveness and of his thrust toward the Other. He has to find a response
appropriately directed to his thrust.

Step 1. At this basic phase of therapy, we will be aiming at the retrieving of the basic *forms and rhythms, and of the order of experiences in
their space/time coordination*, as well as of the basic *forms of transaction
with the Other.* We have to begin with work upon kinesthetic organization in such a way that the basic essential spatio/temporal movements
and sensing articulation begin — if need there be — to fall into the
regular *objective ordering of the life-world-constitution* in experience.
    It is undoubtedly with the proper personal kinesthetic articulation
that the spatio/temporal axis of the life-world functional ordering
should be retrieved. Physical exercises, sports, games, eurythmics, etc.,
would be of crucial significance.

Step 2. Retrieval of the basic *personal life-order in the everyday routine
of life*, that is, of careful grooming and taking care of one's dwelling and
possessions, should be attentively promoted. It has a many-sided significance in re-introducing socio-communal foundational order. Indeed,
the routine grooming activities lay down significant kinesthetic ordering
of the life-world involvement. They pose the first circuit of the objectified intersubjective involvement of the individual, significant as the step
preparatory to the transaction — a pre-condition to meeting the Other in
life's business. These three lines, stimulation of creative potential,
opening toward the transactional relationship with the Other, and the
restoration of a basic personal life-order, will be the object of attention
in a progressive unfolding throughout all the phases of therapy. However, they should be initiated in the first phase, because they belong to
the *inner re-orientation; they also stimulate the restoration of the
functional powers* which by these means will be prepared for the
reconstructive — or re-identifying — progress at a later date.

Step 3. Stimulation and implementation of the patient's *creative "potential".*

*The "Nature-Imitation" Approach in Art Therapy and the Directing of
the Creative Impulse Toward the Proper Re-construction of Experiential
Forms.* With all the above-mentioned means we aim at the basic stimula-

tion of the creative impulse and at the preparation of ways for its implementation in a multidirectional unfolding by the individual. Art therapy
seems to be the main vehicle for this stimulation. Yet, although it is
currently practiced in a great variety of ways, some discussion of its
means, methods, and aims might be useful. In its most popular current
form of "expressional" therapy, it aims at the basic release of the creative
impulse. Simultaneously it aims to accomplish the kinesthetic rhythm/
movement/ear as well as an acoustic/rhythm/color/shape coordination. Consequently it suggests some basic models of expression as
frames within which the personal "thrust into the open" may take place.
As important as this will be at the first phase, I would suggest that at
subsequent phases of therapy it should be expanded into a more *formal*
line. Namely, what is at stake here is the orientation of the *creative
impulse into proper reconstructive channels*. That is, within all the steps
of the "inner-therapeutic" work we must have in mind *the re-ordering of
the life-world pattern by the individual* according to a universal intersubjective schema. We have to aim at the reconstruction of the twisted
formal aspects of experience.

It seems important then to add to the expressive therapy — in which
the focus falls upon the inner impulse — the means for the reorganiza-
tion of the steps on the way along which it might redress the proper
universally valid forms of experiential constitution of the "outer world."
To this effect it seems useful to introduce the "imitation method" in art
therapy. At the second phase of therapy, when the patient is capable of
engaging in a proper teaching procedure, it is the drawing, painting,
sketching of nature by "imitation" which might be of great importance. I
mean the imitation of visual, tactile, etc., aspects of things as they are
being "seen" and "used" in current life-practice, that is, in an artistically
"naturalistic" style. The student should see the "models" for drawing or
painting nature while the teacher draws or paints with him. By showing
him how he himself paints, the teacher should teach how the subject
"should be seen." This imitation-process in the course of which the
teacher gives examples of "how to see" "natural forms" — a method used
over centuries in art academies and ateliers — the teacher, stimulating a
new type of experience, would indicate to the patient the points upon
which his seeing/touching, smelling, etc., experience could and should
be expanded from the form to which it had shrunk.

Furthermore, the effort at "imitation" of the teacher also stimulates the
creative impulse in an enriching fashion. New colors, other than those

previously remarked, appear in the field of experience. Moreover, new forms and colorful spaces different from those in which the object being represented has been hitherto seen not only enrich the enjoyment but also the "surprise" caused by the effort to "see," to "compose," to "transfer forms" differently from the way in which the patient would spontaneously have made it; the "click" that occurred at the juncture of the effort creating this "surprise" is in itself *a click of the inventive/creative mind.* Nothing may be more conducive to stimulating the creative impulse than, first, this bringing forth of aesthetic enjoyment as one of the main vehicles of the life of the spirit which leads the individual out of the "basement" into which he has sunk into the open, and second, expanding creatively his "vision" of colors and forms in their possible variations. Lastly, and most significantly, this imitation of the model nature will allow the individual to break down the shrunken and twisted vision of "external" objectivity, introduce the necessary flexibility of experimentation into his form-oriented functioning, and re-introduce the proper shapes of basic experiences indispensable for life-worldly transaction.

Step 4. On the basis of our foregoing discussion, the creative animation of the patient can be best accomplished and thus render best results in the direction of transactional socialization, if the "creatively oriented" therapy is devised to *transactionally involve other people.* The currently applied "group-therapy" concept is here important; but *to "engage" into transaction* by the concept of "chain-activities", when one fragment of the activity is to be picked up by another person — to be continued until the patient is capable of entering into various types of games — is basically still more useful.

c. The Four Phases of Socio-Communal Therapeutics with Institutional Support[52]

It has been pointed out above that the "rehabilitation" process of the post-hospital patient still calls for a specially prepared nucleus of communal life within which, as in an incubator, he will find the appropriate communal support for the re-identification within the life-world and will thus avoid these types and aspects of the inter-worldly transactions which would hinder his progress. As is well known, in the last fifteen years a drastic change has occurred in the approach to mental health. Because of pharmacological effects upon mental disturbances, there has been a movement to release the patient from the hospital into communal life. It has become, however, apparent that this movement has failed. On the one hand, hospital care deteriorated into a pharmacological treatment alone. The community, on the other hand, did not integrate the released person. Observation of the failures and problems in the then created and hitherto existing institutional support system of the half-way housing, apartment-projects with therapeutic direction, etc., shows that in principle such an institutional support has to be planned for a long-run period.[53] Consequently it has become clear that many distinctions have to be introduced in responding to the stages of progress and in avoiding too hasty capitalizing upon its steps. There seem to be basically four phases of the re-identification or socio-communal re-integration progress to be considered. Lastly, having abandoned the naturalistic attitude, according to which all living individuals may be adequately treated as being basically "beings of natural drives, instincts, etc.," for the view that a person is basically *also a being of a self-invented cultural pattern*, it will be of paramount significance to introduce here a differentiation of persons according to their original moral and cultural commitments. Each person presents his own conundrum of issues and calls for new responses to them.

We are drawing conclusions from the failures of the above-mentioned movement of the "de-institutionalizing" of the patient; a more careful differentiation of the rehabilitation/integration process allows us to distinguish:

(i) *The First Phase*

Socio-communal therapy in its preparatory stage will, as stated above, already begin in the hospital where the patient will enter at an appropriate moment into the therapies discussed above.

(ii) *The Second Phase*

The "extended family home" will already function as the nucleus of communal life. On the basis of current experience, the patient who is living the extremely reduced life-circuit of the hospital is not capable of responding to the demands of current life. The family circle to which he would naturally return is not only too narrow but, once he has outgrown it, is also much too restricted in the life-world involvement to allow for the stimulation and expansion of his reaching out. Lastly, the family circuit is the one within which the communal situation has been obviously unsatisfactory for the individual from the beginning and, consequently, cannot now offer the resources which he needs to recuperate and to expand his potential. On the other hand, however, the natural communal life-order has been, within the original personal life-world pattern, formed with respect to a centralized family life-organization representing the *communal nucleus* of self-identification. The original personal life-world pattern of the social expansion of the person has been, in general, formed within such a constructive *transactional* schema. Consequently, to begin with, a family model offers the unique and best point of departure in reconstructing this communal life-world orientation. It proposes not only the basic schema but, expanded through the number of members, the "reconstituted family group" of eight to ten persons affords an extensive number of stimulation modalities to choose from for such an interrelational practice. The kind and modalities of the thrust and response will be, in fact, appropriately calculated. Indeed, here enters the importance of the cultural orientation of this communal unit. In order that the personal communal life-world pattern be stimulated and then activized, we have to consider the essential — and most "sensitive" — lines along which it has originated. That is, we must consider the cultural life-modalities with respect to which the person identified and gave himself a meaningful structure: morality, ideological orientation (e.g., religious or agnostic), cultural forms of human relations, etc. All of this has to be considered in bringing together members of the "extended family." Yet moral and religious modes come first. The group should be formed on the model of a natural family by a couple especially educated, who would direct the communal and individual occupations. The "expanded family home" would also have a therapeutically guided life-routine and regular psychiatric supervision.

Although involved in a program of duties, entertainment, etc., and in the personal inner-oriented therapy as discussed above, the members of the group should individually hold "day-occupation" programs. No matter how the individual members could be otherwise employed according to their state, there are to be other programs in conjunction with the home support available.

The Occupational Day-program. What is currently called a "day-treat-ment program" can be considered as its tentative example. However, such a day-treatment program has to be differentiated into several phases according to the needs posed by the progressive communal re-identification of the person. For instance, while at the first phase, the so-called "occupational" and "art therapies" may be just — as they are — geared to the basic stimulation of the inventive potential or basic retrieving of kinesthetic orchestration, as well as basic concentration, at the second, that is, transitional phase, marked by a basic transfer from the hospital to the "extended family home," both the occupational as well as the art therapy have to be focused more upon the *grounding of the patient in the functional proficiencies of the capacities for socio-communal transaction.* That is, in the first place although the now prevailing "expressive" style of art therapy has certainly a great role to play at the first phase, at the second phase we could consider giving it a more structured form in the next step, namely, one in which reconstitu-tion of *experiential forms would be almost directly aimed at.*

The Day-Program Aiming at the Retrieving of Basic Occupational Skills With an Orientation Toward Co-operation. There are many aims and methods which are being practiced in the so-called "day-treatment" centers and consequently we do not need to enter into enumeration and discussion of all of them. Within our therapeutic framework it should be, however, brought out that having reached the second phase, the pa-tient may be expected that his therapeutical guidance will lead him into a *life-purposive orientation.* On top of many other therapeutic devices (aiming, e.g., at improving concentration, sharing with others, clarifica-tion of one's problems in discussion with others and attempting to understand them better, etc.), a *purposive skill training may be initiated.* In this respect a priority should be given to *skills that may be practiced in co-ordination with others* (e.g., chain-activities, services, etc.).

The end of phase two should be marked by the patient's progress to the point of his being able to enter into some — as simple as it might be — so-called "professional training", guaranteeing him the capacities to begin, in the third phase, some professional schooling, or "apprentice-ship," or work in the "outside" life-world.

(iii) *The Third Phase*

The re-identification through a "host family" of the mentally disturbed person within the communal life-world would mean in fact that the easing of the person into "regular" life had already been achieved. However, one should assume that such an entry may not be advisable in any abrupt fashion before progress into the "communalization" has reached its peak. The incubation period of the family group being outgrown, there is still a long way to go before the individual will be re-constituted upon a firm and lasting basis. In order to maintain what has been already accomplished and to develop it further, he will attain his complete communal identity only with a full-fledged re-constitution of his communal network. This means his complete transactional immersion within it attributing to him a role and a "situation."

It is necessary, on the one hand, that the person leaves the protective shell of the family group and of day-treatment. Yet it cannot be reasonably expected that one is safe within a ruthlessly competitive world if left to one's own devices. On the contrary, we need a socio-communal support in another type of a quasi "institutional" arrangement. Namely, as experience has shown, in spite of any already acquired progress, even if the individual is well advanced in his "re-identifying" and his attempts to find himself within the communal network with "a role" and a specific situation, lacking them he still needs the support of a nuclear community. Now a "real-life community," which is the best support for his further development, would be a "host-family" that would receive him for a period of a few years until he himself is capable of establishing a nuclear communal unit.

Concurrently with this re-entry into real life, the individual should be capable of undertaking some real-life occupational task by means of which he would, first, enter into the full-fledged transactional communal life. Second, he would unfold his personal network of communal existence. Whether it would be appropriate to undertake special educational programs to secure a future professional life or to pick up again a line of professional work would have to be carefully considered to prevent undue stress and demands which could discourage the thrusts into the life-world and cause them to atrophy.

The "Life-apprenticeship". At this phase of the reconstructive progress of the person the response to his thrust by the community and the communal situation into which he will strive to enter are of paramount importance. The still shaky individual who still has to acquire the communal self-identification is a "life-apprentice." He has to initiate himself into the "regular" roles and functions of the societal group life in which he is to perform in order to become a member of the community; he has himself to seek and find his way for it. He has to meet the incertitudes, vacillations, discouragements, and self-doubts comparable to those of a person passing from adolescence to adulthood. Yet he is not placed within the usual continuity of this transition where normally education and previous life-experience might secure, at least to some degree, a smooth passage.

The skills which he has retrieved need to be tried out in practice as much as further developed; his general system of functioning is still, if not unsettled, at least most sensitive. It is up to the community then to extend to him a helping hand, to receive him with a benevolent acceptance. It is now up to the community to pass a test. It is up to the community to respond to his thrust into its full existence by offering him the possibility of assuming community functions and roles according to his present capacities. To bring this benevolent acceptance into practice three points should be taken into consideration. First, study and work possibilities at a more relaxed pace than the usual intense and competitive tempo should be opened to him. Second, part-time occupations should be available. Third, considering the need of initiation into a profession, which might be already possible, there should be introduced into the regular working modalities, the modality of "apprenticeship."

Indeed, our person is in this phase equipped with the basic skills for life and with potentialities to adapt them and himself to the societal requirements, but he lacks skillfulness, dexterity, precision, accuracy, as well as self-assurance, perseverance, endurance. These all have to develop "on the job." These are not deficiencies but weaknesses and must be accepted as such by the employer or teacher. They must be met with patience, indulgence where appropriate, sympathetic encouragement, and human concerns, that is, with the truly communal spirit. Without such an interchange between the thrust and the response no communal rehabilitation of the person, which would give him an opportunity to lead a human life, can take place.

(iv) *Fourth Phase*

Life-long support can be provided by a "co-operative housing system": As is assumed in contemporary psychiatry, psychic or mental deficiencies cannot be "definitely" overcome; an excessive sensibility accompanies most of them and calls for a slower than usual pace of life and a protected, quiet environment. Furthermore, in our most complex societal life the individual needs a considerable acumen and persistent attention in "taking care of himself." The slackening of this attention at any time will in the case under discussion lead to losing control over one's situation in life. This implies that the already rehabilitated person has to have some guarantee of a life to some degree protected during the *entire* life-period in order to maintain himself in a regular condition. Taking into consideration the various aspects of life which fall into the category of such protective conditions, it would be desirable to provide a basic guarantee of a supportive situation by establishing "cooperative living housing."

Analogical to the already existing and state-operated apartments, the cooperative housing would not only give the necessary stability to the person by being privately owned, but with its ownership would go a partaking in a basic "household community" providing a regular basic maintenance of the apartment privately owned (e.g., cleaning, furnishing, etc.) as well as arrangements for a basic societal involvement of co-operative living (e.g., regular meals in common, socials, etc.). Lastly, there should be the provision of stable and reliable socio-communal therapeutic counselling and help when needed.

All these may now in the present situation of things appear as a farfetched dream. However, so much has been already accomplished by medicine in the biologico-physical sector as well as in communal care for the underdeveloped, handicaped and elderly, that we may reasonably hope for the expansion of care for human beings with nervous weaknesses and hypersensibilities. Hitherto there has been a neglectful and shrunken approach to them in the socio-communal realm. It is the community and society at large that would benefit first from such enhanced care.

The Awakening of the Communal Spirit in the Two-Way Transactional Traffic at the Moral Level between the Community and its More Sensitive Members. As pointed out above, the support of the community is the indispensable condition for the ultimate re-identification of the convalescing person. Having reached the "life apprenticeship" phase, the person enjoys now the "freedom from" his inhibition and inner compulsion indispensable for redressing his life pattern in the resumption of a function and a role within the life-world. He has still to gain the "freedom for" accomplishing it. That is, he has to find appropriate opportunities and the benevolent attitude of his fellow men to encourage him. The need and call for support is shifting now from the artificially prepared setting to the natural community groups.

There are active groups in every community gathering families and single people around a special social aim or ideal: religious, humanitarian, socio-political, sporting, entertainment, etc. Among them the most active are religious groups gathering people belonging to the same religious institution — Christian parishes, humanitarian circles, etc. — bringing together people around the ideal of *human service* (e.g., Rotary) as well as those with a socio-political bent for people with strong concerns with societal issues (e.g., League of Women Voters) and operating basically on a communal level. They all manifest a vivid concern with the welfare of the citizen and a considerable personal generosity of heart. Some of them (e.g., Catholic and Protestant parishes, Jewish, Buddhist religious communities, League of Women Voters) manifested in previous times a direct concern with the welfare of the community which is their base (e.g., the "parish counselling" which has been available for all needing it).

It seems that such humanitarian groupings of the communal life offer the best opportunity for the re-integrative identification of the person. Such groups may mobilize people who will be sensitive to the needs of the person and in their professional establishments offer "apprenticeship" re-integration, with all the understanding which is required. Other groups welcoming participation in their own organization may open a full-fledged communal involvement on the personal and social circuit of interest.

By doing so, the communal groups revive within themselves the communal spirit which seems to have atrophied within our Occidental culture.

TO CONCLUDE:
TOWARD THE REVIVAL OF THE COMMUNAL SPIRIT

When we observe more closely the reasons for this and other initial rejections, it appears that our present-day attitudes in Occidental communal life completely forget the values and modes of "conviviality" with others which can take innumerable forms of mutual sharing in joys and sorrows, and lead to "carrying each other's burden of life" from which no-one is exempt, in reciprocal trust, for the sake of its opposite — a competitive spirit of outdoing others, at going ahead and leaving them ruthlessly behind. These latter attitudes make of life a ruthless struggle; they make the life of the spirit within the community weaken and atrophy. The benevolent attitude toward the Other which is the healthy root of the communal life is extinguished. In order that the person may retrieve in full the sense of his own existence, the community needs to revive its own *sense of the human significance of life.*

The socio-communal guidelines for psychiatric therapeutics along the lines of the moral sense, concern as much the individual who needs to encounter a humane attitude in the Others as the Others who might benefit from this opportunity to unfold their latent capacities for under-standing what is different from themselves and to accept a full-fledged scope of human life to which belong naturally a variety of types of expression and sensibilities, of rhythms of work, efficiencies, degrees of accomplishments, etc. In a truly human socio-communal life all these types should be welcome, finding there both their proper function, instead of being subjected to the law of discrimination, and their right to survive as human, which is now reserved for the most coordinated and one-sidedly striving individual alone.

By reviving its primogenital facet — the benevolent sentiment of the moral sense — which is at the roots of societal/communal living (that is, at the roots of man's "humanity"), Occidental societies may be capable of assuming responsibility and of meeting their obligations toward the living individual to help him fulfill his right to life.

The World Institute for Advanced
Phenomenological Research and Learning

NOTES

[1] Cf. Anna-Teresa Tymieniecka, *Leibniz' Cosmological Synthesis* (Assen: Royal Van Gorcum 1971).
[2] Cf. Anna-Teresa Tymieniecka, "The initial spontaneity: The pessimism-optimism controversy concerning the human condition," in *The Crisis of Culture*, ANALECTA HUSSERLIANA, Vol. V (Dordrecht/Boston: D. Reidel 1976). In this volume we have turned away from the historical approach which from Spengler's *Der Untergang des Abendlandes*, through Toynbee, Husserl, and Weber, has emphasized the transforma-

tions within Occidental culture, to focus upon the disintegration of the human being himself, as manifest in literature, the arts, and social life, etc.

[3] In my early book, *Phenomenology and Science in Contemporary European Thought* (New York: Farrar, Strauss and Giraux 1960), I have shown how phenomenology, stemming from Husserl, Heidegger, Ingarden and Jaspers, has deeply transformed the approach to the human being in numerous realms of scholarly pursuits as well as in the human sciences (sociology, anthropology, psychiatry, etc.), by an attempt to rescue the specifically human element in human nature from the scientific reductionism to which it had been subjected.

[4] Edmund Husserl, *The Crisis of the European Sciences and Transcendental Phenomenology*, trans. David Carr (Evanston, Illinois: Northwestern University Press 1970). Although the conception of "science" has in the last decades greatly changed from the one which Husserl inherited from the nineteenth century, still the depth of his philosophical insights makes this treatise a classic to which reference must be made in any serious philosophical consideration of our present-day "crisis."

[5] I have attempted to show the "nature" of life in a succinct study: "The praise of life: Metaphysics of the human condition and of life," in *Phenomenology Information Bulletin, A review of phenomenological ideas and trends*, Issue 6 (Belmont, MA: World Phenomenological Institute 1983). Cf. also: "Initial spontaneity and self-individualization of life," ANALECTA HUSSERLIANA, Vol. XVII.

[6] Two sections of this study, pp. 29–41, have appeared in my introductory study to: *Foundations of Morality, Human Rights, and the Human Sciences*, ANALECTA HUSSERLIANA, Vol. XV (Dordrecht/Boston: D. Reidel 1983).

[7] The term "vital" is used here to denote the fully developed phase of life and of its conditions.

[8] Cf. the section on Karl Jaspers and Gabriel Marcel, as well as the entire part on phenomenological psychiatry, in my book cited above in note 3.

[9] By 'political' is meant here, in the Greek tradition, the specific feature of man's social nature consisting of constructing a "polis," a state orchestration of social life.

[10] Cf. Calvin S. Hall and Gardner Lindzey, *Theories of Personality* 3d ed. (New York: John Wiley 1978).

[11] Cf. Ludwig Binswanger, *Grundformen und Erkenntnis menschlichen Daseins* (Zurich: Nichans 1953). Ludwig Binswanger, the celebrated Swiss psychiatrist inspired by Husserl and Heidegger, has developed a psychiatric conception of the human being within the "life-world" in which, in contrast to Freudian views, no priority is attributed to a unique driving force within man. Rather, man's entire experiential system as expanded by his interactions with other men within the world becomes the pattern with reference to which psychiatric methods are devised. Binswanger has found numerous followers in several branches of what has become known as "phenomenological psychiatry."

[12] In recent times, a highly developed conception of the specifically human person has come from the famous French psychiatrist Henri Ey. Ey and his school have brought to a culminating point the phenomenological tendency of Binswanger, Buytendyk, E. Minkowski, E. Straus and many others, in vindication of the belief that psychiatric diagnosis and therapy should deal with human nature as a whole. In his famous book, *Consciousness, a Phenomenological Study of Being Conscious and becoming Conscious,*

translated by John H. Flodstrom (Bloomington: Indiana University Press 1978), Henri Ey presents, in its full expanse, the source and the experiential compass of the uniquely human conscious self as a person. Ey, and following him, Lanteri Laura, devise methods of psychiatric diagnosis with reference to the "disintegration of consciousness," i.e., of the person.

[13] It is Max Scheler who has stressed particularly the significance of acting in the understanding of the human person. While emphasizing the autonomy of the person and the differentiation of the person from the individual, Scheler has highlighted particularly the social participation of the "intimate person," endowing it with a specific form of the "social person." Cf. Max Scheler, *Gesamtwerke*, Bd. 2, *Der Formalismus in der Ethik und die Materiale Wertethik* (Bern: Franke Verlag 1966), and by the same author, Bd. 8, *Erkenntnis und Arbeit* (Bern: Franke Verlag 1960).

[14] In this perspective the "autonomy" versus the "conditioning" of the person is the center of the controversy.

[15] Although Max Scheler attempts an inversion of this approach by seeking to show the origin of values in emotions, nevertheless values ultimately emerge in an already constituted form; this could not have occurred without the work of the intellect. Cf. the present writer's monograph, "The moral sense at the foundations of the social world," in *Foundations of Morality, Human Rights and the Human Sciences*, cited in note 6.

[16] By 'sense' is meant here that which "infuses" the linguistic forms with significance so that they may present meanings (it should not be understood in relation to senses, sensory, sensuous etc. referring to sensory organs).

[17] By "substantial persistence" is meant here the way in which the living being "appears" to our senses as a cogent, self-reposing, stable and perduring factor of life and to our actions as a responsive and autonomous partner.

[18] Cf. Anna-Teresa Tymieniecka, "Die phänomenologische Selbstbesinnung, Der Leib and die Transzendentalität in der gegenwärtigen phänomenologischen und psychiatrischen Forschung," ANALECTA HUSSERLIANA, Vol. I (Dordrecht/Boston: D. Reidel 1971).

[19] The distinction between our body as an object and our body as experienced stems from Edmund Husserl. Cf. his *Ideas Pertaining to a Pure Phenomenology and a Phenomenological Philosophy*, Book II (Den Haag: Martinus Nijhoff 1982), Part I.

[20] For Husserl's most careful and masterful analysis of the relationship between our body as experienced and the psyche (or empirical soul), cf. *Ibid.*, Part I.

[21] I have attempted an investigation of the soul in its "essential manifestation" in the third part of my book, *The Three Movements of the Soul*, to appear in Vol. XXII of ANALECTA HUSSERLIANA.

[22] Regarding the concept of the 'human condition,' cf. the present author's monograph, *Poetica Nova . . . a Treatise in the Metaphysics of the Human Condition and of Art*, ANALECTA HUSSERLIANA, Vol. XII (Dordrecht/Boston: D. Reidel 1982).

[23] Cf. Anna-Teresa Tymieniecka, "The initial spontaneity," ANALECTA HUSSERLIANA, Vol. V (Dordrecht/Boston: D. Reidel 1976).

[24] In the above study, I have also indicated that from the very beginning of the individual unfolding of the human being, moral "virtualities" are present.

[25] I have succinctly analyzed this progress, terming it the forging of the "Transnatural Destiny of the Soul" in several of my writings. Cf. "Hope and the present instant," in S.

Matczak, ed., *God in Contemporary Thought* (New York/Louvain: Learned Publications/Nauvelearts 1977).
[26] This progress of spiritual unfolding with reference to the Other, the other self — the "inward witness" — has been the subject of my attention in *The Three Movements of the Soul,* cited in note 21.
[27] The specific "mechanism" of this quest after the ultimate significance of human existence, as conducted with respect to another self, has been shown by the present writer in "Man the creator and his threefold telos," ANALECTA HUSSERLIANA, Vol. IX (Dordrecht/Boston: D. Reidel Publ. Co. 1979). It appears that whoever the Other is, the concrete "encounter" with him or her *in* the significance of life is ever-elusive, because the Other functions merely as a concrete reference point, while it is the "inward witness" that is being consulted by the soul in its innermost depths. I am showing in this study how this comes to light when every supposed "communication" between the soul and the other self necessarily breaks down. With this break, however, the soul is ready for the face-to-face meeting with the Ultimate Witness, who "has been there hidden in the intimate center of the soul all along" (as Teresa of Avila shows also).
[28] Concerning "man's self-interpretation in existence" cf. Anna-Teresa Tymieniecka, "The creative self and the other in man's self-interpretation in existence," ANALECTA HUSSERLIANA, Vol. VI (Dordrecht/Boston: D. Reidel Publ. Co. 1977).
[29] The following analyses have appeared in my monograph, "The moral sense at the foundation of the social world," cited in note 6.
[30] The term "source-experience" — in contradistinction to the classic phenomenological term "originary experience" — has been introduced by the present author precisely to pinpoint the crucial moment within the unfolding of individualizing life at which from the animal action/reaction agency a transition occurs to the specifically human experience simultaneously originating the human subject.
[31] I have outlined the life-progress from the pre-life conditions accomplished by means of the individualisation of the living being in "Natural spontaneity in the translating continuity of beingness," in *The Phenomenology of Man and of the Human Condition: Individualisation of Nature and the Human Being,* ANALECTA HUSSERLIANA, Vol. XIV (Dordrecht/Boston: D. Reidel 1982); and in "Spontaneity, individualisation and life," in *Phenomenology of Life: A Dialogue between Chinese and Occidental Philosophy,* ANALECTA HUSSERLIANA, Vol. XVII, 1983.
[32] The indispensable role of man's inventive/creative function within the virtualities of the Human Condition as the vehicle of the specifically human "orchestration" of all man's faculties, has been stressed by me in a series of published works. Beginning with my book *Eros et Logos* (Louvain: Nauwelearts 1972), I have been developing the phenomenology of creativity in a series of writings. Cf. "Imaginatio creatrix," ANALECTA HUSSERLIANA, Vol. III; "The prototype of action: Ethical or creative?" *ibid.,* Vol. VII; and "Man the creator and his threefold telos," *ibid.,* Vol. IX.
[33] Although I did arrive at the present views independently, they are, nevertheless, in agreement substantially with those of the British moralists of the seventeenth century. Cf. Anthony, Earl of Shaftesbury, "The Moralists," in *The Characteristics of Man.* Vol. 2; and *An Inquiry Concerning Virtue and Merit,* Vol. 2.
[34] Francis Hutcheson asserts that the moral sense is a direct intuition of the Good. That is, he considers the moral sense as the source of the recognition of good and evil in an

immediate perceptive evidence. Cf. Francis Hutcheson, *A System of Moral Philosophy*, Vol. I (London 1755). Although I basically agree with him, because, as Aristotle (*Magna Moralia*, Lib. 2, c. 10, 6a2, chap. 10) says, in discussing "right reason," unless a man bears within himself the sense to distinguish the good from the bad, there is nothing to be done; yet to say that the moral sense is a direct perception of the good or bad is a misleading shortcut through the complexity of human functioning. Hutcheson's view is, however, understandable: he did not investigate the workings of the moral sense within a full-fleldged analysis of the genesis of moral evaluation, nor within the moral orchestration of the transactional intersubjectivity of man. In the present work, this analysis is supplied and the conception of the "moral sense" is expanded.

[35] It has been for the present writer a marvellous surprise to "discover" at the heart of the phenomenological enterprise (as undertaken by Husserl, Hedwig Conrad-Martius, Edith Stein, Roman Ingarden, Max Scheler, etc.) an underlying remnant of the classic philosophical architectonic design, common to Aristotelian and medieval thought in general, which Immanuel Kant — borrowing the expression from the poet Alexander Pope — has called "the great chain of Being." Cf. Anna-Teresa Tymieniecka, *Beyond Ingarden's Idealism/Realism Controversy with Husserl*, ANALECTA HUSSERLIANA, Vol. IV; and Eugene Kaelin,"Man the creator and the prototype of action," *ibid.*, Vol. XI,, where he is presenting the conception of the "great chain of being" proposed by me. Lately, however, I have found a parallel to the conception of the "great chain of being" (which I have elaborated as having its source in man's creative faculties), in the realm of the life of Nature. I have seen it in the phenomenon of the "Unity-of-Everything-there-is-alive." Cf. my essay, "The praise of life: Metaphysics of the human condition and of life," cited in note 5.

[36] For an investigation of the nowadays so popular concept of "human dignity," cf. the present writer's monograph, "The moral sense at the foundations of the social world," Part II: Human rights and Moral Sense, cited in note 15.

[37] I have introduced the notion of 'transaction' as the knot gathering, in a most significant way, the vital functional threads coming from the bio/psychic circuits on the one hand, and from the psychic/moral movements on the other, in the above cited study "Moral Sense . . .". This extremely important notion comes here into its own.

[38] Cf. Ludwig Binswanger, "Über die daseinsanalytische Forschungsrichtung in der Psychiatrie," in *Ausgewählte Vorträge und Aufsätze*, Bd. 1, pp. 190—217, Bern: A. Franke 1947.

[39] "Magna Charta of Clinical Psychiatry".

[40] Ludwig Binswanger, "Freud's Auffassung des Menschen im Lichte der Anthropologie," *ibid.*, pp. 159—190.

[41] Henri Ey, *La destructuration de la conscience*, Paris.

[42] *Annales de Psychiatrie.*

[43] Cf. by the present writer, "The cosmos and the foundations of psychiatry," in *Heidegger and the Path of Thinking*, Duquesne, Duquesne University Press.

[44] Kazimierz Dąbrowski, *Mental Growth through Positive Disintegration*, London: Gryf Publications 1970.

[45] Kazimierz Dąbrowski, *La Psychoneurosis no es Una Enfermedad, Neurosis y Psychoneurisis Desde el Punto de Vista de la Desintegracion Positiva*, Lima: Ediciones UNIFE 1983, p. 12.

[46] Eugène Minkowski, *Le temp vécu,* Paris 1933.

[47] Wolfgang Blankenburg, "Prolegomena to a Psychopathology of Freedom", in *Changing Reality of Modern Man,* ed. Dreyer Kruger, Cape Town -Wetton-Johannesburg: Juta 1984, p. 175.

[48] *Ibid.,* p. 175.

[49] *Ibid.,* p. 176.

[50] *Ibid.,* p. 183.

[51] This failure is to be observed not only in the North American effort at de-institutionalization but also in the Italian implementation of the Psichiatria Democratica. Cf. "The Transformation of Psychiatric Care in Italy: Methodological Premises, Current Status, and Future Prospects", by Paolo Crepet and Agostino Pirella, in the above-cited collection, pp. 162—166.

[52] This practical project owes its stimulus and inspiration to Miss Elisabeth Parkhurst, a Baptist missionary who has devoted a great deal of attention and energy to investigating the way in which the currently available institutions operate. She has inspired me with her innovative insights.

[53] I am aware of the socio-political trend in psychiatry introduced in Italy as "Psichiatria Democratica" by Franco and Franca Basaglia; cf. "The Unfinished Revolution in Italian Psychiatry: An International Perspective," Richard F. Mollica (ed.), *International Journal of Mental Health* **14**: 1—2 (1985). In one of our research seminars, Dr. Wolfgang Jacob of the Heidelberg Medical School gave us a careful account of the methods used in German psychiatry, e.g. at the Rehabilitation Clinic at Heidelberg etc. It seems, however, that a systematic approach to the rehabilitation issue as the aim of psychiatry in which appropriate therapies and communal efforts would be conjointly devised is still missing. Yet these are crucial for the purpose of defining the meaning of rehabilitation and implementing it. It is the aim of the "Pilot Project" outlined below to supply this need.

RICHARD F. MOLLICA

PSYCHIATRY IN QUEST AFTER ORIENTATION

I. PSYCHIATRY NEEDS A CENTER OF ORIENTATION

A. *A Call for Defining the Aim and Direction of Psychiatry*

The traditional gap between scholarly studies of facts and principles on the one hand, and the practical solution of human problems concerning these facts, on the other hand, has in certain fields been considerably narrowed. The mathematical and empirical sciences have been able to directly apply their theories through universal techniques to the needs of human life. The classic discrepancy between theory and practice seems to have been overcome by the latter. Psychiatry, however, remains ambivalent about its ability to integrate psychiatric theory and practice. For example, in spite of enormous advances in the scientific study of human behavior and mental disorders over the past twenty-five years, public mental health services are in disarray.[1] In addition, a proliferation of competing theoretical and treatment programs has emerged. Estimates indicate up to 200 or more "brand names" of psychotherapies.[2] In 1984, an editorial in a leading medical journal, *Lancet,* challenged the psychotherapies to demonstrate their effectiveness.[3] The psychiatric field's difficulties in extending its scientific advances to its practical domain partially lie in its dual nature. Psychiatry simultaneously belongs to the empirical study of mental illness and human behavior as well as to the humanistic study and treatment of the human mind and emotional suffering. In spite of psychiatry's striving to achieve scientific universality in its practical usefulness, it still remains in reality much closer in its applications to art than to technique.[4] [This fact generates considerable criticism and controversy.] The humanistic domain of psychiatry brings the scholar and the scientist into a confrontation between lived human experience and theoretical conjecture about it. In fact, reality keeps the scientific imagination in check. For example, empirical approaches provide inadequate solutions to the suffering of the homeless patients living on our country's streets. Like any technical application of science to the practical needs of human life, psychiatry deals directly with the

101

A-T. Tymieniecka (ed.), Analecta Husserliana, Vol. XX, 101–124.
© 1986 *D. Reidel Publishing Company.*

concrete: in concurrence with the concrete human person. However, while the technical application of science deals with the application of universal principles, psychiatry in radical contrast deals more often than not with unique concrete situations.[5] For example, although depressive illness might exhibit core depressive symptoms, each individual experiences his or her sadness within a unique and highly personal social and psychological context. However, each situation is a unique and unrepeatable occasion for the practitioner to apply whatever the scientific theories may support. Groping for the appropriate attitude toward the psychiatric patient, the practitioner has to have clearly in mind an aim and a direction toward which to direct his treatment.

How does psychiatry at large define its aim? The traditional answer was clear: "to heal the human psyche." The term psychiatry, itself, is derived from the Greek words *psyche*, mind, and *iatreia*, medical treatment. Psychiatry is, therefore, the medical treatment of the human mind (i.e. psyche) with all its attendant emotions, cognitive structures and behavioral manifestations. The question which has plagued the above response has been, "what is the place and role of the human psyche within its circumambient conditions, i.e., the organic body, on the one side, and the societal world, on the other?" It has been apparent that for a psychiatrist to turn his attention solely to the "inner" state of psychic functioning does not suffice. The individual is embedded in an "outer" sphere of an entire societal network of interdependencies and interpersonal relationships. The scientific orientation in reaction to the organic-psychosocial dichotomy has attempted to narrow its investigations by emphasizing the former while the psychological and sociological orientations increasingly expand their realms of supposed expertise. The human psyche — the target of our medical care — is, thereby, lost between too narrow or too broad conceptual schemes that have so far generated limited practical results.

With these comments, we first highlight the still existing lack of orientation in the prevailing trends of psychiatry today. Moreover, we point to the source of this essential deficiency, namely to the persisting discrepancy between theory and practice. Second, we emphasize the need for a clear formulation of the aim, task and direction which psychiatry as a scholarly and scientific medical profession must have in order to be useful and efficient. This postulate can be satisfied only when we find philosophically the central point from which the individual in his/her network of social interdependencies is living and developing. Approaches and therapies abound but they remain disconnected segments forgetting that they should deal with man in his entirety.

B. *Psychiatry Carries the Discrepancy Between Theory and Practice in Its Two Dimensions: Individual Therapeutics and Its Social Condition*

Indeed, psychiatry as a medical specialty which seeks its foundation and principles in scientific research and discovery is involved in a twofold dilemma in its relationship to theory and practice. First, there is the discrepancy between the universal principles proposed by the scientific/ scholarly theories and the concrete practice of psychiatry, on the one hand, and that between the psychotherapies dealing with the higher mental acts and the psychological nature of these acts, on the other hand.

In spite of extensive investigations in the fields of psychology, anthropology, somatic medicine and pharmacology, the treatment of the human psyche remains haphazard. The borderline between organic functioning and the higher mental contents of intellectual acts, as Binswanger puts it — i.e., the interconnectedness between the pharmacological and psychotherapeutic treatment — remains an inscrutable dividing line. Certain physiological malfunctioning and symptoms can be kept in check by means of pharmacological methods; these methods may prepare the ground for the treatment of the psyche. However, the passage from the physiological to the psyche and the inverse, remains a mystery.

Second, as we discover in the conception of the socially inspired psychiatry proposed by Franco Basaglia,[6] there exists the dilemma between the possibility of practicing the art of healing and the availability of mental health services, between what society offers and the actual needs of the psychiatric patient. Since the human individual is dependent upon the social network within which he exists, the psychiatric help which he may or may not receive is dependent upon the network of social services functioning within the given society.

This social help depends upon social ideals, values and regulations which shape the social life and command the way in which the modalities and the distribution of psychiatric help are available. The person as a member of the society is, in this respect, evaluated in a certain way. His/her need for psychiatric care remains dependent upon the attitudes and criteria which the societal values command. Although these attitudes stem from universal principles concerning general conceptions of the society, they still automatically affect the concrete individual. Many types of patients remain defenseless. The more so since they are ignorant of the ultimate reasons for which their call for help is unheeded. In principle and in theory, they keep all the rights of the citizen, whereas in concrete fact, they may be deprived of concrete possibilities to avail themselves of these rights.

Every psychiatric theory carries implicitly or explicitly a set of philosophical assumptions about the human being and his place and role within nature. As is the case with the conception of "life at large" of Freud, so it is with every psychiatric theory. We have already presented the intrinsic need for the determination of the aim and direction of psychiatry and indicated the reason for the lack of it in the discrepancy between theory and practice; a discrepancy which could be brought down to the philosophical assumptions of the main psychiatric approaches. The first part of our paper will engage in a succinct discussion of the philosophical underpinnings of two major psychiatric theories — that of Ludwig Binswanger and Franco Basaglia. Then, in our quest after the philosophical link that would unify the classic separation between bios-psyche-spirit, on the one hand, and bios-person-society on the other, the recent philosophical proposal of Anna-Teresa Tymieniecka will be introduced. Our sociopsychiatric survey of these three major conceptual frameworks will substantiate our common claim for a radical reorientation of the psychiatric profession and of the societal approach to it.

II. PSYCHIATRY IN THE QUEST AFTER THE UNIFIED TERRITORY OF THE "COMPLETE HUMAN BEING": LUDWIG BINSWANGER AND HIS VINDICATION OF "HUMAN EXISTENCE" WITHIN THE "LIFE-WORLD"

As pointed out above, issues involved in the gap between theory and practice are essentially of a philosophical nature. Science itself is compartmentalized into different fields of investigation according to its specific subject matter which entails corresponding methods of research. The subjacent interrelatedness of the fields calls for a special investigation of their respective "natures;" it calls ultimately for principles according to which each of them taken by science separately may be placed within the entire picture of reality, nature and of the human universe. The investigation of such principles of interrelatedness and of the unifying links is what we call "philosophy." Each of the scientific theories is based upon some — most often inexplicit and unclear — philosophical assumptions concerning the nature of its field, as well as of its role within the entire picture. Psychiatry has made a decisive step in clarifying these assumptions directly by interpreting them in terms of some comprehensive philosophical theories. Ludwig Binswanger, for example, with explicit reference to the philosophical anthropology of Edmund Husserl and Martin Heidegger proposed a phenomenological existential theory as the basis for psychiatric diagnosis and practice.[7]

Binswanger, a student and for a long time a follower of Freud, parted with his master on three major points. In the main, he protests against the Freudian conception of the human beings as "*homo natura*," which reduces the entire cultural and social life to biologically oriented life forces, or "drives."[8] Binswanger calls for the acknowledgment of the full-fledged human nature. In Freud's biological view of life, which Binswanger considers to be a myth, the emphasis in the individualization of the human being as soul and psyche is placed upon the dominant role of drives both biological and psychic (*seelisch*). The individual human being is a complex chaos of needs, passions, affects and drives, not a personal agent but a self-unconscious "it" (*es*). As "*homo natura*," man — as the "it" — is a play of driving forces, drives which bring in vital energies and find their manifestation in the psychic life. There is no organizing principle and the "it" lives from moment-to-moment in the satisfaction of the vital needs. It knows no valuation or morality of good and evil.[9] Consequently, in the progress of life, there is no transformation of its forms, but the differences signify merely various ways in which the drives manifest themselves. The "it" remains unqualified and impersonal.[10]

Rooted in this conception of life, Freud's scientific model of the psychobiological mechanism cannot do justice either to the cultural and spiritual existence of the human being or to his unfolding and growth. It is precisely in this unfolding that the positive or pathological transformations of the human psyche occur. For Freud, however, there are no significant transformations, but merely different forms which the elementary drives assume in their dynamic manifestation.

Although Binswanger recognizes with great care the enormous significance of Freud's penetrating analysis of the human bio-psychic mechanisms for psychiatry, he finds it an unjustifiable reduction and simplification of the complete man. In contrast, Binswanger proposes to do justice to man's totality following three new ideas:

(1) the "personal inner biography" (*innere Lebensgeschichte*) contrasted with the "life-functioning" (*Lebensfunktion*) of the human being by means of which instead of the neutral and mythical "it," we are confronted with the concrete human person;
(2) the "*historicity*" of the human psyche;
(3) and lastly, the fundamental ontological structure of the human being as that of "*being-in-the-world*." Within this complex, the idea of the "life-world" as a network of interactive meaning in which the human being exercises his existence while simultaneously experiencing his own personal life acquires a crucial significance.[11]

Pointing out that the bios/soul founded acts of the higher mental/emotional order do not explain the contents of these acts which remain *sui generis* and autonomous from the life forces carrying them, Binswanger distinguishes the "inner biography" (*innere Lebensgeschichte*) of the human person from the functioning of the empirical acts (*Lebensfunktion*). While in the latter, the process of life is being accomplished, the former represents the progress of the personal growth of man in its unique specificity. Universalizing the views of Bonhoffer, Binswanger emphasizes ". . . the difference between the way of the bodily-soul (*körperlich-seelische*) functioning of the organism and its disturbances, on the one hand, and the sequence of the contents of the psychic (*seelisch*) experiences, on the other hand."[12] The states of depression, hallucination, etc., manifest the disturbances of the soul-body functioning, whereas the desire to be ill (*der Wunsch krank zu sein*) is, in contrast, never an expression of the functional disturbances, but the "intentional or spiritual content of a psychic (*seelisch*) experience.[12]

The inner biography which unfolds within the progress of life's functioning, shapes the meaningfulness of the contents by the personal devices of the autonomous human self. There it is that the specificity of the human being as a spiritual being, specificity which cannot be understood in terms of empirical acts, and thus reduced to impersonal drives which remain in the service of the organic progress, resides.

In the capacity to unfold such an autonomous level of life, which Binswanger identifies with a uniquely personal "human existence," is seen the capacity of man to "transcend" (*übersteigen*) the organic, animal life. With the concept of the inner biography, the stress falls upon the historicity of human nature. Human beings distinguish themselves from the animals precisely because they are capable of transcending the automatism of organic functioning with its restricted, merely life-promoting, significance toward the self-devised meaning-fulness of inner experience. Following Höberlin, Binswanger sees the turning point in the "attitude," (*Einstellung*) that the human person assumes toward its life and its inner and outer occurrences in the reaction to which it spins its personal history.[13]

The inner biography brings into view the human person advancing in its progress, undergoing inner transformation and thereby projecting/ establishing itself in its own "history." These transformations relate to the networks of affective-cultural interactions within which human life unfolds and expands the narrow biological realm of nature into that of the specifically human "life-world." The human being is "historical" because his very being is unfolding in the essential interplay with other humans (*Mitmenschen*) within the networks of a shared life.

Here we have the breakthrough to the acknowledgment of the complete human being, which in the basic form of "human existence" or "man's being-in the world" (*Dasein*) constitutes the third crucial point of Binswanger's innovation of psychiatry. Writes Binswanger: "It is without further ado clear that the purely clinical psychiatry . . . is incapable of understanding real human problems, like those of religion or philoso-phy, morality or fine arts, history or education, of genius or of freedom. However, it encounters these problems while turning back to its native soil (*Mutterboden*), that is to the structure of human beingness (*Menschseins*) as such."[14]

Referring to Heraclitus' statement that man in sleep is self-enclosed and abandoned to himself but he awakens to a life with other human beings, within a world of life common to all, the life-world, which he essentially shares with others in his very structure of being a human, Binswanger emphasizes that this world of life is not a physical set of vital conditions, but a meaningful construct which man spins himself while he unfolds his existence. Thereby, Binswanger brings in the basic conception of the *human being as the being-in-the-world*. On this basic structure of being a man which consists in transcending the animal "circumambient world" (V. Uexkuhl) to the emergence of one's inter-worldly existence, Binswanger quotes Heidegger: "Human existence (*Dasein*) transcends means: in the essence of its being the human existence is world constitutive (*weltbildend*) . . .".[15]

The human mode of being, insists Binswanger, going as far as Heraclitus for the elucidation of its roots, but accepting the contemporary systematization of it by Heidegger, resides in man's-being-in-the-world. The meaningful self-constitution of the personal being with respect to the simultaneous constitution within the life-world and of this life-world itself, means that it is with respect to other human beings that we constitute ourselves within our life-world. It is "the basic structure of being human" as being-in-the-world at large, "precisely with-and-for-one-another."[16]

Man, as a being-in-the-world, shapes his historical personality with respect to others. First, the shape of his personal lived world emerges with respect to the intimately personal relatedness of love/hate; second, with reference to the socio-cultural circuits. Evolving within shaping interchange with others, the historical person undergoes transformations, growth or shrinkage, expansion of life effective skills while following the "norm," or their diminishing effects while deviating from the "norm."

Indeed, Binswanger makes these philosophical insights crucially significant for psychiatry when he introduces the spatio-temporal axis of the being-in-the-world of the human individual into psychiatry with reference to the basic interworldly spatio-temporal studies proposed as norms for diagnosis. On the one hand, psychopathological symptoms amount always to some transformation of the patient's basic in-the-world-structure.[17] On the other hand, neither psychology nor psychiatry deal with a "subject" graspable in separation from the "objective" of the world; they confront the "human existence," which underlies such a distinction.

These three innovative concepts transform Freudian theory. They hang intimately together enlarging the notion of the human being in his entire spectrum of existence and manifestation. On the one hand, the life-world existence of the human being manifests its fullness as the person and as a cultural and spiritual being, which the concept of *homo natura* shrunk to the biological realm. On the other hand, the inner biography with its counter-part, the concept of the specifically human factor of life, "existence," switches the emphasis from the role of the drives to that of spiritual forces at work in the process of transcending the narrow automatism of life toward its emergence as human existence (*Dasein*). The abstract mechanical construct of Freud is thus replaced by the concrete fullness of the person situated in its natural network of life-interdependencies.

The shaping of the human person in its historical unfolding marks a spatio-temporal progress in following the norms of the basic spatio-temporal pattern, distorting it or deviating from it. Furthermore, while Freud's "It" conceives the human being as encapsuled and unenmeshed with other persons while knowing neither cultural values nor morality, the conception of the personal fullness of the human being vindicated by Binswanger is based in its unfolding precisely among and with other human beings (*Mitmenschen*).[18]

Turning the philosophical views into a "science of man" (*Wissenschaft vom Menschen*), which may be called "existential anthropology," while maintaining the scientific core of the Freudian mechanical model for psychiatry, Binswanger has caused the narrow limits of the latter to explode. His "existential anthropology" vindicates on the one hand the validity of links within the life-world which maintain the human existence of the individual man; on the other hand, it brings forth the crucial significance of the relationships of the human person with other persons for its psychic development. These two pioneering innovations transform the psychiatric approach to man. The spatio-temporal patterns of the life-worldly existence become norms for psychiatric diagnosis. However, Binswanger's psychiatric theory did not reach beyond the recognition of the symptoms to an appropriate therapy. In his studies devoted to psychotherapy in which he shows the application of the existential anthropological insights to the diagnosis of the concrete patient, Binswanger reveals the still persisting unbridgeable gap between the descriptive recognition of the peculiar situation of the patient and the recognition of the therapeutic means to be used in order to alleviate it. He indicates that the doctor who has to bring help — a healing help — at the crucial zone where body and psyche motivate each other has no technique, no method, no scientific principles. He has for his compass only his inventive imagination.[19] He is like an artist whose skills have to be led by intuition and imagination because both, the artist and the psychotherapist are each time breaking new seemingly unfathomable ground.

Scientific theory dealing with the organism and psychopharmacology may attain a degree of adequacy when applied to concrete cases. However, the "enlightment of existence," that is the clarification of the psychic distortions with reference to the inner biography, may, if carried out favorably bring the psyche out of the entanglements of evaluations and emotions. However, this approach does not carry its work below the borderline of psyche and body where the psychic distortions connect with bodily malfunctioning. In short, the application of

Binswanger's theories to man and the immersion of the person within the life-world within which the bios, psyche and other people are intimately interacting remained at the strictly descriptive stage. The basic structure of man-in-the-world does not reveal the knot operative in these threefold interdependencies. That is, the "mechanism" or the making of this knot in tying up the networks of these interdependencies remained out of sight. Without bringing to light this knot, the philosophical-anthropological theory remains at the static level and cannot meet the demands of concrete therapeutic practice. First, because the latter calls for an "operative" principle. Second, because following the therapy of enlightening the inner biography of the person we remain cut-off from the specific corresponding "outer-biography" of the same person in its actual unfolding within the system of the bios taken in its largest sense as well as within the system of interaction with other persons in the societal circuit of life. Without discovering the principles of these two types of linkage, the hiatus between psychiatric theory and practice is not bridged: it only acquires new form as the gap between diagnosis and therapy.

III. SOCIO-POLITICAL APPROACH TO PSYCHIATRY: FRANCO BASAGLIA'S PSICHIATRIA DEMOCRATICA AS AN ATTEMPT AT A REFORM OF PSYCHIATRY FROM THE SOCIETAL POINT OF VIEW

While opening up the human person through his life-world analysis, Binswanger sought to expand beyond the restrictions of Freud's mechanistic biological circle. Two of his followers in Italy, Franco and Franca Basaglia, taking the latter for granted, focused upon the last consequences of the expansion: the socio-political organization of human life. In 1973, Franco Basaglia pointed out in his famous essay, "Che cos' è la psichiatria?" (What is psychiatry?),[20] Binswanger's warnings of the dangers inherent in the scientific method that, "in moving away from us, proceeds to theoretical conception, observation, examination and dismembering of the real person with a view toward scientifically reconstructing a scientific image of him." The Italian reformers under Franco Basaglia's leadership, proceeded to develop a new methodology toward approaching the mentally ill. In the above essay, Basaglia stated its intention, "so the psychiatrist has to bracket the illness, the diagnosis and the syndrome with which the patient has been labeled. . . . Since the patient has been destroyed more by what the illness has been held to be and by the 'protective measures' imposed by such an interpretation than by the illness itself".

~ The new Italian psychiatry that emerged in the 1960's, which generated the most radical mental health reform seen anywhere in either Europe or the USA, could not comprehend the failure of psychiatry's scientific method to improve the life and social conditions of the asylum patient. Again, Franco Basaglia states:

And so we have, on the one hand, a science ideologically committed to a quest for the origins of an illness it acknowledges to be "incomprehensible" and, on the other hand, a patient who, because of his presumed "incomprehensibility," has been oppressed, mortified, and destroyed by an asylum system that, instead of serving him in its protective role of therapeutic institution, has, on the contrary, contributed to the gradual and often irreversible disintegration of his identity.[21]

The Italian reformers denied the neutral, apolitical role assumed by empirical psychiatry. Although they did not deny the existence of neurologically based illness, they emphasized that any illness had to be considered within a social, political and interpersonal context.

Unfortunately, the socio-political and interpersonal context for the asylum patient in the late 1950s in America and Europe was extremely grim. In 1958, Hollingshead and Redlich elucidated the influence of social characteristics on the type and quality of care assigned patients.[22] For example, the lower the social class of the patient, the greater the use of organic treatments as opposed to psychotherapy, the fewer the number of clinic visits and the shorter the length of the therapy hour. Lower class patients were also found to receive therapists with less training and experience. The lower class patient − organic treatment/ custodial care − inexperienced therapist triad appeared to be a fixed societal relationship as members of Redlich's team re-demonstrated over 25 years later in spite of America's psychiatric reform movement.[23] In contrast to Hollingshead and Redlich, Basaglia and his colleagues offered a socio-political theory that could explain the above realities.[24] *Emargination* was the term used by the Italian psychiatrists to designate the social process by which mental patients were isolated and segregated from society. The Italian reformers believed that the asylums represented the successful management of the displaced poor and served to punish and segregate individuals who were considered socially deviant. *Chronification* was considered by the Italians to be the "second disease" caused by the social exclusion, neglect and mistreatment of patients through institutionalization. Although the Italian reform movement did not deny the neurological or psychological basis of mental illness, they neglected its empirical categories and elected only to treat the "second disease."

As Mollica[25] has revealed, although Basaglia never recognized Gramsci, Basaglia was strongly influenced by the latter's socio-political theory. In 1926, Gramsci who was Italy's leading political theorist and revolutionary thinker, was arrested by Mussolini and subsequently sentenced to a fascist prison where he died in 1937. Following Gramsci, Basaglia also rejected a Marxist theory founded solely on economic conditions. Using Gramsci's concept of hegemony, i.e., that society is a community which accepts patterns of behavior, institutional directions and cultural ideas proposed to large groups of populations by the dominant group — Basaglia rejected those historically based practices in Italian society that repressed the mentally ill. Referring to the "spontaneous consent to accept new ideas," which according to Gramsci lies in the hands of the population, the Italian reformers proposed a social reform that would attack society's hegemonic practices at all levels — cultural values, institutions, social organizations, the educational system, etc.

In their analysis of the relationship between the theory and practice of psychiatry, Basaglia and his colleagues revealed in detail how societal ideals while simultaneously creating positive as well as prejudicial social attitudes, have led to a societal situation in which psychiatry instead of liberating the patient from his inner and outer constraints forces him into repressive institutional structures, such as the asylum. They have brought to light how psychiatry through societal directives actually forces patients into repressive situations in which they remain helpless. Having stressed the specifically social entanglements of the human individual in his life struggle, they have denounced contemporary psychiatry as a socially repressive practice, which hinders the psychologically weaker individual from taking part in regular life. With this emphasis upon the societal ideals and its consequences for the psychiatric patient, they have shifted away from the classic antinomy between the bios versus spirit to that of the individual versus society.

Since it is not a blind vital necessity which in their view is operative in the modalities of social life, but the influence of intellectuals, thinkers and scholars that brings intellectual and moral reforms about, not bloody revolutions, but the intellectual "war of positions," they proposed to Italian society a new set of cultural ideals according to which the repressive system of psychiatry would be dissolved and the patients integrated into regular life. Under the name of *"Psichiatria Democratica,"* a political reform was introduced which was expected to bring about a radical solution to the practice of psychiatry.

However, although the societal ideals represented by the reform have found social acceptance, for example, in the 1978 Public Law No. 180 which closed the asylums, the practical psychiatric results of the dissolution of asylum constraints has varied considerably in different Italian regions and has generated considerable controversy. When we ask why such a drastic turnabout in the recognition of the societal needs of the chronically ill person did not bring about dramatic improvement in the care of the mentally ill, the answer is twofold. In the first place, we may state that the dissolution of the asylum system has not led to the introduction of an appropriately significant system of social support which the psychologically weaker individual might need. In the second place, and foremostly, the old gap between theory and practice reappeared albeit in a new form. Namely, within the reformers' sociopolitical perspective which certainly contributes to our understanding of the human being, the essential link between the societal system of human interaction governed by ideals, ideas and principles, on the one hand, and the concrete human person whose entire being is involved, on the other hand, did not come to light. The societal perspective ignores not only the anthropological-foundational issues addressed by Binswanger, but also those biological realities which underlie the indispensable continuity between clinical treatment and societal integration. The concept of societal reform as proposed by *Psichiatria Democratica* fails to bridge this gap.

Nevertheless, the innovative approach of the Italian psychiatry has considerable merit. It has revealed several points hitherto hidden or ignored of the repressive character of the prevalent psychiatric practice. It has also clarified their prejudicial foundation; just to name the main ones:

1. Several principles upon which the social practice of psychiatry as a system is repressive of individual depriving them of the opportunities to straighten out their lives were revealed;

2. Contradictions between social ideals and actual institutional practices in psychiatry were named;

3. Lastly, those aspects of the socially inspired deformation of the rehabilitative tasks confronting the patient were clarified.

The repressive social structure impairing the psychiatric patient in his rightful effort to normalize his life had been reduced to the following principles:

(1) the dominant role of the empirical assumptions is still orienting the psychiatric institutional approach;
(2) the automatic emargination of the psychiatric patient, that is putting him to the side of the regular current of life;
(3) devaluation of the human capacities of the patient on account of his assumed incurable deficiencies, i.e., his classification as a "chronically ill" person;
(4) social exclusion of the patient, neglect and mistreatment in surroundings that lack due respect for his personal and human dignity which create what the Italian's call "the second disease;"
(5) the "institutionalization" of the staff in the asylum, i.e., the staff assumes the attitude or role of deciding, domineering and paternalizing caretakers instead of exercising their duty in the spirit of help to be given to the patient toward his return to normal existence.

In this way, the theory-practice doctrine of the Italian psychiatric movement not only revealed the striking socio-cultural facts about the situation of the psychiatric practice within the dominating social setting, but also pinpointed its prejudicial leading ideas which are unwarranted by actual facts or distort their meaning. Furthermore, it has brought to light the manipulative strategies of social institutions which keep the mental patient outside of social life, strategies which are grounded in unclarified prejudices stemming from the industrial spirit which gives absolute priority to efficiency, usefulness, self-interest and comfort. Such is the spirit of the contemporary ruthless drive to get ahead of everyone, that whoever is not able to stand up to the competition is abandoned as a hindrance or nuisance. To avoid any claim put by their presence upon the conscience of the society of which they are rightful citizens, they have to be put "out of sight" as an irretrievable commodity, a dead loss by definition.

Psychiatria Democratica aims by radical contrast to make us aware of the moral and legal rights of the psychiatric patient on the one hand, and of the responsibility of society to fulfill them, on the other hand. However, in view of its limited ability to provide the point of junction between societal existence and personal interaction, between clinical treatment and societal integration, it is clear that the final solution to the problems of psychiatry is still to be worked out.

IV. INTRODUCING THE "MORAL SENSE" INTO PSYCHIATRY:
A PHILOSOPHICAL PROGNOSIS FOR THE REORIENTATION
OF AIMS AND METHODS OF PSYCHIATRY PROPOSED BY
ANNA-TERESA TYMIENIECKA

In her recent work, which proposes a "socio-communal integration of the psychiatric patient," Tymieniecka addresses the difficulties posed by the conundrum of issues under discussion.[26] She proposes an "Archimedean point" from which to remap the territory of the scientific and contemporary controversies: the "Creative Condition" of man. The conception of the "moral significance of life," which it entails as pervading the inner workings of existence is the innovative point of this program. This conception is based upon three central ideas:

(1) in a parallel to Freud, it starts out from a basic insight concerning the nature of life; however, avoiding any unwarranted generalizations, it approaches life through its "*individualization*";
(2) in a parallel to Binswanger, the center of attention is the entirety of man; this entirety is circumscribed, however, by the exercise of man's "*creative function*;"
(3) lastly, in a parallel to Basaglia, the specificity of the social network of the life-world is particularly emphasized; its problems are, nevertheless, modified with reference to what Tymieniecka calls the "Moral Sense," which is introduced as motivating simultaneously the origin of both, the human person and the society.[27]

The crucial point, however, in which the practical usefulness of this proposal seems to consist and upon which its theoretical justification appears suspended, lies in the phenomenological conception of the "*human transaction.*" Meant as the prototype of all human relations and operative at the very roots of humanness as the basic sociability of human nature, the theoretical description of its "inner workings" is presented as having practical applications.

In short, by means of what is called "*transactional analysis*," she intends to show the natural passage from the theoretical description of the human person "on the brink of existence" to its metamorphosis into a psychiatric patient, and so demonstrate its concrete practical application.

In order to see how the thorny questions related to the classic antinomies may be resolved by her approach, we have to give a cursory view of the main tenets of the philosophical-phenomenological framework within which the mentally disturbed person is identified as the person "on the brink of existence." To begin with, her identification of life must be pointed out, i.e., with its individualizing differentiation from within life's forces, energies, haphazard strivings, pulsations, drives, etc., relevant to the Freud-Binswanger controversy about nature-spirit. Tymieniecka identifies life with its unfolding progress, this progress with the introduction of order, and order with meaningfulness. The progress of the self-individualizing life is seen as providing, in fact, a *filum Ariadne* along which the complete spread of active forces, energies, strivings, drives, etc., which promote its unfolding, may be appraised in their constructive contribution. They are not placed in a position, but appear in a continuous flow. This latter, unlike a haphazard striving, appears through its constructive outline of ever increasing complexity. The individualizing progress introduces and unfolds life while channeling the cosmic and the so-called pre-life forces into a constructive becoming.[28]

By means of this outline the interplay of operations carrying on the individual process with the circumambient conditions out of which the individual wrings out his identity can be grasped. In addition, the various phases of the growth-transformations which occur in the operational functioning can be distinguished following its evolutionary line of progress. The continuity of the individualizing progress of life not only integrates all forces and virtualities at play without sharp divisions that the objectifying mind introduces while considering them separately. It reveals also the emergence in its advance of new operational factors which introduce transformations of modalities and forms.

Thus the progress of the individualizing life offers the basis for appreciating in concrete the "regular" or "typical" progress in the form, or modalities of transformation and deviation from the "norm," as well as other differences (e.g. shrinkage of some forms or disappearance of some modalities, etc.).[29]

When we now ask after the foremost regulative principle of the operational system which differentiates the individualizing life from its available means, thus introducing networks of existential interdependencies, a principle such that it may simultaneously carry on the inner/outer oriented traffic of the process of growth and account for its progressing complexity, the answer is: *valuation*. The criteria of valuation change with the release of new promptings toward further steps of transformation in growth. In fact, the constructive outline of individualization is punctuated by new types of valuative criteria emergent from within and directed outward. At all phases of the individualizing life preceding the emergence of the unique and unprecedented phase of the Human Condition, the valuative criteria pertain to the relevance of the available means "outside" to the needs of the intrinsic factors of growth of the individual while tying through the inner/outer oriented functioning of the individualizing life the "external" with the "inner" workings; the valuating principle lays down the knots for this network of interdependencies. By the same stroke, a universal ordering of life is emerging in which all the cosmic and life-forces concur. While they acquire a measure of meaningfulness with respect to their role in the origination and growth of life, the individualizing "entity" acquires simultaneously its selfhood. The selfhood is established with reference to the circumambient life-system it projects in doing so. The two, that is the individualized life and the circumambient life-system are existentially convertible; one expresses the other. Together they are elements of this vast network of life's interdependencies and cannot be legitimately considered in any form of separation.[30]

These primeval ties of life cannot be severed without annihilation of life itself and have to be kept in mind if we want to find our way to the intricacies of the situation of man within his own life-world. Thus, an indissociable continuity of elements entering into the life-system of a living being is emphasized; bios and psyche stay within the continuity of life's outline as indissociable.

But what about man? Indeed, at a certain point of the evolutive progress (seen here as an ontological progress of becoming, not to be confounded with the scientific theories of evolution), the prompting virtualities of the constructive advance emerge in an unprecedented fashion and the self-individualizing progress, as well as the ordering of life, takes a drastic new turn. We find here the pivot of life's accomplishments: Man's *creative condition*.[31]

As mentioned above, these emergent novel virtualities and promptings appear endowed with original tendencies: they carry on the inventive/creative criteria for valuation, and mark the advent of the Human Condition. With the new valuative apparatus which the Human Condition carries and with the special creative faculties of consciousness which it establishes in its progress, a new "significance of life" is being devised.

Tymieniecka brings forth the conception of "man — the creator" as basic for characterizing man. She conceives of his creative role as introducing into the meaningfulness of life (projected by the individ-

ualizing progress) a uniquely creative-inventive significance. Thereby
Tymieniecka believes she has overcome the shortcomings of Binswanger's existential anthropology. Showing the progress of life's
interpretation as spread through several phase of constructive complexity, she believes she has dissolved the nature-spirit opposition.
Moreover, the human being stands out from among other types of
individualized life by the unique way of directing his valuative self-
individualization (*self-interpretation-in-existence*). First, in a radical
contrast to the automatized, as Binswanger would concede, selectiveness of the preceding phases of growth, which follow stiff valuative
criteria, but not breaking from them, on the contrary rising from within
them, the human being invests new and unprecedented selectivities.

Second, the creative function of man, in the release of which the
specifically human functional system of the meaning bestowing conscious apparatus comes about, is the instrument by means of which the
living individual immersed into natural functioning invents as its unique
type of fruition his very own universe, together with his very own
beingness, his interpretation-in-existence. As in Heidegger-Binswanger,
it is in the human being interpreting his existential network as being-in-
the-world that Tymieniecka sees the brute force of nature. However, in a
radical contrast to the existential-anthropological view, it is not by this
structure of the human being as such that man transcends nature; the
static structure of human existence is cut out of life. This transcending as
well as this structure remain in a secondary position; they are due to the
creative function of the Human Condition which takes precedence. The
life-world of the living individual consists, as indicated above, of several
circuits of meaningfulness. They are all existentially interdependent.
And yet, it is only with the emergence of the creative/inventive circuit
that the life-world becomes a specifically human world and conversely,
the individual becomes a human person. Within the creative functional
system of the human being, Tymieniecka proposes to find access to the
hidden springs from whence the factors which account for the ties of
the network of the entirety of the life-interdependencies proceed.
Tymieniecka emphasizes in her monograph how her position dissolves
the *homo natura* — spirit opposition.[32] The opposition between the
objectivized quality of forces, acts, energies as being either biological or
spiritual dissolves into an abstraction when we shift from the question of
the origin of the gene, from the quest after the source, to that of the aims

and progress of the functioning of the constructive progress which they serve as well as enact. Taking a resolute step beyond the Binswanger-Heideggerian conception of the ontic substructure of the human individual, *Dasein*, life-world, Tymieniecka rejecting its claim to be the final answer, sees that answer lying instead in the functional system of the Human Condition as the pivotal phase of the constructive individualization of life.

In short, man, that is the human person, is defined as basically the meaning-bestowing functional system. While the vital and gregarious meaningful spheres of the life-world originate in the life-subservient functional modalities, the social and cultural spheres of the life-world, albeit in congenital continuity, take off into an imaginatively "free" sphere on the wings of man's creative function.

Lastly, concerning the societal issues raised by Basaglia, Tymieniecka's system develops in depth the middle ground between the life-world condition of Binswanger and the societal world shaped, but unexplored by *Psichiatria Democratica*. There it is that human transaction and the moral sense make a direct link to the specific issues raised by psychiatry.

In agreement with existential anthropology, Tymieniecka puts forward man's essential being-in-the-world. However, in her framework, the structure of man's inter-worldly beingness extends within the all-comprising field of the phenomenology of life over the existential interdependencies on both sides. These sides include the biological and somato-psychic functional networks, on the one hand, as well as the psychic-social and cultural interdependencies on the other. There is no sharp limit between them. The life-world of man extends over both.

The specificity of the human being is to be stated but as the turning point of life-progress: the emergence of the Human Condition. But there is a specific meaning-bestowing circuit unique among vital strivings, namely the one within which the creative function of man transforms the life-promoting functioning — self-interest oriented gregariousness — into the "other-man oriented" functioning of the specifically "human transaction." This is the emergence in the Human Condition of the "Moral Sense," as Tymieniecka calls it, as one of the factors of the creative function of man that marks the turning point in life's progress: the emergence of the specifically human significance of life.

The organic/vital, vital/sentient, and sentient/psychic relevances of means to needs direct the evaluative operations of the individualizing life-progress (subservient to the progress of life) introducing a meaningful order — which corresponds to V. Uexkuhl's "circumambient world" (*Umwelt*) of the living individual — and is circumscribed by the stretch of his vital interests; the latter is shared by a group but in a "togetherness" which is just gregarious. The emergence of the creative function putting at work the moral sense marks the advent of the Human condition within the vital progress of self-individualizing life; a new turn which this self-individualization takes on.[33]

In fact, according to Tymieniecka, it is this unique factor which introduces into evaluation, so far directed merely by the self-interest of the individual, the "*benevolent sentiment*" in which the interest of the other is being put forward, if not ahead of one's own, at least on par with it. The emergence of the moral sense as the criterion of valuation in the relations between individuals allows the living being to enter into a specifically human inter-relation, which is basic to any human contact as a human: "*transaction*." We arrive here at the heart of her entire argument.[34]

The human transaction orients living individuals one toward the other. They both become "human" searching beyond passive acceptance of the unifying life-situation toward sharing and weighing mutual interests. Thus, the significance of what we consider as the social world and of the living being originates simultaneously. The transaction is the link between man and "fellow-man" which "links" the steps of their specific growth and transformation within the societal life-world. It pinpoints the radical significance of human interaction for human unfolding on the one hand, and of the communal/societal role which the individual has to assume and perform in order to unfold within the network of these interactions — as well as to maintain himself in existence — as the human person, on the other. Let us stress here Tymieniecka's point that it is in human transaction that the living being establishes step by step his human sphere, that is his personal selfhood in the reaching-out-response with others.

Human transaction draws upon the individual as well as communal modalities of valuation operative in the communal setting in which it takes place. Human personality unfolds its specific features through the role it assumes within concrete communal ties with others. As it becomes invigorated in its growth by the expansion of these ties, so it weakens and shrinks in their loss. With their loss the personality beyond a point falls out of communal inter-relations while it simultaneously disintegrates in its specifically human functioning.

In this philosophical-anthropological analysis of life the quest for diagnosis and therapy of the mentally disturbed person is naturally tied up with the crucial role which "human transaction" plays in the growth and transformations of the human person. Identifying the situation of the mentally disturbed person with the person's finding itself "on the brink" of regular human life-affairs and demonstrating that the recurrence of crises in the so-called chronic condition of the psychiatric patient amounts to the shrinkage of his faculties with respect to his life-involvement with others, Tymieniecka emphasizes the role of the human "transaction" as the discrete thread to be picked-up and followed-up; first, in pursuit of the inner biography toward a diagnosis; second, as a thread toward a meaningful therapy.

The linkage between the regular unfolding of the human person within its societal inter-relations and the members of the community with whom these relations are tied is revealed as the moral linkage. This moral linkage of the underlying benevolent sentiment of man to man in the vigor and equipoise of the maxim, "everyone his due share," is kept alive by the way in which social attitudes and practices treat every member of the society. The human life worthy of a man depends upon it but also the life and quality of society. Along these lines, Tymieniecka proposes a "Pilot Project for the socio-communal re-integration of the psychiatric patient." The main aim of psychiatry (theory and practice) being now defined as the re-integration of the patient into the societal life-world, the principles for a chain-like therapeutic orientation are grounded in the philosophical-anthropological conception of the individual sketched above.

Some of her ideas for a "Pilot Project" include these:

1. The functional continuity of individualized life does not allow for a sharp division between "inner directed" clinical therapy of the psychiatric patient (whether pharmacological, that is, concerned primarily with the organic functioning, or psychotherapy, as concerned chiefly with the psyche) and the "outer directed" socio-communal therapy dealing chiefly with the retrieval of social capacities and skills. The gap or antinomy between the treatment of the psyche and of its bodily manifestations (Binswanger), as well as that between the clinical treatment within the asylum and the therapy still needed but missing (Basaglia) is here naturally dissolved. In her life-functional perspective an inner/

outer oriented therapy to be applied from the start is devised to the contrary in a continuous line consisting of several discrete but intermotivated steps.

2. Therapeutics means change. The same line of orientation and the same goal are to be kept in mind.

3. She stresses the continuing functional line of personal expansion. The so-called "bodily functions" are, in fact, an indispensable conductor to the retrieval of the life-worldly functioning, inasmuch as the impulse or desire to practice them — which is one of the direct aims of inner directed therapies is originary as much from the awakening of the moral senses as the direct moral/orientation of the outer-oriented psychic functioning leading toward the societal involvement and its practices.[35]

Yet, the basic point Tymieniecka emphasizes is the moral readiness of people within communities to allow — and help — the psychiatrically weaker person to share in the communal life: their birthright. With the factor of the moral sense as accounting for human ties and for the societal circuit of the life-world, Tymieniecka offers this as a way of integrating the natural modalities of life on the one hand, and the specifically human social ones on the other. With this special instrument, human transaction, Tymieniecka believes psychiatric theory and practice become two faces of the same coin.

V. CONCLUSION

This essay has presented the theoretical framework of three psychiatric orientations aimed at reforming psychiatric practices. All three attempt not only to protect the mentally ill from "reductionisms" — both psychological and biological — but also they advance psychiatric treatment as a project sensitive to the patient's unique human subjectivity. Of course, each approach, no matter how daring, has failed to successfully bridge the gap between psychiatric theory and practice. Binswanger's methodology has not demonstrated capability of extending its reform to the majority of the mentally ill — the so-called "unworthy poor" who suffer from serious mental illness. *Psichiatria Democratica,* although advancing a radical political transformation in Italy, has not produced effective clinical practices. Gramsci's emphasis on intellectuals (including psychiatric practitioners) being *specialista e politico* has been reduced almost entirely to the latter. Finally, Tymieniecka offers a still untested psychiatric phenomenology of rehabilitation, i.e., a sort of phenomenologically based, truly human, integrated community. Her major contribution, however, is to recognize that psychiatry must offer as its primary orientation those therapeutics which return the mentally ill to the world of social relations. As Redlich,[36] Astrachan et al.,[37] Mollica,[38] and others have revealed, rehabilitation is in fact the most ignored of all psychiatric tasks because it confronts directly the essential moral conflicts of our society. At the heart of the theory/practice dilemma lies the inability of psychiatry to accept as its major goal, the return of the mentally ill to their proper place in society.

Department of Psychiatry,
Harvard Medical School

NOTES

[1] Mollica, R. F. "From asylum to community: The threatened disintegration of public psychiatry," *New England Journal of Medicine* **308**: 367—373, 1983.
[2] Parloff, M. B. "Shopping for the right therapy," *Saturday Review* **3**: 14—20, 1976.
[3] Editorial. "Psychotherapy: Effective treatment or expensive placebo?," *Lancet* **14**: 83—84, 1984.
[4] Toulmin, S. (ed). "Mental health," *The Journal of Medicine and Philosophy* **2**: 1977.
[5] Toulmin, S. "On the nature of the physician's understanding," *The Journal of Medicine and Philosophy* **1**: 32—50, 1977.
[6] Mollica, R. F. (ed). "The unfinished revolution in Italian psychiatry: An international perspective," *International Journal of Mental Health* **14**: 1985.
[7] Tymieniecka, A-T. *Phenomenology and Science in Contemporary European Thought.* New York, Farrar, Strauss and Giroux, 1961.
[8] Binswanger, L. "Lebensfunktion und innere Lebensgeschichte in Ausgewählte Vorträge and Aufsätze," Band I Zur phänomenologischen Anthropologie. Bern, A. Franke A. G. Verlag, pp. 52—53, 1947.
[9] Binswanger, L. The Study cited above, p. 72, quotes Höberlin's *Der Charakter,* Basel, p. 413, 1925.
[10] Binswanger, L. "Freud's Auffassung des Menschen im Lichte der Anthropologie," p. 170 in the above cited collection.
[11] Cf. Tymieniecka's book cited above, pp. 48—81, 118—164.
[12] Binswanger, L. "Lebensfunktion und die innere Lebensgeschichte," in the collection quoted above, p. 55.
[13] *Ibid.,* p. 72.
[14] Binswanger, L. "Heraklits Auffassung des Menschen," in the above collection, p. 98.
[15] Quoted by Binswanger in "Ueber die daseinsanalytische Forschungsrichtung in der Psychiatrie," p. 194; Martin Heidegger, "Vom Wesen des Grundes," *Husserl-Festschrift,* p. 97, 1929.
[16] Binswanger, L. "Ueber Psychotherapie," p. 134.

[17] Cf. Binswanger, L. "Ueber die daseinsanalytische Forschungsrichtung in der Psychiatrie," pp. 202–216.

[18] Binswanger, L. "Heraklits Auffassung des Menschen," p. 103.

[19] Writes Binswanger, "It is (inspiration) the first condition of each psychotherapeutic, as of each medical, yes, of each art in general." Binswanger, L., "Ueber Psychotherapie," p. 137.

[20] Basaglia, F. "What is psychiatry?" *International Journal of Mental Health* **14**: 42–51, 1985.

[21] *Ibid.*

[22] Hollingshead, A. B. and Redlich, F. C. *Social Class and Mental Illness* New York, J. Wiley and Sons, 1958.

[23] Mollica, R. F. and Milic, M. "Social class and psychiatric practice: A revision of the Hollingshead and Redlich model," *American Journal of Psychiatry*, **143**:12–17, 1986.

[24] Mollica, R. F. "From Antonio Gramsci to Franco Basaglia: The theory and practice of Italian psychiatric reform." *International Journal of Mental Health* **14**: 21–41, 1985.

[25] *Ibid.*

[26] Tymieniecka, A-T. "The Moral Sense and the Human Person within the Fabric of Communal Life," a monograph in *Analecta Husserliana, the Yearbook of Phenomenological Research* XX: 3–100, 1986.

[27] This monograph continues her study, "Moral Sense in the Foundations of the Social World," *Analecta Husserliana* XV, 1983.

[28] The conception of the "individualizing life" was introduced by Tymieniecka in her earlier work, *Why is There Something Rather than Nothing, Prologema to the Phenomenology of Cosmic Creation*, (Assen, Royal Van Gorcum, 1972) and corroborated further in numerous writings some of which have appeared in *Analecta Husserliana*.

[29] Cf. the above-cited "Moral Sense in the Foundations of the Social World"; also "The Moral Sense and the Human Person" herein: 29–44.

[30] Tymieniecka, A-T. "Die phänomenologische Selbstbesinnung: Der Leib und die Transzendentalität in der gegenwärtigen phänomenologischen und psychiatrischen Forschung," *Analecta Husserliana* I: 1–10, 1971; also by the same author, "Cosmos, Nature and Man and the Foundations of Psychiatry," in *Heidegger and the Path of Thinking*. Pittsburgh, Duquesne U. Press, 1970; and "Initial Spontaneity and the Translacing Continuity of Beingness," *Analecta Husserliana* XIV: 132–151, 1983.

[31] Cf. "The Moral Sense and the Human Person within the Fabric of Communal Life," 48–54.

[32] *Ibid.*, 29–42.

[33] *Ibid.*, 45–48.

[34] *Ibid.*, 80–93.

[35] *Ibid.*, 40.

[36] Redlich, F. C. "Medical rehabilitation and psychiatric rehabilitation," *Psychiatric Annals* **13**: 564–511, 1983.

[37] Astrachan, B. M., Levinson, D. J. and Adler, D. A. "The impact of national health insurance on the tasks and practices of psychiatry," *Archives of General Psychiatry* **33**: 785–794, 1976.

[38] Mollica, R. F. "Resisting reform: Acknowledging the effects of good psychiatry/bad psychiatry," *International Journal of Mental Health* **14**: 1–8, 1985.

JOHN R. SCUDDER, JR. AND ANNE H. BISHOP

THE MORAL SENSE AND HEALTH CARE

Medical ethics has become a major concern of philosophical ethicists, theorists and practitioners in the health sciences, and the general public. Spectacular cases such as Baby Fay, Karen Ann Quinlan, and Barney Clark have drawn much attention to moral problems in medical practice. Generally, attention has been focused on those cases which appear to result from recent advances in medical science and technology. To these problems ethicists have brought their expertise in moral analysis, making moral judgments, and applying traditional ethical norms to moral problems. A new area of ethical study has arisen — biomedical ethics. But this welcomed interest in medical ethics often rests on three questionable assumptions. (1) The practice of medicine itself is an adjunct to science and technology. (2) Ethical issues arise from the advances in medical science and technology. (3) These ethical problems are best resolved by applying the expertise of traditional philosophical ethics to these problems.

The biomedical ethics approach tends to ignore that health care itself is a moral enterprise in that it aims at the physical and psychological well-being of the ill. As such, the moral issues in health care are found within that care itself as it is practiced in the day-to-day relationships of physicians, nurses, patients and others. Put differently, medical ethical problems arise out of the intersubjective relationships of those involved in health care and are present even in the most mundane cases. These relationships are focal to all moral issues in health care and the advances in medical science and technology merely magnify the moral issues inherently present in health care itself. Since the moral norm of health care practice is inherent in that practice (promoting the physical and psychological well-being of the ill), medical ethics focuses on how medical practice can best fulfill the good it is intended to accomplish.

Such an approach to medical ethics requires recovering the moral sense present in medical practice. For the authors, this recovery was initiated by hearing Anna-Teresa Tymieniecka deliver a paper on the moral sense and the human sciences.[1] This paper spoke forcefully to the paper which we read at the same conference concerning medical science

125

A-T. Tymieniecka (ed.), Analecta Husserliana, Vol. XX, 125–158.
© 1986 by D. Reidel Publishing Company.

as a human science.[2] In addition, it provided a context for a more adequate understanding of the implications of research we had previously done concerning fulfillment in nursing. Further study of Tymieniecka's treatment of the moral sense also generated other insights into the meaning of health care.

We will begin this paper with a brief treatment of Tymieniecka's moral sense, especially as it relates to the human sciences. Then we will argue that medical science is a human science in which the moral sense is embedded in medical practice. Further, we will show that ethical issues occur when this moral sense is inhibited by self-interest, by conflicts between the medically correct treatment and moral good of the patient, and by tensions resulting from conflicting intentionalities in health care practice. After focusing on the physician and medical practice, we will consider the moral sense and nursing practice. We will concentrate on how fulfillment comes from the convergence of the sense of nursing practice with the moral sense. Finally, we will explore the moral sense from the patient's point of view, an area which has been very much neglected. In general, we will show that Tymieniecka's treatment of the moral sense makes possible a more adequate understanding of health care practice which combines the professional sense with the moral sense, thus providing a much sounder treatment of medical ethics than the bioethical approach.

I. THE MORAL SENSE

The moral sense, according to Anna-Teresa Tymieniecka, is the highest and final sense which makes man unique. She contends that the moral sense appears to complete man's self-interpretive development which begins with the life processes and continues through a conscious meaningful relationship to the world. When, in the self-interpretive existence of man, self-interest is understood intersubjectively, requiring mutual agreement and consensus, the moral sense appears. However, it does not evolve out of this stage, but rather appears as a specific irreducible new factor. This new factor — the spontaneous benevolent sentiment — has "been 'lying there in waiting' as the virtual element for the *specifically human meaningfulness of the life progress* and of the life-world."[3]

The moral sense is not to be identified with rationalistic ethics which attempts to deal with this phase of human existence through "neutral deliberative analysis" based on moral principle.[4] Tymieniecka argues

that the moral sense of right and wrong must come before moral judgments. Indeed, the benevolent sentiment of the moral sense endows
neutral moral judgments with "justifying reason." Nor is the moral sense
to be confused with ethical emotivism which understands moral evaluation as expression of feeling. Instead, she contends that "morality surges
into the interpretation's subjective realm as an *interpersonal affair.*"[5]

The moral valuation is a process within the intersubjective context — with an
interworldly extension — and yet operated by the innermost core of the individual's
conjoined meaning-bestowing faculties (intellectual, aesthetic, moral). As such, it is an
intersubjective and interworldly instance by which the human being basically established
himself within the life-context, his own *social* realm. Indeed, the moral valuation introduces a "moral point of view" into the interpretation of life-events, actions, and their
interrelations. This moral point of view consists precisely in approaching an interpretation of the life-significance of personal feelings, qualities of judgment and actions, from
the stand-point of benevolence. In objectifying intellectually the valuative contexts,
circumstances and their conflicting moments on the one hand, and, on the other, conditions for applying further the benevolent spontaneity while it pervades the tendencies
and propensities of the human being which enter into the valuative and selecting
process, we derive moral "principles," "standards" of conduct, and "rules" of behavior.
These constitute an entire apparatus for meaningful interpretation, which is necessary
for thinking and judging in moral categories. Hence the progressive intellectual extension of the spontaneous moral sense into a directive power of the social life and world.
Its principal interpretative propulsion is: Everyone ought to valuate and act "benevolently."[6]

Benevolence is not an emotion even when emotion is directed toward
the other because emotions are "not co-determined by the 'for.'" For
example, if we apply this to health care, the person giving health care
actually cares for someone who is ill. This does not mean merely that
they have an inner emotion called care directed at another person. In
health care, caring is concretely caring for the ill. Such care is benevolence in the sense that it places the good of the other over our own but
also allows us to understand "the 'good' of the Other," as "our own
good, or the Good in general."[7]

The benevolent sentiment has "a directive from its 'center,' a 'center'
of qualitative orientation 'which resides in its sense.'" This directive,
while not giving an explicit sense of what benevolence (for the sake of
the other) concretely requires, bears it "germinally as a proficiency to be
crystallized in the experiential exercise." This exercise reveals a universal principle: "the good of the Other."[8]

A person becomes a moral subject either by exercising the moral
sense as "benevolence oriented for the 'good' of the other . . . or by

malefic deviation from the promptings of benevolence."[9] A person's moral being is formed by good or evil actions. A person first turns to the good or evil mode and next is molded by it. Acting morally is accompanied by enjoyment — good actions produce positive enjoyment and evil actions, negative enjoyment.

In the case of a morally benevolent sentiment, the human self is penetrated by what in self-reflection he crystallizes into "self-contentment", "self-satisfaction", "self-respect." These moral sentiments generated by the exercise of the moral sense are amplified by releasing the sentiments of "self-reliance," "security," and self-"worthiness" deposited in one's own inaccessible and autonomous, indestructible *moral force*. Furthermore, the benevolent sentiment enables the subject to expand his self-interpretative reach into the life-world circumstances, relations, influence upon others, etc. In contrast, the malefic sentiment which penetrates the functional system with inimical, envious stirrings, hate of the Other, and suspicion, turns the subject from his natural give-and-take prompting into an effort at *rational calculation*. Interpreting the Other by calculation of his motives of interacting, with respect to losses or gains for his life-interests, makes the wealth of his interpretative resources shrink, and this automatically shrinks the network of his morally significant intersubjective life arteries.[10]

Those who choose the malefic orientation are "invariably 'sooner' or 'later' shattered by 'regret' and 'remorse.'" In contrast those who choose the benevolent orientation find self-satisfaction and confirmation in the self-interpretive course. They "choose and keep choosing: in short, for the *meaning of life.*" This self-interpretation is the ultimate source of human happiness. "The moral subject, in benevolent acting finds his own reward! In evil-doing, his own punishment."[11]

Moral valuation is foundational for the human enterprise in that to be human "means an *intersubjective, social* significance of the self-interpretation of man, which brings about the meaningfulness of his own uniquely personal existence." "Self-interpretation according to the moral sense" requires "the outwardly exercised, intersubjective life of morally interpreted actions (of the social world) on the one hand, and the moral molding of the subject, who performs the moral actions (feelings, emotions, experiences, etc.), on the other. When this self-interpretation indicates that one has acted for the good of the other, one experiences moral enjoyment as self-contentment and self-worthiness. Thus, one finds happiness. "Since happiness is considered the fulfillment of life, we may say that the cultivation of benevolence and its exercise in life-engagements means a 'fulfillment' of the human being."[12]

Fulfillment in the moral sense requires more than attending to the prompting of the moral sense. It requires moral valuation in which the conflicting interest of "the benevolent sentiment prevails over the life-tendencies of self-interest." This occurs through the "*valuative-preference mode*" in which we attribute "to each of the conflicting elements its *due*." In attributing each conflicting element its due, it is essential that the appreciating subject expresses "*his own* stand, *his own* perspective" which calls for the mode of appreciative sincerity. In addition, fairness is required in that "the active implementer should follow all the appreciative aspects of the moral judgment with detachment of all circumstantial pressures." In following the moral sense in acting for the good of others one does not usually go through the whole process of moral valuation. Instead one repeats "the modal moral 'attitudes' in repetitive kinds of situations."[13] In short, one acts out of moral conscience.

Acting out of moral conscience has important implications for health science practice. For example, a surgeon, faced with an important moral choice while operating, could hardly repeat the whole process of valuation before acting. All that can be required of him is that he act "in good conscience."

In the practice of benevolence, we are much affected by how our actions toward others are received. The "receptive interpretation of the significance of our deeds by the Other . . . motivates directly our conscience toward a reflectively critical *revaluation* of this significance and beyond." Also positive approval and appreciation of such acts "enhances our moral sentiment." It calls forth emotive power and gives certainty to our moral convictions. Further, appreciation by the other reaffirms the appropriateness of our actions as practical contributions to the well-being of the other. Also, when the moral sense fails to empower us adequately to act benevolently in a given situation, the approbation of the other can give us sufficient power to act rightly. Attunement to the appreciative response of others stimulates "within ourselves the spontaneity of the moral sense" and leads "us to repeat it and thus will mold accordingly our psychic functions and our conscience."[14]

The interactions of moral subjects lead to the institutionalization of moral attitudes in the progressive unfolding of human culture. These attitudes differentiate between benevolent and malefic tendencies and are crystallized into objectified valuative processes which constitute a people's moral values.

Thus, the objective status of a moral value rests neither on being an "autonomous 'entity' functioning as an *a priori* for the moral experience, nor on being a subjective product of the act of moral valuation." A "moral value is 'objective' insofar as . . . it emerges from, and in turn expresses a *universal type of configuration* of the life context." In a stronger sense, it is objective "insofar as the *significance* of the interpretive step advanced by the performance of a moral valuation (and judgment-action) . . . stems . . . from the universally valid moral sense," rather than "from repetitively subjective psychic functioning." The moral sense works through particular subjective valuations so that the moral value "reaches beyond any single, concrete life-context into each and every instance of the same types of transactional situations in which any individual might find himself," and hence has "universal significance."[15]

The moral sense obviously has important implications for the human sciences. The social world on which the human sciences focus "resides in the working of the moral sense." Consequently, any adequate human science must consider the workings of the moral sense in human relations. Of course, the human sciences have another major concern — articulating a particular intersubjective relationship. For example, in psychiatry and psychology this consists of "capturing, describing, diagnosing and devising therapeutic approaches" to mental illness. In so doing, the way in which mental illness deviates from and destroys normal consciousness is articulated. However, it is important to "appreciate the constructive role of the moral sense in the existentially viable (normal) state of the individual's intersocial transactions and the consequences of the extinction of the moral sense in their degeneration."[16]

The exercise of the moral sense directly relates to our ability or inability to project ourselves into a network of intersocial relationships. Involvement in the intersubjective societal network is decisive even for matters concerned with vital functioning. Therefore, "the investigation of the life significance of the moral sense as the foundation of intersubjectivity enters into the deepest concerns of psychology and psychiatry."[17]

Those human sciences which specifically investigate the social world such as anthropology, sociology, social history, etc. all reveal the importance of the moral sense. "The social world, being an intersubjective product . . . would never have emerged and the life-world would have remained an anonymous flux had not the benevolent sense offered us the lever to lift the anonymous life cycle to a societal life order."[18]

Therefore, utilitarian and pragmatic treatments of social life are inadequate because they ignore how the moral sense affects our communal existence.

Finally, investigations of the evolution of culture articulate "the strivings of the human spirit expressing itself in the human (societal) interactions, toward an intersubjective harmony" which is "indispensable for man to accomplish in his individual *self-interpretation in existence*, the significance of his life." This movement of the spirit can be accounted for only by "the sentiment of benevolence, universally shared and institutionally implemented in social forms." In fact, this cultural striving is evident in the development of "such sociocultural institutions as law, government, political agreements, commitments for respecting human rights, social movements for freedom of conscience in religion, and others."[19] Articulation of these cultural strivings which are embedded in all socio-cultural institutions, including health care, is a primary goal of the human sciences.

II. HEALTH CARE, HUMAN SCIENCE AND THE MORAL SENSE

The human sciences, according to Tymieniecka, have two primary functions. One is to articulate the sense of a particular intersubjective relationship and the other is to explore how the moral sense is evident in that relationship. We will show that health science is a human science in both senses. We also will argue that medical practice is not founded on the natural sciences but rather in a caring relationship and that the moral sense is embedded in medical practice at its foundation in the physician-patient relationship. Second, we will explore how the moral sense is embedded in nursing practice and how it is related to personal and professional fulfillment in health care. And finally, we will explore how the moral sense is evident in what patients expect of the professionals who care for them and in how they respond to that care. In our examination of the relationship of physician, nurse and patient, we will not only show how the moral sense is foundational to that relationship but also how the moral sense discloses the meaning of health care.

III. MEDICAL PRACTICE

Philosophers and physicians have recently become very much concerned about medical ethics. The spectacular advances in medical science and technology have led to the creation of biomedical ethics. Ironically, biomedical ethics, while claiming to be a humanistic contribution to medicine, is unwittingly acknowledging that the health sciences are natural sciences by assuming that biomedicine is the foundation of the health sciences and that moral problems result primarily from advance in medical science and technology. Actually, medical care itself has traditionally been rooted in the care of human beings who are ill and need professional help. Recent advances in biomedicine merely contribute to the efficacy and efficiency of this care but do not change the nature of the health care professions. Therefore, medicine is appropriately studied as a human science.

Husserl helps us distinguish between a natural science and a human science. They are alike in that they are both human ways of being in the world and both theoretical ways of being in the world. They are different in that a natural science examines that which is not specifically human and a human science studies that which is specifically human. Husserl thinks the natural sciences are naive in that they do not recognize that science itself is a human activity and that the meaning of scientific theory and explanation rests ultimately on the meanings of the lived world. Scientists who try to treat human beings naturalistically are even more naive in that their approach eliminates that which is specifically human, especially meaning. For instance, Husserl points out that scientists who use the psychophysical approach to psychology fail to see that they must grasp the essence of a human experience, i.e., anxiety, before they can study it physiologically.[20]

Husserl not only objects to naturalistic approaches to the human sciences because they are naive but because they put the cart before the horse. He defines human science as follows:

In these sciences theoretical interest is directed exclusively to human beings as persons, to their personal life and activity, as also correlatively to the concrete results of this activity. To live as a person is to live in a social framework, wherein I and we live together in community and have the community as horizon. . . . Here the word 'live' is not to be taken in a physiological sense but rather as signifying purposeful living, manifesting spiritual creativity — in the broadest sense, creating culture within historical continuity. It is this that forms the theme of various humanistic sciences.[21]

From the fact that the human science should study the specifically human, it does not follow for Husserl that human scientists can ignore what the natural science can disclose about man. He says that the study of human science would be much simpler if "the world were constructed of two, so to speak, equal spheres of reality — nature and spirit — neither with a preferential position methodologically and factually." But this is not the case.

The practitioners of the humanistic science consider not only the spirit as spirit but must also go back to its bodily foundations.[22]

These bodily foundations are understood through the natural sciences. In short, Husserl contends that the human sciences study man as a spiritual being and that the corporeal is included as it affects man as a spiritual being.

Certainly in the health sciences the corporeal is more focal than in the other human sciences. The health sciences are primarily concerned with how the living body affects human well-being. They are concerned with practices which attempt to heal, cure, assist and support the ill. The study of the health sciences focuses on these practices. The health sciences are like education, political science and economics in that they study practices as they take place in the lived world. But it is evident that the corporeal is more focal in the health sciences when compared to other human sciences which focus on practice, like education. In education teachers are primarily concerned with broadening the horizon of meaning in which students understand and act in the world. Students attend school primarily to enhance their meaningful relationship to the world. For example, the student who is taught $4 + 4 = 8$ but does not know the meaning of 4 cannot be said to have learned that arithmetic function. On the other hand, when a sore throat is diagnosed as a strep throat and treated successfully with penicillin, the patient need not know that he had a strep throat or that it was treated with penicillin for the patient to be cured. Our point is certainly not that sound medical practice should not include sharing with the patient the meaning of his condition and his treatment, but simply that healing can occur without this meaning. The fact that cure can be effected without meaning fosters the belief that modern medicine is founded on the natural sciences. Even Husserl, himself, came to the conclusion that folk medicine had been transformed into modern medicine by the natural sciences.

We can illustrate this in terms of the well-known distinction between scientific medicine and 'naturopathy.' Just as in the common life of peoples the latter derives from naive experience and tradition, so scientific medicine results from the utilization of insights belonging to purely theoretical sciences concerned with the human body, primarily anatomy and physiology. These in turn are based on those fundamental sciences that seek a universal explanation of nature as such, physics and chemistry.[23]

Actually, Husserl made this distinction in 1935 which was at the beginning of modern scientific medicine as Edmund Pellegrino points out. The ability to cure in a radical sense, utilizing science, has occurred within the lifetime of physicians who began their practice at about the time Husserl was writing. Pellegrino defines cure as "the eradication of the cause of an illness or disease, to the radical interruption, and reversal of the natural history of a disorder." The spectacular successes in the ability to cure especially since World War II, according to Pellegrino, have inclined many physicans to think of medical practice as curing through scientific means and understanding rather than as the traditional practice of caring for ill patients.[24]

Before World War II, physicians were more engaged in healing than in curing. By healing we simply mean that physicians were able to enhance the functioning of the healing powers of the body. Although now cure is possible in some cases, much of medical practice still concerns healing. Doubtless, in the future, with the advance of medical science, cure in the radical sense will become increasingly a function of medicine. But this will not fundamentally change the practice of medicine since, according to Pellegrino, medical practice "is a special moral enterprise because it is grounded in a special personal relationship — between one who is ill and another who professes to heal. . . . Healing is a mutual act that aims to repair the defects created by the experience of illness."[25]

Medical practice begins with the illness of the patient. Illness concerns a disruption in the lived world of the person, not a pathological description of disease. As Tristram Engelhardt points out, illness is what is experienced by the person who becomes a patient. Disease is the pathological definition of that illness used in diagnosis by a physician.[26] In everyday life people often confuse these two meanings. For example, when a husband was asked about his wife's illness, he replied that he did not know what it was. In fact, he had had the same illness a few days prior and knew well what she was experiencing. What he did not know was the pathological category to which it belonged. Also, he believed that the illness was of short duration and did not require medical attention. He was sure that medical attention was not required because in his case the body had healed itself in a short time without medical treatment. Much illness is still handled without professional help and often with folk medicine.

When the disruption by illness is severe enough that a person needs assistance, the person seeks medical help and becomes a patient. Both illness and being a patient are human experiences which have been well described phenomenologically. Illness primarily affects the lived body. According to Sally Gadow:

The immediacy of that primary being-in-the-world is ruptured by incapacity, the experience of being unable to act as desired or to escape being acted upon in ways that are not desired. Immediacy, in short, is shattered by constraint.[27]

Mary Rawlinson describes illness as a rupture in the man-world relationship.

Illness names that experience in which our own everyday embodied capacities fail us. Illness obstructs our ordinary access to the world and presents the body as a signifier for the way in which we are limited and can be impeded in our encounter with the world.[28]

She contends that illness alters this relationship in four ways. First, our body becomes the center of our concerns. Usually we are not aware of our bodies.

When our embodiment fails, we discover that embodiment ordinarily means reaching for, going toward, attending to what is present, and in short, enjoying the capacity to encounter what is other.[29]

When we become ill — almost in direct proportion to the severity of our illness — the body fills our consciousness.

Whereas our embodied capacities ordinarily provide the background to the figure of our worldly involvements, in illness our body, and particularly that aspect which pains, becomes itself the figure of our intention against which all else is merely background.[30]

Second, illness "confounds our capacity to expect. Our embodiment seems unreliable and unpredictable." This unreliability ranges from temporary disruptions in our everyday activities, to adjusting to a permanent loss of some capacity, to the termination of all possibilities in the near future. Third, illness makes us aware that our concerns are not merely determined by our own choosing. The ill person does not decide to become absorbed with illness and pain. This imposition brings an acute awareness of our own finiteness and of the possible and actual loss of self. And finally, "illness distorts our ordinary relationship with others insofar as it debilitates, humiliates, and isolates."[31] Illness isolates us from others because of our absorption with our own illness and suffering and because others tend to think of the ill as not like me. In addition, illness

results in a surrender of one's autonomy and integrity of person out of necessity or in hope that this surrender will be in the end useful in the effort to recover those capacities which the illness obstructs and threatens. This surrender makes one vulnerable and leaves one at the mercy of others in significant ways."[32]

In summary, Rawlinson states why we disvalue illness:

(1) the obstruction of our capacity to 'possibilize' and take up involvements in the world, (2) the way in which illness disrupts and derails our direction of our own histories, and (3) the extraordinary dependence and lack of self-sufficiency and self-control which illness brings.[33]

A person becomes a patient when illness leads him to seek professional help. As Pellegrino points out, a person experiences "a pain in the chest, the finding of a lump, the loss of appetite, morning nausea, dizziness on bending over . . . which leads us to seek help."[34] According to Pellegrino, seeking professional help is the first element in the patient-physician relationship. The act of profession is the second element in the patient-physician relationship. When the physician accepts the patient, he or she makes two implicit promises, according to Pellegrino. First, the physician is competent and possesses the knowledge which one needs. Second, he or she promises to use that knowledge in the patient's interest. This relationship is therefore based on profession and trust. Thus, it is different from a commercial relationship which is based on mutual self-interest or a legal relationship which is based on contract.

An ill patient becomes extremely vulnerable when he entrusts his life to a physician's professional care. The gap between patient vulnerability and professional knowledge and skill is closed during the act of medicine. This is the third element in the physician-patient relationship, according to Pellegrino. In the act of medicine the physician and patient come to a decision about what to do about the illness — a decision which is technically right and morally good.

Medical science can determine what is physically wrong, what can be done about it and what is likely to be the outcome. But it cannot tell what ought to be done for the good of a particular patient. If you follow that cure model, the biomedical model, then what is medically good, what is medically indicated, what is scientifically correct becomes what is good for the patient. But, reflect for a moment and you'll see the two are not the same. The good decision must also fit this particular person's concept of the good life and of the way he or she wants to live.[35]

Since medical decisions aim ultimately at the good of a particular patient, medical practice cannot be determined or founded primarily on natural science. But, if we follow Husserl's interpretation of a human science, it is obvious that natural science plays a more significant role in medicine than in other human sciences. However, it should be noted that Pellegrino's "phenomenological" description of the physician-patient relationship is not drawn from science but from the articulation of the essence of medical practice.

The study of medicine should not only be a human science because it comes from descriptions of the essence of medical practice but also because the physician-patient relationship is one of caring. According to Pellegrino, care has four meanings in the health professions. The first sense of care is compassion for a fellow human being who is ill. The second is doing for another what he cannot do for himself due to his illness. The third sense of care is when the physician or nurse takes responsibility for treating the patient's medical problem using professional knowledge and skill. The fourth sense of caring is taking care of the patient using the craftsmanship of medicine and nursing.[36]

Those health care professionals who use the biomedical model of the patient-physician relationship understand the physician's role to concern only the third and fourth senses of care. Care in the first two senses is relegated to other health professionals. This bifurcation of health care can be seen most clearly in nursing practice. Traditionally nurses have been involved in patient care in the first two senses. Indeed, nurses traditionally expressed compassion in direct care for the patient such as bathing. Recently, some nurses have seen their role as primarily assisting the physician in caring in the third and fourth sense, while others continue to favor the traditional forms of care. Currently, nurses are trying to overcome this bifurcation by including all four senses of care in their practice as evidenced by primary nursing.

Rather than bifurcation of health care, all five contributors (Pellegrino, Richard Zaner, Sally Gadow, Mila Aroskar, and Tristram Engelhardt) to a philosophical treatment of the relationship of physicians, nurses, and patients advocate integral care in which *all* health care professionals care in all four senses.[37] All of the above authorities on the philosophy of health care are concerned that recent advances in medical science are leading health professionals away from viewing their practice as caring for persons who are ill toward curing illnesses through application of scientific knowledge and technical skills. So-called biomedical ethics, insofar as it claims that moral issues arise out of advances in medical science and technology, supports the curing and the technical model of medical practice.

Most bioethicists take a "humanities" approach to health care rather than the human science approach. Those who take the humanities approach say, in effect, that the biomedical model is the correct one but that physicians face moral choices which require the assistance of ethicists. These ethicists contribute their skills to the resolution of "bioethical" problems through knowledge of classical moral norms and skill in analyzing moral problems. In contrast, those who use the human science approach, like Pellegrino, search for the essence of health practice as it occurs in the world. The study of health care practice indicates that health care is primarily a relationship of caring. Certainly, adequate health care requires use of the contributions of biological and chemical science and medical technology. Unfortunately, this reliance on the natural sciences leads some students of medicine to conclude that medicine is founded on natural science. Indeed, this is a comfortable position because that aspect of medical practice which is clearly related to natural science is both foundational and demonstrable. A shot of penicillin can, in fact, be shown to cure strep throat and, in this case, biochemistry is the foundation of the cure. Unfortunately, human science is neither demonstrable nor foundational in the same conclusive way. The human scientists can only investigate the essence of practice and say to other investigators, "This is the sense of that practice. Isn't it obvious to you that it is?" The reason for this is that the human sciences aim at that which is specifically human and, therefore, not causal. Instead, the human sciences aim at meaning, and in this case, the meaning of health care practice. Health care is essentially a caring relationship concerned with the curing of illness when possible. If cure is not possible, health care is concerned with care that promotes the natural healing of the body. If that is not possible, it supports, comforts, and aids those who must live or die with illness.

It should be obvious that health care in the above sense is the benevolent care of the moral sense as Tymienicka interprets it. This moral sense operates within health care practice itself and is not brought in from the outside as in the case of the moral norms or moral reasoning of biomedical ethics. This implies that a physician or a nurse acts morally when the benevolence of the moral sense encompasses professional health care.

If physicians and nurses act morally by following the moral sense embedded in their professional practice, why are there so many moral issues and problems in medical practice? The bioethicists would have us believe that moral problems result from the new possibilities of medical treatment afforded health care professionals by advances in medical science and technology. But is this the case? Although it is true that our ability to prolong life has been greatly increased, the problem of when to let die has been in medical practice as long as there have been physicians and, indeed, before that in the cultural mores of primitive people. Modern advances have, in fact, dramatically brought to our attention the moral issues endemic in medical practice. However, focusing on the more spectacular cases may blind us to such perennial moral issues as the one raised by Cain: "Am I my brother's keeper?" Or as Tymieniecka puts it, do I act out of benevolent concern for the other or out of my own self-interest? Even when self-interest becomes "enlightened" in the sense that I understand that my well-being is bound up in the well-being of a community, as Tymieniecka points out, this is an inadequate basis for community and certainly for health care within that community.

A second reason for moral problems in medical care is evident in Pellegrino's distinction between what is medically right as determined by the physician and what is morally good as decided by the patient. Traditionally, many physicians have held that only the medically right treatment should be considered. As a consequence of recent recognition of patients' rights, the patient has a greater role in determining the nature of treatment. Even when the patient's right is clearly acknowledged, however, the patient is still very dependent on the physician's expert knowledge in making a sound choice. Just as the patient is dependent on the physician, the physician is in turn dependent on the patient for his understanding of the patient's wishes. When neither physician nor patient understands each other's position, the moral issue is apt to become one of rights — who has the moral (and legal) right to choose. However, a conflict over rights often obscures the fact that both physician and patient are moral agents. The physician no less than the patient has to concern himself or herself with what is morally right. As Pellegrino contends, this requires a dialogical relationship between physician and patient in which they decide together on the treatment which best fulfills the moral sense of benevolent concern for each other.

A third reason for moral problems in health care is the nature of professions. As Tristram Engelhardt points out, the medical profession, like all professions, is a conglomerate of intentionalities which have developed over time.[38] Although the dominant sense of medical practice is the moral sense (caring for the ill), it includes other intentionalities. For example, the economic sense of health care comes from the intention of the free enterprise system, namely the "bottom line." This intentionality is clearly in conflict with the moral sense of the medical profession. Also, the medical profession includes the sense of promoting the well-being of the members of the profession. For example, this intention did indeed work against the moral sense of benevolence in the case of the AMA limiting the number and size of medical schools at a time when many qualified applicants were denied admission and there was a great shortage of physicians in small towns, rural areas and inner cities. And, one might add, this is a violation of the intention of the free enterprise system in that it restricts trade, thus keeping prices high. However, the medical profession, like all professions, must promote the good of the profession in order to fulfill its mission. For example, one thinks of how the low salaries and prestige of teachers have affected negatively the quality of education in the public schools. Tymieniecka suggests how the conglomeration of intentionalities in the medical profession can be dealt with while keeping the focus on the moral sense of benevolence. This can be accomplished through the "*value-preference mode* in which we attribute to each of the conflicting elements its *due*."[39]

Giving "each of the conflicting elements its due" is often very difficult in health care because the moral sense embedded in medical practice must work through a network of social and legal rights and privileges which regulate the relationships of physician, nurse and patient. For example, physicians have exclusive legal rights to diagnose and prescribe treatment to patients. What happens when these rights conflict with the moral obligations of the nurse is evident in the case of *Tuma vs Board of Nursing, State of Idaho*.[40] Tuma was an instructor in a college nursing program. She and a nursing student were caring for a patient who had a diagnosis of leukemia and had an order for chemotherapy. The patient related to Tuma that she had had success in treating her disease for twelve years through the practice of religion. Tuma subsequently discussed with the patient alternative treatments to chemotherapy, including natural products. In the end, the patient continued with her chemotherapy and later died. The physician believed that Tuma had overstepped the bounds of her practice and filed a complaint with the board of nursing. It was determined that Tuma's action did not contribute to the patient's death. However, the Board of Nursing in the State of Idaho suspended Tuma's license for unprofessional conduct. The *Tuma* case made clear the legal right of the physician to prescribe treatment. But what of the patient's right to seek medical advice elsewhere? And what of the nurse's right to give counsel when requested which does not prescribe treatment but suggests other possible treatments?

To focus this issue more clearly, consider the following imaginative variation. A patient does not want to follow the *only* treatment her physician will prescribe for her incurable cancer. At best, following that, or any, treatment, she can live from a few months to a year. The treatment her physician prescribes would be very costly and would involve spending most of her remaining life in a hospital. Morally she does not want to burden her family financially and personally she wants to spend as much time as possible at home with her family. She asks her nurse, who knows well her wishes, if there are other types of treatment available. From her experience, the nurse knows that other competent physicians have, in fact, prescribed other treatments and shares this information with the patient. The patient consults with another physician and chooses a treatment which is more in accord with her moral sense and her personal desires. Clearly, this is a case in which the physician's legal right to prescribe the professionally "right" treatment is in conflict with the good of the patient and the moral sense of the nurse.

Another intentionality in the conglomerate medical practice is the objective intentionality of science and technology which, after all, is very different from that of interpersonal care based on benevolence. Sally Gadow puts this issue very clearly.

> The reason, then, that technology poses a greater threat to dignity than does less complex care is related to the experience of otherness. Mundane care and simple apparatus involve measures that persons usually can manage for themselves. But complicated measures and machinery are more disruptive; they can remove the focus of control and of meaning from the individual by imposing otherness in two forms, the machine and the professional: (1) the apparatus asserts an otherness that cannot be ignored or easily integrated into the physical or psychological being of the person, and (2) complex techniques require greater expertise than many persons possess, and professionals may be called in to manage the procedure.[41]

Gadow contends that this objectification of the patient does not simply result from the use of technology but from the consciousness that informs professional medicine. The patient is regarded as an instance of a disease categorically defined and as a problem to be solved. For example, Chris Sizemore whose case was treated in the movie, "The Three Faces of Eve," reports that although her mental illness was important enough "to make a motion picture about," she felt that she was "so unimportant I didn't even tell people who I was."[42] Gadow contends that the resolution of this problem requires understanding care in a new way. For her "caring is attending to the 'objectness' of persons without reducing them to the moral status of objects."[43] She contends that treating patients as subjects rather than objects requires speaking with them as fellow human beings concerning the meaning and implication of their treatment and illness and touching them as fellow human beings.

Gadow also contends that objectivity enters health care as professional methodology. Medical professionals have developed routine procedures for handling certain cases and accomplishing certain tasks. Thus, health care comes to have two meanings. The first is routine professional care learned in professional schools of medicine and nursing. The second is personal care for this particular person who is ill. In the *Tuma* case these two senses of care came into conflict — the physician demanding that routine professional procedures be followed and Tuma responding from the spontaneous moral sense to the specific request of a particular ill person.

Achieving unity of the professional sense and the moral sense is difficult because the professional sense has been fractured by the various specialties within medicine and nursing. Recent attempts at "holistic medicine" often miss the nature of this splintering in that they call for physicians, nurses, social workers, psychologists, and clergymen to work together in one comprehensive program of health care. Holistic attempts at unity often fail because, as Tymieniecka has pointed out, they fail to treat the divisions and conflicts within the various human sciences. She contends that true unity comes from recapturing the moral sense implicit in these sciences.[44] We will show how recovering the moral sense restores unity to the divisions within nursing practice and in so doing brings fulfillment to nurses.

IV. NURSES AND FULFILLMENT IN HEALTH CARE

We will show that fulfillment comes to nurses, and presumably to other health professionals, when two senses of care converge. When applied to contemporary nursing, these are the two senses which Tymieniecka contends concern the human scientists: (1) articulation of the sense of various intersubjective relationships and (2) disclosing the moral sense evident in those relationships. Interestingly, Huston Smith has pointed out that "meaning" itself has these two essential meanings: one referring to the intelligibility of an activity and the other to the moral worth of that activity.[45] In nursing and other health care professions, we have designated these senses as the professional sense and the moral sense. We will show that nurses find the fulfillment, as Tymieniecka suggests, when the professional sense and the moral sense become one.

Before the professionalization of nursing, the two senses of nursing were clearly in harmony. For example, on a nursing poster under the painting, "The Sick Child," by Gabriel Metsu, a caption reads, "Nursing: The Oldest Art." Regardless of the accuracy of the claim of "oldest art," it is evident from the painting that caring for the ill is a fundamental human phenomena in which the art of care and the moral sense are conjoined in a personal relationship.

Nursing as a profession was originally merely an extension of personal, direct care for the ill. Historically, nursing has been distinguished by being that health care profession most concerned with direct care. That is, nurses have been assumed to act out of compassion for the ill by doing those things which the ill cannot do for themselves, such as bathing, feeding and other activities of daily living. But with the advance of medical science and technology since World War II, the function of nurses has been distinguished from that of physicians by contending that nurses care and physicians cure. The stress on cure has led nurses, in addition to direct care, to become the physician's assistants in curing.

If we apply Pellegrino's four senses of care to nursing, it becomes evident that one problem in contemporary nursing is to see the relationship between care for the ill in the moral sense and in the professional sense. His first sense of care as "compassion" is obviously care in the moral sense of caring for an ill person out of benevolence. His second sense, direct care — doing for the ill what they cannot do for themselves — can readily be a concrete expression of the first sense since it requires direct contact between persons. But when done strictly as prescribed routine, it becomes professional care. The third and fourth sense of care are obviously professional care: using knowledge and skill and exercising competence in eliminating or alleviating the patient's problem. The advancement of professional knowledge and skill and the advances in medical science and technology *all* have led to stress on professional cure at the expense of the moral and direct senses. In nursing care, the primary struggle has focused on whether the professional sense of the nurse should come from within the nursing profession (care) or from the medical profession (cure). This professional issue is set within the wider context of the relationship of professional care to direct and moral care. Of course, this wider context is the human intersubjective context of the moral sense. The logic of Pellegrino's treatment of care in medicine clearly assigns the priority to the moral sense of care over both professional and direct care but incorporates both within it. If this is true for medicine even with its current stress on cure, it is certainly true for nursing with its traditional focus on care.

The finding of a recent experiment we conducted concerning fulfillment in nurses supports the contention that both the professional sense and the direct sense of care must be incorporated within the moral sense.

V. INVESTIGATION

Our investigation attempted to discover whether practicing nurses regarded nursing as direct, personal care out of compassion or performance of professional roles aimed at cure utilizing medical science and technology. Also, we attempted to discover how nurses understood the relationship of caring in Pellegrino's first and second senses to the third and fourth senses. We especially wanted to know whether the stress on medical science and technology had led nursing away from traditional care out of compassion.

We knew that we would encounter at least two difficulties in attempting to elicit the meaning of care from the direct experience of nurses. In previous attempts to arrive at the sense of professional care from personal experiences of nurses (Bishop) or teachers (Scudder), we had encountered two pitfalls. First, when asked to treat their experience by descriptive methods, some teachers and nurses merely describe what they typically do. Second, some nurses and teachers who had learned in theory classes that good practice includes x, y and z, merely determined whether x, y and z were actually done in their practice.

To avoid these pitfalls, we selected a phenomenological procedure advocated by Barritt, Beekman, Bleeker and Muldering.[46] We asked sixty practicing nurses and senior nursing students to respond to the following:

I. Write an account of a single experience, something simple and straightforward, from your nursing practice when you felt most fulfilled as a nurse and thought you were most completely being a nurse. As much as possible stick to a descriptive language and do not include interpretations and attributions of causality to your writing. Don't lose yourself in factual details. It all begins with the lived experience and that is what you should strive to describe.

II. Write an account of a single experience, something simple and straightforward, from your nursing practice when you felt least fulfilled as a nurse and thought you were least being a nurse. As much as possible stick to a descriptive language and do not include interpretations and attributions of causality to your writing. Don't lose yourself in factual details. It all begins with the lived experience and that is what you should strive to describe.

We asked for these concrete descriptions of actual experience in order to avoid generalized descriptions of routine behavior or evaluation of professional activity in terms of theoretically derived criteria. We reasoned that in telling of a concrete experience in which the nurses felt that they had been most clearly and completely a nurse or least fully a nurse, we could arrive at the experiential meaning of care for nurses.

VI. MOST FULFILLING EXPERIENCES

The responses of these nurses made evident that their most fulfilling experiences were those in which care was experienced as personal benevolence which subsumed professional care and/or direct care. All of their descriptions focused on personal caring relationships with patients, regardless of whether they involved either/or direct care and/or care which required a high degree of professional knowledge and skill and professional language. The following description illustrates the point:

Approximately 25 years ago I was assigned to work with a female patient who had suffered third degree burns over 60% of her body. I was assigned to her every working day for approximately one year. She required intensive nursing care as well as psychological support. Initially, she was on complete isolation. Her care was time-consuming and challenging. She was on a Styrker Frame, receiving intravenous fluids, had dressings over 60% of her body and was completely dependent for all physical needs. I spent an average of three to four hours per day carrying out the doctor's orders as well as providing essential nursing care. We came to know each other as I have never known another patient. She trusted me and talked of her husband's alcoholism, her concern for the welfare and safety of three small children, financial worries and her physical pain and depression. She needed several operations for skin grafting, intensive rehabilitative services and the emotional support from nursing staff to cope. It was such a pleasure to be a part of the healing process and to see her leave the hospital after a year's hospitalization, even though emotionally a part of me went home with her.

In the above case, in which care is present in all four senses, it is evident that the sense of fulfillment came from care in the first sense, i.e. personal relationships in which professional activities are done out of compassion. Fulfillment from these professional activities seems to bear little or no relationship to whether caring was done in the second, third, or fourth sense. One nurse who responded put this very well in the following description:

I felt most fulfilled as a nurse when one day a patient at my place of employment told me. "Things are just not right when you are not here." I had been giving her bedside care, medications, treatments, and conversation. I oriented her to date and time, the events going on in the world. This statement made me feel I had accomplished more than "just nursing."

Note in the above description that "just nursing" refers to caring in the second sense — direct care. She seems to be making a distinction between that which can be learned as a role and that which is uniquely a part of her personality and of her relationship with this person, her patient. However, "just nursing" does not imply that role should be separated from personal relationships, but that fulfillment as a nurse requires that the role be encompassed in personal care for the ill person.

Personal care also encompassed professional care when done in third and fourth senses. Consider the following description in which compassion is expressed almost entirely in the third and fourth sense.

My most fulfilling experience deals with my first and only cord prolapse of a woman progressing with the aid of Pitocin Stimulation and how a normally Progressing labor can suddenly change into a critical episode of life saving measures for her unborn child. What made this experience memorable was the fact that I detected what felt like a cord at a vaginal check and the physician was called, but misdiagnosed his vaginal check. My suspicions called for another check and fetal heart monitoring. Without further doubt of what was occurring, the physician was recalled and an emergency C-Section took place with a healthy baby boy born. It took place quickly but was particularly self-satisfying — especially since I see the child and his Mom and Dad upon occasion and see his growth progress.

Even when the second, third and fourth senses of care are well integrated, fulfillment for the nurses comes from the moral sense becoming apparent in the professional sense as in the following example.

Two years ago I had the opportunity to deliver total patient care to a 25 year old girl with end stage congestive cardiomyopathy. She was in congestive heart failure with many ventricular life threatening arrhythmias. She was well aware of the fact that she was going to die and admitted her fright to me and also asked me point blank if she was going to die. We were able to discuss such problems as how could this be explained to her 7 year old daughter? Who would care for her 7 year old daughter and her own 16 year old retarded sister? I also discussed with her family, some of the fears she had acknowledged and encouraged them to discuss these things with her. Her main request of me was that I sit by her bed during the night and simply hold her hand. In addition to ministering to these needs, I also monitored her vital signs, changes in physical assessment, titrated various vasopressors and vasodilators to maintain optimum cardiac output. I overrode our strict visiting policies to allow her husband and daughter to sit at her bedside as they wished with the understanding that they would promptly leave if asked to do so by any of us.This patient remained in my unit for 4–5 weeks in critical condition before dying and though we all felt the hurt of losing her, we also felt the joy obtained by providing emotional and physical support along with patient teaching to both patient and family and helped both patient and family to accept and begin to deal with her inevitable death.

In the above example, the second, third and fourth senses of care are well integrated into benevolent care which gives fulfillment to the nurse. Also in the above case, as in the other cases, this care goes beyond the immediate clinical situation to include family, orientation to the outside world, and/or continued relationship after the illness.

In the nurses' descriptions of their most fulfilling experiences, the stress was on the nurse-patient relationship rather than profession-nurse relationship, the hospital-nurse relationship or the physician-nurse relationship. Even in the few descriptions which included cooperation between nurse and physician, the fulfillment came not from the good nurse-physician relationship but from the better care for the patient which resulted from their cooperation.

VII. LEAST FULFILLING EXPERIENCES

Although the least fulfilling experiences were varied, the majority of the factors causing concern were those inhibiting patient care. The most frequent inhibiting factors were uncooperative patients, patient overloads, lack of skill or knowledge, and lack of contact or cooperation with physicians. For example, one nurse stated, "I felt least like a nurse when I had twelve patients and only enough time to change the beds and bathe the patients" and another nurse wrote "I felt least fulfilled when an elderly man wanted me to talk with him and I had too many beds to make, too many baths to give." Another nurse wrote that her least fulfilling experience was her unsuccessful attempts to locate the physician of a patient experiencing chest pain. Time lapsed and the patient died. The nurse expressed feelings of inadequacy, helplessness, and sadness.

Some of the unfulfilling experiences *seemed* not to focus on patient care, however. For instance, some nurses resent physicians not respecting their professional judgment.

I was never allowed to use my nursing judgment and I had to perform every task strictly by the rules or by doctor's orders. The times when I did attempt to make a judgment and communicate that to others, I was generally made to feel that I had made an error or I was disciplined verbally that it was not my place to make these kinds of decisions.

Others resent doing maid-type activities, especially when patients treat them as servants. "I felt unfulfilled when I had a patient who treated me as a maid instead of a nurse. She constantly was ordering me around asking me to empty her trash can and clean up her room" and the "patient requested services of me that were more like a housekeeper." Nursing students especially resented rigid, cold supervisors. One resented a patient's flirtations. "The patient made many 'flirty' comments which I ignored. 'My you're a pretty little thing.' 'If I had you, I sure would treat you like a princess.' I pretended not to by fazed by his behavior, but caring for this type of patient was definitely not fulfilling." Although the above resentment of affronts to personal dignity and to professional judgement and status seem not to result from inhibiting care in the moral sense, they actually do. As Tymieniecka points out, one's benevolent care of another needs to be recognized and appreciated by the other. In all of the above cases not only was that recognition and appreciation not given, but the response made such care difficult, if not impossible.

The problem of how to relate to dying patients was often mentioned. Some nurses found that they were able to develop an intense personal relationship with dying patients. They were able to discuss aspects of living and dying for the first time. Patients and families were very appreciative. In two cases, the dying patients through their personal relationships with nursing students contributed to the students' confidence in themselves as nurses and as caring persons. "As a nursing student I was assigned to a dying patient. In the midst of all her pain and suffering, she constantly praised everything I did for her. Just to experience the presence of a person so at peace with herself was a terrific highlight in my life." In contrast, some nurses found caring for dying patients least fulfilling because they were unable to help them and as a consequence felt hopelessness and depression. "When I had my first experience with a dying patient, who had throat cancer, I watched this patient slowly suffocate and I felt utterly useless." Another patient "had no family and was a hemiplegic due to multiple trauma. He could only communicate with notewriting and he scribbled a note, 'Please let me die.' I felt inadequate and confused."

One surprising finding from examining these descriptions was that the response of the patient to the nurse's professional and personal care was the most important ingredient in determining whether or not an experience was fulfilling or unfulfilling. We had expected that nurses would find relationships with physicians and hospital bureaucratic requirements least fulfilling. But hostile and uncooperative patients con-tributed most often to lack of fulfillment. "I felt least fulfilled when the 13 year old boy's first words were cussing at us, to leave him alone. How dare he! We had saved his life." In contrast appreciative, suppor-tive, and cooperative patients contributed most to fulfillment. "I felt most fulfilled while caring for a forty-three year old woman with meta-static cancer. We cried together, laughed together, and conspired to-gether to meet her needs even if it meant bending hospital policy."

Our conclusion from this exploratory investigation is that nurses feel most fulfilled when they experience their professional care as personal benevolent caring for the ill. The way of caring for the ill, whether it was in the second sense of direct care or in the third and fourth sense of professional care, did not seem to be essential. The nurses seemed most fulfilled when caring in any of the above senses was done out of the moral sense and in a personal relationship with ill persons who appre-ciated and confirmed the moral worth of their care.

These nurses felt fulfilled when the professional sense of care ac-complished what its moral sense of care intended and the patients recognized and appreciated the nursing care. But how often does this occur in a nursing career filled with routine roles, a multitude of tasks, nursing care plans and complicated equipment? A cynical nurse might say, "Sure there are exceptionally fulfilling moments but for the most part, nursing is doing the routine tasks of patient care." But such cynicism fails to recognize that professional roles, knowledge and skills are not necessarily separate from or in opposition to compassion and personal involvement. In fact, Paul Ricoeur shows us why the two are necessarily related to each other. In an essay Ricoeur contrasts being a neighbor with providing a service. For Ricoeur, one is a neighbor in the first person. The meaning of being a neighbor is being available to another person when you are needed as you are needed. In contrast, a social role is performed in the third person. One performs specific services by virtue of having a certain social status and the service is provided to those in certain categories. Does this distinction require one to choose between being a neighbor and performing a social service? Ricoeur thinks not. Both being a neighbor and providing a social service are united in having the common intentionality of charity.[47]

This intentionality is often obscured by the nature of social function and institutional services.

The ultimate meaning of institution is the service which they render to persons. If no one draws profit from them they are useless. But this ultimate meaning remains hidden.[48]

When intentionality is hidden we often blame social roles and technology for lack of personal meaning in institutions. But Ricoeur warns of the folly of such accusation.

The theme of the neighbor is primarily an appeal to the awakening of consciousness. It would be absurd to condemn machines, technocracy, administrative apparatus, social security, etc. Technical procedures and, in general, all 'technicity,' have the innocence of the instrument.[49]

Although technology and social roles may obscure personal meaning, they do not necessarily oppose it. Personal meaning appears when the neighbor awakens my moral sense of benevolence and I respond with appropriate care for him. If I respond with routine professional care, then I need confirmation in a first person encounter from *this* person who has benefited from *my* professional service. The following description is an excellent example.

The most fulfilling experience I ever had was when a child I was caring for arrested but was successfully rescuscitated. I had written my notice that day — I wanted out of nursing — it was killing me. The baby stopped breathing while we were on the elevator coming back from x-ray. I did mouth to mouth on her until we got back to the room and the code team arrived. The baby responded beautifully. Naturally I felt good. But when the mother praised me for "saving" her baby, I tried to tell her that what I did was not so special; anyone can do mouth to mouth. "But it was you," she said. "You were there. If you hadn't wanted to be a nurse in the first place and been working that day, I wouldn't have my baby."

In the above case, it is evident that the nurse understood that she had merely performed a professional role. In the past she, like most professionals, carried out her roles assuming that she helped patients but often without much direct evidence or confirmation. But her most fulfilling moment occurred when the personal moral sense of performing her roles was made manifest to her in a concrete personal relationship through the appreciative response of the mother.

Caring for the ill is generally done by nurses through professional functions and activities. If we had asked nurses to give the meaning of being a nurse directly, many would probably have listed professional functions and activities. Usually routine functions and activities are carried out in faith because there is little direct confirmation of their moral worth. The events and situations described by the nurses were those in which they encountered the moral sense of their professional care. Then the meaning of the routine, the roles, the technological function appeared and the moral sense of their vocation as nurse was confirmed. It made little difference whether the neighbor was encountered in direct care or technical care; the moral sense of their profession came from direct encounter with a person for whom they had cared. This deeply felt confirmation of the moral sense of being a nurse occurred when the intentionality of benevolence embedded in the role of the nurse became manifest in a personal relationship in which an ill person was helped. Then the professional sense of being a nurse and the moral sense of being a nurse converged in an integration of personal and professional care rooted in benevolence.

VIII. THE PATIENT

Our treatment of the nurse's relationship to the patient has made evident that some patients respond to nurses from the moral sense. For example, one nursing student was given confidence in her professional ability, a greater sense of her worth as a person, and a greater recognition of the potential for compassion in human beings by an elderly patient dying of cancer who responded to her treatment with benevolence. This indicates, as Tymieniecka suggests in her treatment of the moral sense, that patients have a moral obligation to appreciate and confirm those who give them health care.

Oddly, the moral obligation of patients to physicians, nurses and others has not been adequately treated, if at all. In an article exploring the relation of the patient to physician and nurse, Richard Zaner implies that patients often relate to those who give them health care out of the moral sense. Zaner points out that patients are unusually patient in their dealings with physicians. They continue to trust physicians even when their treatment indicates that such trust may be unwarranted. Put differently, they tend to give the physicians the benefit of the doubt concerning their treatment. Only when the evidence is overwhelming that they have not been treated adequately or fairly, do most patients confront their physician. Then they merely expect the physician to acknowledge his/her mistakes and not to charge them for inadequate or harmful treatment and to do their best to remedy the hurt and incapacity they have brought about.

Patients are remarkably resilient and forgiving — published statistics to the contrary notwithstanding. These (and other) patients seem reluctant to pursue legal redress, even where it may not be unreasonable. Unexpectedly, this woman, like so many others, was able to understand that physicians are human, too; that they make mistakes; and that only at times are they culpable. Often, what's important for these patients seems not the mistakes, but the willingness (or unwillingness) of the physician to own up to a mistake and be ready to make amends in some reasonable and caring way.[50]

In short, patients expect, as a minimum, fair treatment from their physician.

Patients, however, do expect more than fair treatment from their physician. As Zaner points out, "sick people 'want *to know* and they want to know that the people who are taking care of them *really* care.'"[51]

The moral sense is evident in the expectation of patients that physicians will honestly apprise them of their condition. That patients seek to understand their condition even when not given adequate information by their physicians amazed physician Robert S. Mendelsohn. He wrote, "a remarkable finding of these interviews is the consistently profound base of information patients manifest about their condition. . . . patients are aware of the major aspects as well as the subtleties of causation, diagnosis, treatment alternatives and prediction of outcomes." Mendelsohn admonishes his fellow physicians, "Stop talking down to your patients; stop patronizing them as if they were children or stupid or retarded or all three."[52] Certainly, this implies that patients expect their physicians to have a moral obligation to treat them as adults capable of understanding their condition. Further, they believe that physicians have a moral obligation to help them understand their condition and prospects. Zaner puts this very well.

Patients want and seek to know, and in the interest of proper and appropriate dealing with them, the norm that obliges that accurate, adequate, and understandable information be promptly and continuously given to them seems unquestionably demanded.[53]

Patients need to know because knowledge of their condition and treatment often spares them from unnecessary anxiety. But also knowledge makes it possible for them to accept the moral responsibility of caring for themselves insofar as they are able. Physicians and nurses can help evoke this sense of responsibility by including the patient as a member of the health care team. As one patient, a heart attack victim, put it, "we were like a team and this was a campaign. I was a member of the team. I was the cause of all the trouble but I was also a member of the team, We were holding hands." As a result of this experience, one would expect greater cooperation between patient, physician, and nurse because they had achieved a "communality" in which the patient/physician and patient/nurse relationship had become "a *covenant* and not a contract."[54] This moral covenant should hold even after the patient has left the hospital and assumed the moral responsibility for his own care with the guidance and advice of his physician.

Patients expect their physicians to fulfill the primary moral sense of their profession in that they expect them to really care for them. Zaner forcefully states this moral imperative.

Patients want to know that those who care for them really care. To be sick or injured is to experience ourselves as diminished, as afflicted to various degrees and in precisely those ways which mark us as distinctively human: in our freedom to choose and act, our ability to think and imagine and plan, the intimacy of our relatedness to our own bodies and minds, and our relatedness to other persons. To want to know and to be really cared for, as afflicted, is a uniquely demanding moral phenomenon.[55]

Patients do expect physicians to act from the moral sense in care for the ill, but do patients have an obligation to care for others who are ill? For example, one patient suffering from multiple fractures in an automobile accident with an extremely serious injury to her back, during her rehabilitation and therapy, made rounds in her wheelchair to comfort other patients and organized group singing for other patients. Certainly, when someone who is suffering from illness or accident comforts and entertains other ill persons, this is acting from the moral sense in its highest form.

The above example of care for the ill would not be regarded as exceptional for a well person, however. We expect well persons to visit and comfort the ill. The moral sense of benevolence is evident in this cultural expectation. But we make exceptions for the ill and injured. We expect them to be self-centered in their preoccupation with their pain, suffering, treatment, and prognosis. If a patient stoically endures pain, cooperates with physicians and nurses, and assumes responsibility for fostering his or her own healing, we normally consider that patient to have fulfilled the expectations for a morally good patient. Thus our moral expectations of an ill person are different from those of a well person.

When patients who cannot recover completely are faced with living with chronic illness or debilitation, our moral expectation of them also changes. But what of their moral expectations of themselves? Certainly, a person who was formerly the family bread winner, but can no longer be, faces a moral crisis concerning the worth of his or her life. Then the moral sense operates in a very different context. How the moral sense can operate in even very restricted contexts was evident in the case of the woman dying with cancer who gave comfort and assurance to the young nursing student who was caring for her. Thus, debilitation and chronic illness do not suspend the moral sense but alter its direction. For example, in the American movie, "Whose Life Is It Anyway?," the quadraplegic who could no longer be a sculptor decided he had the right to discontinue the technical support that kept him living. But his relationship to other persons implied that he had much to give because of his intelligence and ability to communicate with others. In contrast to Richard Dreyfus' portrayal of the lead character, the same character in the British version seemed to have only the choice between death and continuing a meaningless life. Thus, the British version of the film raised the question of the right to take one's life. In contrast, the American version poses the question, "Does one have the right to take his life when he obviously has much to give to others, even in his limited condition?" Put differently, is the moral sense suspended when one can no longer pursue the chosen direction of his life and must work in very restrictive conditions? Or does the moral sense require that the disabled and chronically ill ask, "What can I give to others in the limited condition of my life now?"

When patients do recover from illness, Zaner believes that they have a moral obligation to those who are ill. He quotes Albert Schweitzer who contends that illness "seems uniquely capable . . . of awakening 'a moral sense that is usually dormant but that on special occasions can be brought to the surface.'"[56] In commenting on the reflections of a forty-two year old coronary by-pass patient who had recovered a new zest for life during his illness and treatment, Zaner observed that there is something else here which also is remarkable: gratitude. He is, of course (as is plain in his earlier remarks), grateful to his wife and family for "putting up" with him and sticking by him, to the physicians and surgeons and nurses (for the most part), and to the technologies that made it possible for him to recover. But he is also grateful for his "total experience," for having recovered "new values" and a new sense of priorities — of what is truly important for him.[57]

This sense of gratitude coming after the experience of pain, uncertainty and frustration unites former patients together as Schweitzer clearly saw.

All through the world, there is a special league of those who have known anxiety and physical suffering. A mysterious bond connects those marked by pain. They know the terrible things man can undergo; they know the longing to be free of pain. Those who have been liberated from pain must not think they are now completely free again and can calmly return to life as it was before.[58]

Having recovered from anxiety, physical suffering and pain, former patients, according to Zaner, have an obligation to those who will undergo the misfortunes they have recently endured.

Being marked by the gratefulness of recovery, seems to bear a moral meaning that is often undetected and unencouraged. Our good fortune in being enabled by others to recover obligates: this good luck which befalls the patient, no less than the illness itself, must not (as Schweitzer says) be taken for granted, its burden muted or forgotten. For now, having been ill and having recovered, patients are obligated to give something in return to others, especially to those who now find themselves marked by pain, anxiety, and suffering and long to be free of them.[59]

That patients are responding to this moral obligation is evident in the number of help groups such as Alcoholics Anonymous, Reach to Recovery, Colostomy Clubs, and others.

IX. CONCLUSION

We have shown that Tymieniecka's treatment of the moral sense does, in fact, help disclose the benevolent intentionality of health care. Medical care begins with illness which inhibits or prevents a person from living in the world in accustomed and desired ways. When this person seeks professional help from the physician and the physician agrees to give that help, a caring relationship is formed. In this relationship, the physician agrees to use his or her knowledge and skill diligently for the good of the patient.

The greater knowledge and skill resulting from advances in medical science and technology rather than changing the fundamental caring relationship, contributes to making the medically right decisions concerning diagnosis and treatment. But the physician agrees not only to use his knowledge and skill to come to the right medical resolution but also to work with the patient to arrive at the good decision for the patient. Moral problems arise (1) when the physician considers only the medically right decision and not the good decision from the patient's point of view; (2) when the physician puts his self-interest above the moral good; and (3) when tensions between the various intentionalities embedded in medical practice inhibit that practice from achieving its moral end.

We have shown that fulfillment in nursing, and presumably other health professions, comes from the convergence of the professional sense of practice with its moral sense. Then professionally established procedures cease to be ends in themselves and become means of achieving the benevolent goal of nursing. However, for this fulfillment to be achieved, patient and nurse must develop a personal relationship in which patients appreciate and acknowledge their good care.

The primary expression of the moral sense by patients is appreciation and acknowlegement of good care. However, the moral sense is implicit in the patient's trusting and patient attitude toward the physician. Also, it is apparent in their expectation that the physician will treat them fairly and honestly concerning their diagnosis, prognosis, and treatment and will care for them personally as well as professionally. Although the moral sense of patients has been inadequately developed as a cultural expectation, the moral sense has important implications for the way in which patients relate to each other while ill, for persons who return to the everyday world with chronic illness or debilitation, and for the relationship of persons who have recovered from illness to persons who become ill.

Tymieniecka's moral sense discloses *the* sense of health care. Health care, when examined as it is practiced clearly indicates that it should be studied as a human science rather than as a natural science. However, full articulation of medical practice shows that it is not only intersubjective but that its ultimate sense is benevolent. Further, fulfillment by medical practitioners requires a convergence of the professional sense and the moral sense into a caring relationship involving mutual appreciation and confirmation.

Lynchburg College

NOTES

[1] Anna-Teresa Tymieniecka, "The Moral Sense in the Phenomenological Praxeology of the Human Science," an unpublished lecture delivered at the Third Annual Human Science Conference, 16–19 May, 1984, West Georgia College, Carrollton, Ga. and available on tape from the Department of Psychology, West Georgia College.

[2] John R. Scudder, Jr. and Anne H. Bishop, "Health Science as a Human Science: Disclosing Health Care Practice." Much of this paper which was read at the Third Annual Human Science Conference has been included in this article under the heading, "Health Care, Human Science, and the Moral Sense."

[3] Anna-Teresa Tymieniecka, "The Moral Sense," *Foundations of Morality, Human Rights, and the Human Sciences: Phenomenology in a Foundational Dialogue with the Human Sciences, Analecta Husserliana*, Vol. XV, ed. Anna-Teresa Tymieniecka and Calvin O. Schrag (Dordrecht-Boston: D. Reidel, 1983), p. 21.

[4] Ibid., p. 19.

[5] Ibid., p. 22.

[6] Ibid., p. 23.

[7] Ibid., p. 26.

[8] Ibid., p. 27.

[9] Ibid., p. 28.

[10] Ibid., p. 29

[11] Ibid., pp. 29–31.

[12] Ibid., pp. 31–32.

[13] Ibid., pp. 32–33.

[14] Ibid., p. 35.

[15] Ibid., pp. 39–40.

[16] Ibid., p. 41.

[17] Ibid., p. 42.

[18] Ibid., p. 42.

[19] Ibid., pp. 42–43.

[20] Edmund Husserl, *Phenomenology and the Crisis of Philosophy*, trans. Quentin Lauer (New York: Harper and Row, 1965), p. 117.

[21] Ibid., p. 150.

[22] Ibid., p. 152.

[23] Ibid., p. 149.

[24] Edmund D. Pellegrino, "The Caring Ethic: The Relation of Physician to Patient," in *Caring, Curing, Coping: Nurse, Physician, Patient Relationships*, ed. Anne H. Bishop and John R. Scudder, Jr. (University, Ala.: University of Alabama Press, 1985), p. 9.

[25] Edmund D. Pellegrino, "Being Ill and Being Healed" in *The Humanity of the Ill*, ed. Victor Kestenbaum (Knoxville: University of Tennessee Press, 1982), p. 157.

[26] H. Tristram Engelhardt, Jr., "Being Ill and Being Healed" in *The Humanity*, p. 142.

[27] Sally Gadow, "Body and Self," in *The Humanity*, p. 88.

[28] Mary C. Rawlinson, "Medicine's Discourse and the Practice of Medicine," in *The Humanity*, p. 74.

[29] Ibid., p. 75.

[30] Ibid., p. 75.

[31] Ibid., pp. 75–76.
[32] Ibid., p. 77.
[33] Ibid., p. 78.
[34] Pellegrino, "The Caring Ethic," p. 14.
[35] Ibid., pp. 15–16
[36] Ibid., pp. 11–12.
[37] Anne H. Bishop and John R. Scudder, Jr. "Introduction," in *Caring, Curing, Coping*, p. 5.
[38] H. Tristram Engelhardt, Jr., "Physicians, Patients, Health Care Institutions – and the People in Between: Nurses," in *Caring, Curing, Coping*, pp. 62–63.
[39] Tymieniecka, "The Moral Sense," p. 32.
[40] *Tuma v. Board of Nursing of State of Idaho*, 100 Idaho 74, 593 P.2d 711 (1979).
[41] Sally Gadow, "Nurse and Patient: The Caring Relationship" in *Caring, Curing, Coping*, p. 35.
[42] *Lynchburg* (Virginia) *The Daily Advance*, 26 November, 1984.
[43] Gadow, "Nurse and Patient," pp. 33–34
[44] Tymieniecka, "The Moral Sense in the Phenomenological Praxeology of the Human Sciences."
[45] Huston Smith, *Condemned to Meaning* (New York: Harper and Row, 1965), pp. 13–40.
[46] Loren Barritt et al, *A Handbook of Phenomenological Research in Education* (Ann Arbor: School of Education, University of Michigan, 1983).
[47] Paul Ricoeur, *History and Truth*, trans. Charles A. Kelbley (Evanston: Northwestern University Press, 1965), pp. 98–109.
[48] Ibid., p. 109.
[49] Ibid., p. 107.
[50] Richard Zaner, "'How the Hell Did I get Here?': Reflections on Being a Patient," in *Caring, Curing, Coping*, p. 86.
[51] Ibid., p. 92.
[52] Robert C. Hardy, *Sick: How People Feel About Being Sick and What They Think of Those Who Care for Them* (Chicago: Teach 'Em, 1978), p. viii.
[53] Zaner, "Being a Patient," p. 95.
[54] Ibid., p. 88.
[55] Ibid., p. 98.
[56] Ibid., p. 102.
[57] Ibid., p. 103.
[58] Ibid.
[59] Ibid., p. 104.

GODWIN SOGOLO

ON A SOCIOCULTURAL CONCEPTION OF
HEALTH AND DISEASE

In November 1984 surgeon William DeVries performed what is now referred to as the latest in medical miracles; he kept his patient, William Schroeder, alive with an artificial heart. DeVries' first attempt with artificial heart extended another patient's life by 112 days. These feats in medical technology are regarded as a step forward over the transplant of human organs — heart, pancreas, liver, and kidney — for which physicians in the West have now attained a high degree of expertise.

Such developments (normally received with applause and a sense of achievement by both medical professionals and lay men) raise questions concerning the changing goals of medicine and the varying conceptions of health and disease across cultures. For several reasons, in most developing parts of the world, particularly in Africa,[1] developments in organ transplant/implant are things of distant importance. First, medical research in these areas are yet to attain the level of sophistication in which attention is directed toward the replacement of impaired parts of the human body. Second, the main problem is infectious diseases, whose solution lies more at the level of prevention than at the level of cure. Third, the conception of health and disease in these parts of the world differs in major respects from the principles of modern medical research in the West.

In this paper I shall argue for a conception of health and illness that is culture-bound. While admitting that there are areas of agreement between this conception and that on which Western clinical practice is based, I intend to show, against this latter clinical conception, that what constitutes health and disease is socioculturally determined. With specific reference to Africa, I shall attempt to establish that modern medical practice in this part of world has focused attention on problems other than those that are central to the needs of the people. I attribute this misdirection of efforts to the superimposition of an alien conception of health and disease which, because it is value-neutral, fails to reckon with important sociocultural factors. Finally, I shall suggest a reorientation in Third World medical practice based on the conception of health as the general ability to function in a sociocultural milieu. The attainment

159

A-T. Tymieniecka (ed.), Analecta Husserliana, Vol. XX, 159–173.

of health in this sense leads to productivity and hence human social well-being, which in this paper is considered to be the final goal of the medical profession.

There are varying conceptions of health, even within the medical profession or within a given community of men. For this reason, perhaps, the World Health Organization (WHO) has provided a definition of health which, because it is so general, seems to absorb all possible conceptions. "Health," according to this body, "is a state of complete physical, mental, and social well-being and not merely the absence of disease or infirmity." Many see this definition as one fraught with difficulties mainly because of its generality. It could be argued, for instance, that the WHO definition has the untoward implication of rendering normative the concept of health and its cognate terms such as "well," "normalcy," "sanity," and even antithetically related concepts such as "disease," "illness," "sickness," etc. But I am inclined to believe that this very implication is the intent of the WHO definition and therefore its strength.

One other major point of objection against the WHO definition is that it broadens the scope of health into the general sphere of "social well-being" and therefore fails to draw a boundary between the responsibilities of the medical institutions and those of other organs of society. It is argued further that some of society's greatest evils stem not from inadequacy in health care but from matters of political injustice, unfavorable environmental conditions, scarce resources, and economic decisions. This is undoubtedly true. But when one adds to the list lack of adequate shelter, shortage of food, and malnutrition, the line between the concerns of medicine and those of other institutions becomes significantly blurred.

Malnutrition, for instance, may not in strict clinical taxonomy be regarded as a disease. Yet it would be wrong to see it as a mere predisposing factor. Malnutrition, properly conceived, is a diseased state whose responsibility falls within the medical profession insofar as medicine is concerned with the business of preventing disease or preserving health. Even when not regarded as a disease itself, malnutrition has been established to be the main cause of several diseases. Ralph Pettman describes the consequences of nutritional deficiencies and how they lead to irreparable incapacitations. Concerning lack of protein, for instance, Pettman writes:

Scientific evidence suggests fairly conclusively however, that not enough protein (whatever the limit is) at decisive points in the diet of a pregnant mother and or in the first years of a baby's life is enough to impair intellectual capacities and both the ability and willingness to learn thereafter. . . . Subsequent undernutrition has the effect of further increasing the individual's susceptibility to infectious diseases like tuberculosis, pneumonia and acute diarrhoea and promoting . . . a general condition of apathy, lethargy, and irritability.[2]

Beyond susceptibility to clinically defined diseases, Pettman emphasizes the physical and mental handicap that is biologically imposed on an individual through malnutrition. A victim, he says, is debilitated and incapacitated to the extent that he cannot afford to perform more than a few basic and limited human activities. Writing in the same vein, but with particular reference to psychiatry, J. C. Carother claims that a state of chronic undernutrition and anemia predisposes a person to the development of neuroses and psychoses such that a short period of famine will precipitate pellagra. In fact, Carother thinks that any mental conditions resulting from deficiency in food intake is liable to lead to pellagra.[3] The point of emphasizing malnutrition as a diseased state or at least of showing its close affinity with disease is to support the WHO definition of health which incorporates human social well-being.

In the strict clinical definition of the term, health means unimpaired function of an organism or part of an organism and disease its functional impairment. However, in defining disease as the cause of organic dysfunction, it should be noted that not all diseases are accompanied by dysfunctional manifestations nor by manifest symptoms such as pain and discomfort. This is the basis of the distinction that Joseph Margolis draws between the concept of disease and that of illness: "A disease is either what is apt to cause a diseased state or that diseased state itself. Illness is simply a diseased state manifest to an agent through that agent's symptoms — sensations, introspective cognition, proprioceptive awareness and the like . . . a temporary condition of ailing (or complaint) not caused by a diseased state at all."[4] By this distinction between disease and illness Margolis draws attention to tanthanic diseases, those that are clinically evident but escape the patient's detection because there are no symptoms accessible to him. Where symptoms are accessible to the patient and he is a complainant, the term illness, according to Margolis, becomes a more appropriate description than disease. It is therefore possible to say of a person that he has a disease but that he is not ill, if the symptoms are not recognized. However, this nonmanifestation of symptoms does not mean that the disease is not disruptive to the person's normal functioning.

René Dubos is therefore right in defining health as the ability to function.[5] Dubos argues against the conception of health as a positive ideal state "which can be achieved by eradicating all diseases from a utopian world." Instead, he sees health as an unattainable "mirage because man in the real world must face the physical, biological and social forces of his environment, which are forever changing. . . ."[6]

Dubos argues that any acceptable conception of health must relate to this struggle, which varies from individual to individual and from culture to culture. Health must therefore be conceived in relation to the activities of a person or group of persons in the pursuit of particular goals. Dubos gives illuminating examples of how health is differently conceived by people of different professions, how, for instance, "a cloistered monk rising during the night to pray and chant" conceives of health differently from "the pilot of a supersonic combat plane" and how "a farmer's wife with several children and a New York fashion model of the same age have very different physical requirements and therefore different concepts of health."[7]

Thus health and disease in a particular culture are determined by varying factors, some of which have to do with the people's level of economic and sociocultural development, occupation, and degree of intellectual awareness. It is thus possible, for instance, for a defective state such as myopia to be regarded as a form of illness by a predominantly literate society whose occupation has to do with reading and writing; whereas it may not be so regarded by a community of illiterates. As Leon Kass rehearses the argument, "If various functions and activities are the measure of health, and if functions are affected by and relative to circumstances, then health too . . . is relative."[8]

This relativist conception of illness is sometimes expressed at cross-cultural levels. Peter Sedgwick quotes instances in which by Western standards an entire community may be ill even though the community itself feels healthy. He refers to the Rockefeller Sanitary Commission on Hookworm which in 1911 found that "this disease was regarded as part of normal health in some areas of North America," and a South American Indian tribe in which "the disease of dyschromic spirochetosis which is marked by appearance of coloured spots on the skin was so 'normal' that those who did not have it were regarded as pathological and excluded from marriage."[9] Examples parallel to these abound in many parts of Africa. For example, a person afflicted by epilepsy, schizophrenia, or hypertension, insofar as that person is not disabled in his daily activities, is (contrary to the expected pronouncements of Western physicians) regarded as healthy.

This is what Dubos means by saying that health is an "expression of environment . . . a state of adaptedness." Disease is thus the inability of a person or community to adapt to changes in the environmental conditions, and health can only be understood in relation to such conditions and the history of a particular people:

The criteria of health are conditioned even more by the aspirations and the values that govern individual lives. For this reason, the words health and disease are meaningful only when defined in terms of a given person functioning in a given physical and social environment. The nearest approach to health is a physical and mental state free of discomfort and pain, which permits the individual concerned to function as effectively and as long as possible in the environment where chance or choice has placed him.[10]

Considering health in relation to factors of function and adaptation to environment, Kass argues that "what is healthy is dependent not only on time and circumstance but even more on customs and convention, on human evaluation . . . without human judgment, there is no health and no illness."[11] This view contrasts with the theoretical notion of health as an objective and value-free phenomenon in which the health of an organism consists in the ability of its parts to perform certain natural functions. Strictly speaking, these are biological functions — the heart pumping blood, the stomach digesting food, the kidney eliminating wastes, etc. Within the whole it is possible that some of these parts will be deficient. But as a self-regulatory system the organism may still be said to be healthy, if it maintains a certain degree of equilibrium and stability despite the failure of some of its parts.

However, the indicator for this stability is judged in relation to the organism's ability to act in the pursuit of its goals. Particularly in man, an ability to engage in activities in relation to his aspirations constitutes his fundamental health-indicator. The point has already been made about the differences in kind and in degree between the pursuits of individuals. These differences throw in the element of subjectivity in what, for an individual, is regarded as a healthy state. In group-oriented activities, health even becomes a more normative concept since it is determined by group goals in relation to environmental circumstances.

This is the sense in which Dubos conceives of health as the ability to function and disease or illness a state of mental or physical incapacitation, all in relation to the problems of adaptability.

HEALTH FROM A CULTURAL VIEWPOINT

"Being healthy," according to Robert Brown, is a relational predicate of degree, similar to "being hard" and "being small." Just as a thing is hard or small, hard or small with respect to some area (as judged in relation to similar things), so we may say of a person that he is healthy (among his fellows) or that the person is "healthy with respect to the performance needed."[12] Now, since the level of performance is normally set by each individual for himself, health so conceived would depend on how high or low the individual places himself on the performance scale. Might we not say the same of different culture? Could it not be said that what constitutes health is culture-bound, that is, that it is determined by a people's social expectations and level of performance in relation to demands placed on them by environmental circumstances?

Recent achievements with artificial heart implantation and organ transplantation follow a conception of illness that has to do with disordered function of human parts. This conception in which illness is a result of structural changes in the cells of parts of the body and the consequent failure of those parts to function is not shared by all communities of men. Its main determining criterion seems to be the level of medical technology and its attendant high level of specialization. In many parts of the world, particularly in Third World nations, medical attention is yet to be directed to such restricted ailments unless, of course, they lead to disturbances in the individual's general capacity to meet social demands, e.g., the ability to work or to perform other social activities. In other words, disease or illness in such societies is holistic in conception — a man is considered ill if he displays a state of unusual feeling, suffering pain or incapacity and in danger of death or mutilation. Once his day-to-day life activities are affected by this general feeling, such a man is said to be ill whether or not the causes are traceable to specific structural changes in the cells of the body.

This conception of health and illness may be considered unorthodox in modern medical practice, but it comes out clearly among the Yoruba community in Nigeria. The Yoruba word *alafia*, which translates health, according to Z. A. Ademuwagun, "embraces the totality of an individual's physical, social, psychological and spiritual well-being in his total environmental setting."[13] Contrary to the claim by Lewis that "it is the presence of disease that can be recognized, not the presence of health,"[14] the Yoruba believe that both states are recognizable and in a negative terminology they conceive of illness as *aisan,* which translates the absence of health. Again, their holistic conception of health and sickness is reinforced when the Yoruba speak of the former state as "when *ilera* (body) is strong and active" and of the latter as "when *ilera* (body) is broken down."[15] The main indicator of health or disease in Yoruba is thus the ability or inability to perform one's routine work, adequate or inadequate performance.

There is in most African cultures even a wider conception of health and illness which incorporates both the physical state of a person and his sociopsychological dispositions. To be healthy according to this conception is to be in total harmony with all the forces that assail man's well-being. Some of it may sound metaphorical, but a person is said to be ill in Africa when he is afflicted by forces such as hunger, unemployment, laziness, strained human relationship, lack of money, infertility, domestic problems, etc. He is considered ill insofar as these factors impair his productive abilities and therefore his overall capacity to fulfill his aspirations in life. One possible interpretation of this conception is to see these factors as causes or manifestations of illness rather than states of illness themselves. But this interpretation is immediately rendered absurd when the traditional African speaks of restoring a patient's health by curing him of his poverty, laziness, hunger, infertility, strained relation, etc., implying that these are illnesses themselves rather than causes of illness.

An important aspect of the African conception of health and illness is that it is the whole human body that is considered either well or in a diseased state. It is not merely some part of the body. Unlike in the West where a patient, when consulting the physician, often throws some hint as to what part of the body he thinks there is affliction, the traditional African (except in a few cases of ailments such as headache, stomach ache, pains in some easily identifiable anatomical parts of the body or external injuries due to accident) is generally nonspecific as to the part of the body afflicted by disease. Even the healer whom he consults does not press for such specific information. This nonspecificity in associating diseases with parts of the body is clear from the fact that, generally, traditional healers do not start their diagnosis of illness by a physical examination of the patient's body. Their primary concern is with the patient's background in sociocultural and in divine/supernatural relations.

It is tempting to explain this holistic conception of health and illness in Africa by saying that both the patient and the healer are ignorant of the detailed anatomical components of the human body. Another possible explanation could be that they know about the physiological components of the body but that they remain ignorant about the causal relation between a given disease and the specific parts of the body it afflicts. These possible explanations are tenable if only the African believes in a mono-causal conception of disease. In traditional African culture a given illness or disease is generally explained by reference to several causes, some of which in modern scientific thought appear to be logically incompatible. An African healer may attribute a disease to a scientific/natural cause not too dissimilar to the germ theory of modern medicine. Yet he may also believe that the same disease is "caused" by supernatural forces. He would then proceed to cure the disease in these two seemingly incompatible directions.

Normally, any such conception of illness that appeals to supernatural forces, deities, spirits, witchcraft, etc., is classified as a form of animism that is common in the history of every society. For example, early medical practice in Scotland took this form in which, according to M. Clough, "healing lay in propitiating the powers (supernatural) against which the patient might have offended. . . ."[16] Supernatural factors play an important role in almost all preliterate (ancient and contemporary) societies of the world. It is common for modern scientific thinkers to read irrationality into all supernatural approaches to medical healing, but I would argue that in relation to the African conception of health and illness, this impression is misleading. Although animistic in outlook, the traditional African concept of disease or illness conforms, at least in part, with the basic norms of modern medical practice. To a traditional African, the causes of illness fall into two major categories, the *primary* and the *secondary*.

Primary causes of illness are those predisposing factors not directly explicable in causal terms. Some of these take the form of supernatural entities such as deities, spirits, and witches; others are stresses due to either the victim's contravention of communal morality or his strained relationship with other persons within the community. Secondary causes

involve direct causal connection similar to the cause-effect relation of
the germ theory in orthodox modern medicine. If, for instance, a man is
suffering from stomach ache and acute diarrhoea and is vomiting, he is
suspected to have eaten "poisoned" food. It has been reported that in
Yoruba, *ete* (leprosy) is spread either by spiders, through chewing sticks
on which flies have landed, or by drinking local gin.[17] The Yoruba
concept of *kòkòrò*, synonymous with "germ" in English, suggests that
there is in the culture a nonmetaphysical/causal explanation of disease.
These explanations may lack the theoretical details of modern medicine,
but they are in principle similar to diagnoses in modern medicine, their
truth or falsity being irrelevant. Our main concern here is with primary
causes of illness and their possible relations with secondary causes.

Primary causes of illness are subdivided into two interrelated sets:
those involving supernatural forces and those of stress due mainly to
strained social relations. Of the two, the former appears to be more
incompatible with the principles of modern medical explanations. For
most scientific minds, any explanation couched in supernatural terms is
a direct negation of the principles of science. But Robin Horton has
argued that this apparent incompatibility is an illusion. According to
him, the traditional African healer whose diagnosis of disease refers to
supernatural forces is not doing anything categorially different from
what the modern scientist or physician does.

Horton points out the similarity between the diagnosis of the modern
physician and that of the traditional healer by arguing that they are both
invoking theoretical entities of different kinds for explanatory purposes.
They have a common goal which "involves the elaboration of a scheme
of entities or forces operating 'behind' or 'within' the world of common-
sense observations."[18] They are both in search of theories that bring
order and regularity into an apparent world of disorder and irregularity.
By use of an analogy, Horton compares the principles involved in
scientific theory with those of a traditional diviner who appeals to
nonvisible, nontangible forces:

The situation here is not very different from that in which a puzzled American layman,
seeing a large mushroom cloud on the horizon, consults a friend who happens to be a
physicist. On the one hand, the physicist may refer him to theoretical entities. "Why this
cloud?" "Well, a massive fusion of hydrogen nuclei has just taken place." Pushed further,
however, the physicist is likely to refer to the assemblage and dropping of a bomb containing certain special substances. Substitute "disease" for "mushroom cloud" "spirit

anger" for massive fusion of hydrogen nuclei and "breach of kinship morality" for "assemblage and dropping of a bomb," and we are back again with the diviner. In both cases, reference to theoretical entities is used to link events in the visible, tangible (natural effects) to their antecedents in the same world (natural causes).[19]

If this analogy is to stand, there must be some causal connection between the "spirit anger" and the illness. This connection is provided by the second arm of our primary causes, namely, stress due to the contravention of communal morality or by strained social relations. Here, however, the distinction between primary and secondary causes seems to disappear. But the point is that modern medical practice is replete with stress-induced diseases such as rheumatoid arthritis, psoriasis, essential hypertension, coronary-artery disease, hypertensive haemorrhage, peptic ulcer, etc. Vaguely speaking, the explanation is that stress reduces the natural resistance of the body against these diseases such that people in a state of stress are more susceptible to their affliction than those who are not socially disturbed.

It is important to distinguish the African conception of stress from the Western conception. A business executive in the West might suffer from stress if his business was on the verge of collapse; a heavy day's work without rest might provoke a state of stress, or his anxiety for possible contingencies might make him suffer from stress. In traditional Africa, stress is due mainly to strained relationship either with one's spiritual agents or with other persons within one's community. It may also be due to a feeling of guilt arising from a breach of communal norms. For example, if an African is involved in an adulterous act with his brother's wife, whether or not this act is detected, the person undergoes stress, having disturbed the social harmony. If he cheats his neighbor, has been cruel to his family, or has offended his community, the anxiety that follows may take the forms of phobias, either of bewitchment or of the affliction of diseases. He feels, and in fact is, vulnerable.

A parallel case can be made between the traditional African conception of medicine and that of orthodox medicine. Just as the modern Western medical scientist would explain certain diseases by a conjunction of the germ theory and reduced resistance due to stress, so may the traditional African healer conjoin primary and secondary causes in his explanation of diseases. The difference, though, is that the former has at his disposal a well systematized body of theories to follow while the latter works on a piecemeal basis of trial and error. Note, however, that not all orthodox medical physicians are theoreticians in the scientific sense of the word. There are many whose practice is based on trial and error — they follow the germ theory without knowing its mechanisms. In the same way, it could be said that the traditional African healer follows certain principles although he is unable to say exactly what these principles are.

The apparent animism of the traditional African concept of health and disease is more manifest in the methods of diagnosis. Unlike the modern physician who has to rely almost entirely on the pharmacological efficacy of his drugs, cure for the traditional African healer is directed toward the two targets of primary and secondary causes. The traditional African healer may be confident of the pharmacological activities of his herbs, but that is not all. The herbs are efficacious, so he believes, only if the primary cause has been taken care of. The herbalist is thus also a diviner, which gives his profession a metaphysical outlook. But, again, this is a mistaken impression. As we said earlier, the primary causes result in the weakening of the defense mechanisms of the body. Cure in this respect simply means restoring the body to a state of increased capacity to heal itself, a state in which the pharmacological efficacy of the drugs is maximized.

Again, there is a parallel of this kind of integrated approach in modern medical practice. The well-known placebo effect in orthodox medicine, in which confidence and positive belief on either the part of the physician or the patient produce favorable effect, is well-nigh indistinguishable from the dual approach of the African healer. Belief here must be distinguished from mere unquestionable faith of the religious type. It is a psychological state of confidence that leads to physically effective results. In it (either African or modern medicine) the conviction that his physician is competent and that the drug works helps to restore the body to a state of harmony between it and the applied drug.

Psychological states, attitudes, and beliefs have been known to play significant roles in traditional African medicine; they now provide acceptable explanations for some of the ailments that have in the past been attributed mainly to supernatural forces. J. C. Carother claims that anxiety, for instance, which in Africa is believed to be an outcome of bewitchment, leads to phobias "whose physical symptoms take predominantly the forms of gastric and cardiac neuroses and of impotence.

Anorexia nervosa or something akin to this, from time to time may be fatal. Fears that the food is poisoned may initiate the syndrome, but its continuance is governed by a feeling (a disguised depression) that the unusual struggle has been lost and the time has come to die."[20] Again, it is clear from this why the diagnostic method is such that the primary cause (in this case, bewitchment, believed to be the cause of anxiety) must be counteracted first or simultaneously with the secondary cause.

Whether in witchcraft, in sorcery, or in spiritual possession, these beliefs are too submerged in traditional African culture to be ignored in matter of health and disease. This is not to say they are true, meaningful, or even rational. The important point is that the beliefs are, as a matter of fact, held, that they play an important role in the diagnosis of illness, and that they affect the pharmacological activities of drugs. Dorothy Rowe stresses the importance of such beliefs in her critique of the orthodox approach in the psychiatric administration of psychotropic drugs. According to her, it can be established "that if a person believes that he has good reason to be anxious or depressed (this 'reason' may not be rational or even expressible in words) the drug does not change his belief and the effect of the belief overrides the effect of the drug."[21] Psychotropic drugs, according to her, are like aspirin, which takes away the pain of toothache without healing the tooth.

There are conceptual difficulties with any account of illness that draws simultaneously on both natural and nonnatural forces. Where the nonnatural forces are social or psychological factors, the problems may be taken care of by psychoanalysts. But in Africa, where the causes of illness are a blend of supernatural forces (gods, deities, spirits, etc.) and natural forces (germs, parasites, kòkòrò, etc.) the apparent difficulty that emerges is similar to the body/mind problem, a subspecies of the general issue of how a nonphysical entity can possibly interact with a physical entity.

This problem is more assailing from the point of view of Western culture where there is a deep-seated bifurcation between the natural and the supernatural. Despite the general belief in the West that the universe is governed by ineluctable laws of nature, there remains a presupposition that these laws work side by side with supernatural forces (at least, it is so for Western religious believers). But in traditional African world-view, there is no clear distinction between the natural and the supernatural, and it is believed that existential entities are arranged in a hierarchical order within a single continuum.[22] This may sound incompatible with our earlier distinction between primary and secondary causes of illness, but it explains why in the mind of an African conflict does not arise between the two sets of causes.

Even if it were assumed that the dichotomy between the natural and the supernatural exists in traditional African thought, the apparent conflict in the people's explanation of illness may still be resolved by invoking the difference in principle between primary and secondary causes. It could be said that a healer in tropical Africa, attending to a patient suffering from (say) severe cerebral malaria, is aware (if only vaguely) that his patient's ailment is caused by a parasite (secondary cause). But in a culture where almost everybody suffers repeatedly from bouts of malaria and where the disease is normally not severe, it is obvious that the patient's consultation is bound to move beyond the "how" question to the "why." "Why such a severe attack and why me and not someone else?" These are quests for primary causes beyond the level of the physical. Note that in searching for answers to these questions, unlike in Western cultures, the concept of chance hardly plays a significant role.[23]

The issues raised at the level of primary causes cannot be resolved by applying the canons of scientific reasoning. Indeed, viewed from the paradigm of science some of the claims made are likely to sound meaningless, irrational, and false, if these terms are ever applicable. The crux of the matter is that the apparent conflict that exists between primary and secondary causes can be shown to be unreal. There is therefore no absurdity involved in an integrated diagnostic process that blends the natural with the supernatural nor in a curative process involving the pharmacological activities of herbs and the appeasement of supernatural entities.

It has to be granted that the language of discourse in traditional African culture in matters of health and illness is different from that of orthodox medicine. Attempts to analyze the former from the standpoint of the latter have either been inconsistent or produced some of the most bizarre results in medical literature. Observations have shown, for instance, that in traditional African medical practice a single drug is sometimes believed to be pharmacologically potent in curing such antithetical ailments as diarrhoea and constipation. In others, such as U. Maclean reports of some Yoruba herbalists, one herb is believed to be able to cure several diseases, most of which are unrelated. Reporting on the herb, *Kòròpò* (*Crotalaria retusa*), Maclean writes:

While some people seemed to prescribe it primarily as an analgesic for rubbing on swollen joints or any painful area, there were numerous advocates of its use for "eye trouble," for dysentery, and for applying to the swollen breasts of pregnant women. In addition it was regarded by some as an aid to conception, being said "to retain sperm in the vagina," and it was reputed to bring on labour when this was overdue . . . that it would induce a divorced wife to return to her husband; that it would guard a house and its occupants against dangerous medicine; that it would assist in the arrest of "evil-doers and lunatics." It was even regarded by one herbalist as a valuable aid to school children's memory, when taken along with fish.[24]

That one drug is capable of curing so many diseases may not even be as striking as the blend between its physical potency and its magical powers. Yet it would be hasty to dismiss the claims for these reasons. They may, after all, not be as incredible or as irrational as they appear when seen in relation to the people's conception of health and disease and in the context of the culture of the practitioners.

University of Ibadan

NOTES

[1] The author is aware of research successes in this direction in South Africa.
[2] Ralph Pettman, *Biopolitics and International Values: Investigating Liberal Norms* (New York: Pergamon Press, 1981), p. 107.
[3] J. C. Carother, *The African Mind in Health and Disease: A Study in Ethnopsychiatry,* WHO Monograph Series, no. 17 (Geneva, 1953), p. 121.
[4] Joseph Margolis, "The Concept of Disease," *Journal of Medicine and Philosophy* **1**, no. 3 (September 1976):243.
[5] See René Dubos, "Health as Ability to Function," *Contemporary Issues in Bioethics,* ed. Beauchamp and Walters, pp. 96–99.
[6] Ibid., p. 98.
[7] Ibid.
[8] Leon R. Kass, "Regarding the End of Medicine and the Pursuit of Health," in *Contemporary Issues in Bioethics,* p. 105.
[9] Peter Sedgwick, "What is 'Illness'," in *Contemporary Issues in Bioethics,* p. 116.
[10] Ibid., p. 99.
[11] Leon R. Kass, "Regarding the End of Medicine and the Pursuit of Health," in *Contemporary Issues in Bioethics,* p. 105.
[12] Robert Brown, "Physical Illness and Mental Health," *Philosophy and Public Affairs* **7**, no. 1 (1977):22.
[13] Zacchaeus A. Ademuwagun, "'Alafia' — The Yoruba Concept of Health: Implications for Health Education,"*International Journal of Health Education* **21**, no. 2 (April—June 1978):89.
[14] Andrew Lewis, "Health as a Social Concept," *British Journal of Sociology* **4** (1953): 111.

[15] Ademuwagun, "'Alafia'," p. 90.

[16] Monica Clough, "Early Healing," in *Proceeding of the Royal College of Physicians of Edinburgh Tercentenary Congress,* ed. R. Passmore (1981), p. 183.

[17] U. Maclean, *Magical Medicine: A Nigerian Case Study* (Penguin Press, 1971), p. 87.

[18] R. Horton, "African Thought and Western Science," in *Rationality,* ed. B. Wilson (Oxford: Blackwell, 1974), p. 132.

[19] Ibid., p. 136.

[20] Carother, *The African Mind.*

[21] Dorothy Rowe, "Philosophy and Psychiatry," *Philosophy* 55, no. 211 (1980):110.

[22] Okot P'Bitek, *African Religions in Western Scholarship* (Kampala: East African Literature Bureau, 1970), pp. 46—69.

[23] See J. O. Sodipo, "Notes on the Concept of Cause and Chance in Yoruba Traditional Thought," *Second Order: An African Journal of Philosophy* (University of Lfe) 11, no. 2 (1973);12—20.

[24] Maclean, *Magical Medicine,* pp. 84—85.

H. TELLENBACH

THE EDUCATION OF A MEDICAL STUDENT

If one wants to talk about "the education of a physician," one can hardly do so without a clear concept of what a physician is, or without at least an essential idea of what constitutes the practice of medicine; because it is such concepts that must serve as a model of a physician's education. Present-day medical training is not aimed at such a conceptual idea; on the contrary, even the most contemporary educational systems are completely devoid of such an ideal and the imagination to carry it out. The education offered is confined to the dimension of scientific research and its practical diagnostic and therapeutic application to an ill person. Because a medical student is expected to do no more than to acquire specialized knowledge and to learn its technical applications, his qualifications for the profession are assessed only on the basis of certain intellectual talents and learning abilities. As an individual, he may be whatever he is or whatever he happens to become — possibly even a misanthrope! The result can hardly be anything else but the picture of the physician that Otto Dix has drawn and Büchner shows in *Woyczek*. Even the commendable intention to augment medical training by medical psychology and medical sociology will not decisively change this situation. Nothing will be changed if this training is carried out in terms of physiological psychology, and only little will be changed if the psychoanalytic school of yesterday is called upon to promote the notion that the psyche is an organ (which has been unduly neglected by physical medicine), a "psychic apparatus" to which analogous scientific methods of investigation can be applied. This school considers the psyche as such to be radically determined by natural human drives, and it thus emphasizes the training in certain techniques of examination and treatment and statistical verification of the results; in other words, psychoanalysis without a new *mediation* of the relationship between drives and the mind.

The basic reason why we have to discuss the problem of educating a physician is that Western medicine does not have an inherent self-understanding from which it could develop a concept of the practice of medicine. Western medicine is based on the laws of nature, and its

A-T. Tymieniecka (ed.), Analecta Husserliana, Vol. XX, 175—184.
© 1986 *by D. Reidel Publishing Company.*

application is a manipulation of the system of cause and effect. The manipulation makes the glory of its success — but it is equally the reason for what seems so hard to admit, namely, its renunciation of any ambition to heal in the sense of restoring a natural life-situation. We may learn what healing really means by looking to Egyptian, Greek, Arabic, medieval, Chinese, or Tibetan medicine, or we may let philosophers tell us — as Gadamer does in his paper "An Apologia for the Art of Healing."[1] These historic concepts should arouse an enthusiasm, because they have — with the inner consistency of all genuine creations of the human spirit — grown out of the substratum of myths, and from this they comprehend disease, and astonishingly enough health also — as arising from the selfsame existential basis. This is the profound reason for studying the history of medicine, a subject grossly underestimated by the educational planners of today. We find in these ideals models for the medical profession and for great spiritual concepts of health and disease; we even find practical therapeutic implications which — in high cultures such as the Islamic — gave rise to institutions that we today can only contemplate with envy.[2]

Within *our* concept of medicine, however, we cannot even determine what disease actually is, because to us both health and disease are notions of value that do not occur in the realm of the natural sciences. That incredible process of nature's assuming power over us, of its exerting this power in a dictatorial fashion, of its disengaging itself from the web of interrelations by which, in times of health, it cultivates "us" as "we" can cultivate and shape it — such things are, in our medicine, reduced to the category of pathological symptoms. There is no room for the existential aspect of disease, for the basic change of being, which is the essential trait of illness.

Present-day medical training does not offer the student any experience of the structured aims that determine his relationship to the patient, nor any awareness of how he ought to educate himself or perhaps even transform himself, so that he may be able to really do justice to his profession of healing. What is missing is the development of an inner authority which — while knowing how to employ special knowledge — performs the function of what Aristotle called the phronesis: that circumspect recognition of what can responsibly be done in the face of the established biographical situation of the individual and one's own knowledge of the nature of Man; in other words, to find the pragmatically most beneficial way of restoring the central balance of which Hippocrates speaks and which we call "health."

But we are not merely concerned about the fact that we are lacking an education toward a specificity of personal relationship; we equally perceive the deficiency as being a certain instability within the medical profession as part of the framework of political and social life.

Since we are reflecting upon the physician of tomorrow, let us try to view two possible further developments and their inherent inadequacies that result from a general rejection of a more comprehensive way of forming the Hippocratic attitude of the doctor. The first possibility is based on the proposition that the physiognomy of our age is generally distinguished by the pragmatic application of its scientific advances. Since we depend on this for our very existence, we tend to overlook the fact that — side by side with this process of transposing the results of pure research to the area of practical operations — the shadow of a destructive and irrational element inevitably enters the arena of life. However helpful and even liberating the possibilities are that physics, chemistry, pharmacology, bacteriology, genetics, and even psychology have bestowed on us, their usefulness has long since been outstripped by their destructive aspects. This is obviously the case with atomic, biological, and chemical weapons; this is equally true of analytical psychology, which immodestly claims to understand everything, only to cancel out all norms of value; for analytical psychology regards man's spiritual powers as being nothing more than emanations from an unconscious and irresponsible source, rather than being the servants of truth, or even the primal powers of myths.

Let us take the example of genetics to illustrate the possible consequences. We all know that its scientific results seem to indicate that the selective breeding of man is possible. Genetics itself, however, does not have a compelling reason for carrying out such an experiment, because pure science is not primarily directed toward practical ends. But it is not difficult to imagine political or social power groups that would gladly make use of such scientific results for their own particular purposes, for instance, by claiming that the research and domination of space is necessary and beneficial for human society. Without a doubt, such power groups would seek to have the expediency of their intentions legitimized scientifically and, for this purpose, would turn to the geneticists. What would *their* decision be like? And how would they substantiate it? In no respect could they draw upon experience gathered during their medical training. Very likely their opinion would be based on that kind of consciousness that regards the sciences as the legitimate heir to the cybernetes-office of philosophy, and that accordingly "The development of scientific steering methods for the life of society" is destined to "shape the profile of our epoch."[3] Such a possible decision would be characteristic of the situation that produced it, like our own situation is characterized by an increasing emphasis in favor of the sciences over the aesthetic, ethical, and religious avocations of human society. A hundred years ago, when for the first time the physician no longer needed to take qualifying examinations in philosophy but only in the natural sciences, the dignity of man was still a living and shaping force in his profession, so that he could confidently leave such decisions, as just simulated, to the authorities of the aesthetic, moral, and religious culture to which he belonged. All this has long since undergone a fundamental change. The inadequacy appears in the fact that now one can no longer assume as a matter of course such self-evident sensitivity toward the norms and connotations of life. For a long time — but most urgently since the medical profession's concept of itself during the Nazi period — this situation has called for a corrective. But such a corrective can only be achieved if the medical student's awareness is expanded toward the insight that what ultimately distinguishes man is his spiritual existence; that this existence is based on tangible and irrevocable conditions that are inherent in him as a human being — and that as such determine the area outside of which even the most ambitious plans do not make any sense.

Today, at every step, we are being confronted with indications of the second aspect within a possible future development. At our universities all manner of radical activism, in league with liberal tolerance right up to anarchic utopianism, is pressing toward its realization. We are thus referring to the possibility of a radical change in our total context of society brought about by any one of the conceivable nondemocratic life-practices and life-interpretations. As lightning, certain situations enter our field of vision that reveal how inadequate and essentially unaccountable the training of our physicians has been as regards its influence upon their actual practice of medicine. If a dictatorial power were to force regulations upon them, with the argument that such regulations were instituted for the benefit of society, then this weakness would quickly be revealed. His training, of course, has in no way prepared the physician to deal with such directives, for which the authorities advance sufficiently engrossing arguments. Based on forceful political ideologies, such arguments can easily bend the mere feeling for what is right as a treatment of a human being, particularly when, argued in the guise of legitimacy, they confront the medical field from outside; we have had bitter evidence for this. In Nuremberg, physicians were being prosecuted who used ideological reasons for justifying their deadly experimentations with man. V. v. Weizsäcker was right when he said that in Nuremberg the spirit of scientific medicine was being inculpated. In the face of such experiences, there should be profound concern over our failure to offer the kind of medical education that — although it may not protect the physician against a breakdown of character — may still raise his power of resistance against indoctrinations devoid of or hostile to values. It would at least make him aware of the fact and the reasons for his own failure or his refusal to obey. Isn't it obvious that an education which is solely directed toward conveying positive knowledge, cannot possibly arm the individual against believing things to be true and right when in fact they are merely profitable to society and industry; nor can it help to protect him against becoming an instrument that can be manipulated by purposes extraneous to his professional obligations?

The memory of Nuremberg, however, may also direct us toward that primary region where we have to vindicate our discourse on the training and education toward medical responsibility. In my opinion, to have indicted these physicians for murder confused the facts of the case. The actual implications were much more vicious; having been prevailed upon to adopt an experimental relationship toward man, such as one previously only had toward animals, these physicians were brought to a point where they had to abandon one of the most deeply rooted presuppositions of Western culture: man as made in the image and likeness of God. It was Dostoevski who, in Rodion Raskolnikov, created a forerunner of such abandonment. After having killed the old usuress, Raskolnikov says: "It was only a louse I killed . . . a useless, disgusting, mischievous louse"; "among all of them I chose the most useless louse."[4]

This does indeed deny the very center where reposes the essence of education. In German, we use the word *Bildung* ("forming") for education; it has its origin in the mystical concept of the Middle Ages: the soul bears within it an image of God in whose likeness man was created, and the meaning of man's life lies in educating himself; in other words, molding, imagining forming himself toward this likeness.[5]

Before we continue with the meaning of *Bildung* or "formation," I want to make it unmistakably clear that nothing is further from my mind than a devaluation or disavowal of scientific research. It may be difficult to look upon the fact that even the psyche is made out to be a part of purely physical nature and forced into the perspective of natural laws, while its objective configuration cannot possibly correspond to such a way of thinking, nor to the methods that belong to it. But even to declare unpretentiously this approach to psychic research the *only* scientific one would not give us the right to disavow research. Scientific research has no choice but to follow its path, however utopian the regions may be to which it leads. The only problem it presents to us, as physicians, is what status we confer upon the results of this research and to what degree of exclusiveness we apply them. This means that the slowly maturing physician must be shown that these results do not form the most worthwhile knowledge, but an awareness of the highest aim which all domination of nature within the field of medicine ought to serve; and that this highest aim of medical treatment consists of nothing but the restoration of that balance which we must learn to see as the essence of health. It is the task of a specifically directed education to demonstrate this and to recognize that the loss of balance in a state of disease is not only an objective medical-biological fact, "but also a biographical and social process."[6]

As such, education has no goal outside of itself. The education of a physician is always the education of the self, of a spiritual essence, that can enable one to correspond to the demands of his profession. Education evolves within the interactions of a community, in language and customs, where traditions are fruitfully modulated to innovations. In such an atmosphere a practical power of discernment can develop; a sensitivity for values and taste, a sense of tact as regards human intercourse as well as the most adequate approach to another human being, which might do him justice and might be commensurate and appropriate for him. Thus it is not a matter of storing knowledge for the purpose of transposing it into a function: "in education the knowledge absorbed is not like means that have lost their function; nothing disappears in an acquired education, rather everything is preserved within it. Education is a genuinely historic concept."[7] According to this definition, education becomes a determining factor in life, even from a physical aspect. Perhaps this "unloseability" is nowhere more clearly demonstrated than in certain processes of brain atrophy, which largely robs the patient of his power of discernment in the sense of his no longer being able to render an account according to logical reasons. But it does not to the same degree, rob him of proper behavior patterns, in the sense of formal manners which, through self-education, have become an integral part of the self. Such patients can still radiate the atmosphere and exhibit the behavior patterns of the society in which they grew up. Without reflection or effort, they are able to do what "one" does in a given situation. They can still largely grasp the meaning of a certain situation for themselves and for another person. To my mind, one of the most essential aims of medical education ought to be that the physician learns to comprehend and assess the individual from within the context of those meanings — and their innate affective powers — which shape that person's relationship to the world. It follows that the physician — with respect to the human being — ought not only to pay attention to the onto- and phylogenetic development of the living organism, but above all to a genesis of the interrelations between meanings and effects, such as they are formed in the individual by way of his personal experience within the social order and purposes of his life-environment. To transcend our understanding of man as an object of medical natural science, without context and without a personal history, in favor of seeing him within the context of earthly existence and history, seems to be an essential aim of medical education.

Under such circumstances we may no longer avoid the consequences as they regard the education of medical students. Therefore, we must resolutely oppose the tendency of those people to whom the threatening disintegration of the university is but a signal for turning medical departments into specialists' schools. The exact opposite is what is needed! It is necessary to bring into the field of medical science the spiritual essence of the philosophical disciplines, which are concerned with the totality of man and thus may give us deeper insight into his being. Such a demand is not new. It has become ever more apparent that we will never get beyond the barbarism of actualities if we do not firmly establish medicine within the continuity of our knowledge of man. Therefore, we should not elude the problem of what is involved in the way of theoretical and practical instruction if we are to realize this endeavor.

The first and foremost requirement would be that, in the first semesters, the medical student be offered a preseminar that will disclose to him the foundations of anthropology. This might begin with unmasking the clandestine metaphysics of modern medicine, which keeps the student ignorant of the general philosophical presuppositions of his field. The philosophical premises of the various concepts of the nature of man in the field of medicine would have to be explained in order to elucidate how they result in the respective methods of treatment; for instance, it would have to be shown in what way modern medicine is based on the thinking of Descartes; how modern medicine reduces man to a functional aggregate that obeys natural laws; that the processes investigated and established in this fashion may not by any means be mistaken for existential reality as such, but that this kind of knowledge was arrived at by our thinking of man in the same way as we think about nature, which can be made the object of natural science and its methods of investigation. It would have to be demonstrated that biographical medicine has its roots in Dilthey's work — that psychoanalysis ought to look for its basic anthropological concepts in Schopenhauer, Schelling, Kierkegaard, Nietzsche, and others. Furthermore the student will have to be shown that, for medicine, there is not only inductive experience but also hermeneutic, phenomenological, physiognomical, and, not least of all, atmospheric experience. The teaching of philosophical anthropology would also have to consider those types of human existence that are the special concern of the physician, as, for instance, it would have to interpret the meaning of anxiety, exhaustion, hunger, thirst, and emotional involvement, etc. In addition, there ought to be an introduction into the essential forms of human coexistence — for instance, a phenomenology of love, as Binswanger has suggested — and not only a theory of sexuality. Beyond that a knowledge about man's proclivity toward transcendence should be imparted, as, for instance, a phenomenology of hope, without which an understanding of melancholy is impossible. And what does philosophy — for instance, Schelling and Kierkegaard — tell us about the essential idea of illness and health? And should not also the therapeutic dimension of a student of medicine include that interpretation of disease which to modern man seems utterly absurd: disease seen as a potential possibility for creative transformation, which was the primary cause for the actualization of the religious genius of St. Augustine who, in turn, had such an impact on Nietzsche and Novalis. Along with all of this will have to go a discussion of the curtailment of our therapeutic aim: we must make it clear to the student that the much admired and basically admirable progress in modern therapeutic methods also represents a progression-from-something, an eccentric moving away from the personality of the patient and from an understanding of his disease as a form of his particular state of being, the treatment of which must aim toward a transformation of this state. The medications made available through modern medicine frequently are decisive healing aids toward this end, but their contribution toward the actual healing process is, by their very nature, strictly limited. Beyond that, what is conducive to his convalescence can be established only through analyzing the patient's life-history, through recognizing what biographical meanings and effects changed his situation.

This will bear fruit only — and this is the second pragmatic conclusion — if one has learned to create a common therapeutic situation with the patient, which alone allows one to experience and understand the patient's state as it differs from one's own. In the process of developing this common level the particular individuality of a person exceeds itself and enters upon a higher unity. Thus, in the education of a physician, it is of ultimate importance that he be led to see that "understanding is always the horizon where presumably separate horizons merge."[8] How a patient's "function in life" is interlocked with and patterned after an always completely different inner biography, which is extremely difficult to grasp and may only be approached hermeneutically (L. Binswanger). Only a thorough acquaintance with this type of experience will enable the student to adopt attitude that will prove to be the proper medical approach in a tangible situation. In practice this means: courses must be given in exploring biographical data, in inducing the patient to communicate, and in apprehending the decisive factors of his life. The student must be taught how to keep this process constantly in view in what Freud emphasized as "suspended attentiveness," so that he can work out what Scheler calls a "grammar of expression" in order to explore, above all, the specific weight of certain interrelated meanings that have their origin in decisive events, and therefore not only form the center of meaning of a person's life but constitute the interrelated effects which, as such, can be the cause of disease.

As a third and perhaps most decisive educative factor there must be an opportunity of choosing and attaching oneself to a teacher or tutor. "One really only learns from the person one loves" (Goethe). I mean an attachment to a protagonist and a molding of the self toward a living image — the sphere of genuinely pedagogic Eros that should again become an integral part of the study process. Experience shows a readiness on the part of the students to seek for alternatives to the constantly and incessantly expanding externalization and objectivization of medicine. The majority of students are receptive to the interplay of discursive empiricism and of the essence of the human being. They know that the workings of causality are of elemental importance for the ordering of nature; but they also feel that the operations of causality can be no more than one element in the aim of a physician's task. It is the possibility to integrate into oneself what is inherent in this conception of v. Gebsattel: we must safeguard this possibility for the education of a medical student — beyond all scientific didactics.

Emerit. Ärztlicher Direktor
der Abteilung Klinische Psychopathologie
der Psychiatrischen Klinik Heidelberg

NOTES

[1] H. G. Gadamer, "Apologie der Heilkunst," in *Kleine Schriften* (Tübingen: I. Mohr, 1967).
[2] I. Schaarschmidt, "Der Bedeutungswandel der Worte 'bilden' und 'Bildung' in der Literaturepoche von Gottsched bis Herder" (Ph.D. diss., Königsberg, 1931).
[3] H. G. Gadamer, "Über die Planung der Zukunft," in *Kleine Schriften*, p. 161.
[4] F. Dostoevski, *Rodion Raskolnikoff* (Münich: Piper, n.d.), pp. 558, 368.
[5] Schaarschmidt, "Der Bedeutungswandel der Worte 'bilden.'"
[6] Gadamer, "Über die Planung der Zukunft," p. 218.
[7] H. G. Gadamer, *Wahrheit und Methode* (Tübingen: I. Mohr, 1960), p. 9.
[8] Gadamer, *Wahrheit und Methode*.

PART II

THE MORAL SENSE IN PSYCHIATRY: THE SWITCH FROM THE ISOLATING APPROACH TO THAT OF "TRANSACTING" WITH THE OTHER

CHARLES E. SCOTT

THE MORAL SENSE AND THE INVISIBLE
OBJECT

I accept Professor Tymieniecka's departure from Husserl and Scheler in
her account of the moral sense. I also accept her claims that the moral
sense is found in being-with-others and that the moral sense does not
originate in constitutive acts by the intellect. I speak within her emphasis
on life-development and on "the intimately individual character of
ethical practice." I shall look at one aspect of the issue she raises
regarding the moral sense, and that is its relation to psychotherapeutic
processes. I must deal with therapeutic processes broadly in these short
remarks, ignoring the important differences in different types of therapy
particularly in relation to different types of disturbance. I shall not deal,
for example, with issues of insanity or schizophrenia or with issues of
society and culture in relation to madness and therapy. My remarks lead
to questions about the language within which we conceive the issue of
moral sense. I want to further the discussion of psychotherapy as a self-
interpretive process that engenders moral sensibility, one that begins in
a type of relation that is not an act of consciousness, a process that need
not be the object of intelligent focus to be effective and generative of
moral sensibility. Whether we are dealing with the origin of values when
we deal with psychotherapy in relation to the human condition is an
issue I will not be able to address. If that question of origin were raised,
we would need to deal with the discourse within which concern for
origin and universality arises. I shall suggest throughout these remarks
that a central question for us is how we speak interpretively of the
psychotherapeutic process. Can we speak of it without the dominance of
constitutive action by the intellect? Can the process of which we speak
speak through our language about it?

What is the experiential situation in therapy which gives a basis for
perceiving what is good? Life-transformation is the only aspect of
perceiving good in therapy that we shall consider among many possible
ones. Perception of good, as well as evil, occurs in the processes of life-
transformation. Life-transformations give or yield the perception of
good and evil as people undergo those processes. The image we want to
avoid is one of a person in the presence of something and who is acting

187

A-T. Tymieniecka (ed.), Analecta Husserliana, Vol. XX, 187–192.
© 1986 *by D. Reidel Publishing Company.*

in a way called perceiving. The transformations which involve, in our case, two persons, are themselves awarenesses in which the two people are a part. Values emerge from the transformative awareness. Values are ongoing instances of the transformation.

In a therapeutic process a person ordinarily experiences allowance and freedom for self-expression and for being as he or she is. In the language of Professor Tymieniecka the therapist's benevolence for the sake of the other prompts the other's life-interest. Instead of shrinking or calculation, life-expanding self-expression and enjoyment of coming to life takes place. Something cramping and hindering or something fearful and threatening falls away. Perhaps an enervating anxiety passes away. One experiences a new force of life, a quality of insight that comprises a different way to be. Prior to this happy part of the process there have been probably anger and blockage, perhaps a brittleness of psyche, perhaps childish infatuation, and very likely an incapacity to deal productively with certain kinds of things. The two people working together often have not felt a depth of understanding between them. They have felt disinterest in spite of intentions to the contrary. One or the other has been bored or irritated. At times nothing worthwhile would happen. One has probably talked and talked to no end or lapsed into dead silence. There have been direct attacks, flanking actions, thrusts and retreats, elaborate camouflaging, parrying, bluffing, and above all diverting and deferring actions. But throughout a simple awareness has been developing, a sense for each other, a quiet sensibility of acceptance that holds nothing back. And this sensibility emerges more strongly now and then and gives a strange clarity for the two people. I say *strange* because probably neither person can speak about it easily without lapsing into jargon and stammering. On the other hand, this clarity of sensibility is far from mystical or romantic. When it happens it is defining and clarifying. It becomes too ordinary in the relationship for extraordinary language. The two people are becoming a part of this sensibility and are feeling its ordering power too clearly to doubt it or, frequently, even to notice it. On occasions this sensibility pervades everything — the way the curtains move in the wind, the pen's lying on the desk's blotter, the curl of rug by the brown chair, the slight ripple of the other's facial muscles on the left side, and the silence of the room's corners.

Lawrence Raab's poem "The Invisible Object" helps to address what we are talking about:

Held toward light
in the shape of your hands
it's clearer than water,
clearer than glass,
than air.
It holds nothing back.

Against the sky, sometimes
it's red, sometimes it's blue.
Set in the grass in your garden
the bees stumble through it.
Looking closely
inside there's always another day
where your life could be.
Yet it has nothing to do
with dreams that change
when you change your mind.

On a table at night it becomes
the center of your house,
a small fire in a darkened room,
and the silence bends toward it
and touches it with one finger.

Lost for days it appears
suddenly
in a pile of old hats.

No reason to be there.
no reason to be anywhere else.
(From Lawrence Raab, *The Collector of Cold Weather*,
New York: Ecco Press, 1976)

The "Invisible Object," this unconstituted occurrence that happens with the two people, is in some instances at least like another life. In it rejection, for example, does not happen when one would ordinarily expect rejection. Nor does censoring. The unconstituted invisible object accompanies the boredom, anger, inertia, guilt, self-infatuation, and immaturity like a different way of being. Raab's phrase is "like another story." In this case it is like a story in which a person's life is fully allowed: allowed in this sense: one person who has many problems, who today, let us say, has an aching back and who is mildly depressed and who has tried unsuccessfully to write three paragraphs over a period of two miserable hours, this person sitting there looking deceptively secure and comfortable finds himself disposed toward this other person. He does not particularly like or dislike the other one. He is actually fairly nonplussed. He certainly would not seek out this other one at a party or on a weekend. But he knows this other one through long exposure and knows himself with the other. He has intended as best he can to be affirmative and to sense as well as possible his own feelings, thoughts, reactions, and adumbrations. He has tried to be honest, tried to work on some hard problems, and has tried to feel and think openly. For all the trying's worth, it has been an iffy process. More failure than success. But the two have come to know each other in the failures, in the partial efforts, in the feelings and occasional passion, in the attempts. And through the efforts this "invisible object" has developed at times and both have experienced some hope, an unusual kind of affirmativeness that has predisposed both people to sensitivities and interests that are different from the ones that usually characterize them. Perhaps the slackness of skin with its slight cast of yellow no longer harbors vague disgust for the other, but now seems alive and peculiar and worthwhile. Perhaps one of the them feels pity and sorrow for his parents' suffering, parents whom he would also like to push into oblivion.

This invisible object has accompanied a change in life. Look at another of Raab's poems, "Accidents of the Air":

> Like nothing in this world.
>
> Not the sound of the shirts
> being sewn, the ladders
> lifted, the water poured,
> or the tree
> filled with wind.
>
> Another story, then. Another life.
>
> Where trees become
> more like single leaves, more like
> the legs of tables, and you
> discover the reports
> of witnesses.

They say *It's not the chipped*
skeletons in their cages
or the cold
arrival of the lists.
And it's not the silence.

But another life
altogether.

Where light
at evening crosses
The lawn
like nothing you had imagined.

in this life or any other.
    (From *The Collector of Cold Weather*, Ecco Press)

"Another story, then. Another life." *In* this change a person also undergoes a revised sense for goodness. Raab's poem suggests that the concreteness of things, their peculiar significances and ways of being, might exercise their influence on how we speak:

"But another life.
Altogether.

Where light
at evening crosses
The lawn
like nothing you had imagined.

in this life or another."

Certainly a revised perceptiveness emerges in which the way things are carries a sense for life; fewer things are distorted by fear, denial, or imaginative flight. Things, perhaps, are experienced as they are in their differences and in *their* happenings. As Stevens says in "Of Mere Being": "You know then that it is not reason/That makes us happy or unhappy." It is something else, quite different from intelligence or the powers of consciousness to constitute objects.

We are looking for ways to speak about this "something else." Its difference and its generative power, which are particularly evident in psychotherapy, are not easily said in most available philosophical traditions. Our recent traditions have not been particularly sensitive to the occurrences of this something else. Our abstract languages seem to lose the concreteness that we want. The considerable difference between philosophical and poetic ways of thinking may itself be symptomatic of the problem we are discussing. It may be that the discourses by which we conceive goodness and value lack the sensibility, the mind, about which we wish to speak. How are we to aim philosophically at the human being in such a way that its difference from cognition, constitution, and intuition of essences is thinkable and speakable? How are we to think of human self-interpretation so that the concreteness of self-interpretations are held and born in our language? My guess is that we need to bring to expression what we talk about in how we talk about it.

If we take the process of therapy as an example in which a moral sense develops, we will find that the values emerging from being free and open with each other address our discourses, and we will be motivated, perhaps, to speak the goodness of therapy in our speech about it and about the ways things are. This moral sense will have the value of looking for what is most alive in our speaking relations and will ask after the invisible object in the midst of our ordinary boredom, irritation, fear, and obsession. How will our speech value the living process which as Raab says becomes the center of our house? Is this center found in the decentering life-transformation that can occur in therapy, for example? My guess is that we should say yes, that we should find out how to speak from and about transformations as sense-forming and as being an awareness that is able to recast our ways of speaking about moral sense in relation to intelligible forms. Is the absence of form, the transformation of form, at the "center" of our moral sense? If that is true, transformation, not personal formation, is the object we need for understanding the moral sense.

Vanderbilt University

NOTE

The poems by Lawrence Raab, "The Invisible Object" and "Accidents of the Air" from *The Collector of Cold Weather*, Ecco Press, 1976, are reprinted by permission. © 1976 by Lawrence Raab.

PATRICK DE GRAMONT

THE GENESIS OF A PURPOSEFUL SELF

I. PURPOSE AND THE NATURE OF REALITY

When having a sense of purpose is considered to be an aspect of human behavior by our psychological theories, it has not, in our cognitive, behavioral, and psychoanalytic view of man been given much of a role. Nor has purposefulness been addressed in terms of the full implications of what an act of purpose must imply. What has been lacking has been an approach that places purpose within the broader issue of the relationship of meaning to reality. This broader issue might be described as follows: does meaning reflect an independent reality that is then reassembled by our nervous system into some faithful facsimile of a reality that is already there? Or do we somehow create our reality, and hence our meaning into something that could not have existed prior to that creation?

For those who believe that we reflect some ready made, independent reality as spectators, it then becomes necessary to explain how we become involved with that reality. Those who hold to a spectator view of reality generally say that this involvement occurs on the basis of desire or fear, since our involvement is said to be prompted by the need to reduce these states of arousal. If we may be said to create our reality, and hence our meaning, then meaning and reality become the same thing; and rather than reflecting a ready made reality, we would have to say that we *are* that reality. We need not then evoke a principle of tension reduction to account for how this works. But we would need some sense of how purpose becomes an aspect of this process, since creating a reality suggests a sense of purpose, which must refer to more than a need to reduce states of arousal. In what follows, "having a sense of purpose" will refer to our capacity to promote a kind of experience, because the experience promoted enables us to realize a meaning. And what I will be leading up to, as I elaborate on the meaning of what was just said, is a view of how this capacity for purpose originates in our development.

193

A-T. Tymieniecka (ed.), Analecta Husserliana, Vol. XX, 193–205.
© 1986 *by D. Reidel Publishing Company.*

II. EARLY COMMUNICATION AS INDUCED AFFECT

Since the issue I have described boils down to how we come to have a sense of meaning, I might start with a preliminary definition of meaning. Meaning is defined as that which is communicated from one person to another. This definition would appear to set aside the question of the nature of our reality by focusing upon meaning as an interpersonal event; and yet we will soon see that this is not the case. The advantage of this provisional definition is that it will lead us to a sense of how meaning is transmitted interpersonally, as well as to what gets transmitted; and it will also provide us with a sense of how meaning might be created in this process. A further advantage of defining meaning as a interpersonal event is that it will enable us to look at some recent research on preverbal communication between infants and their caregivers, and to further examine and question what is involved in what we call meaning, and how that meaning is communicated from one person to another. That the meaning we will be looking at is preverbal will permit me to avoid the complications of what a language system brings to this process. And that this preverbal meaning is of necessity experienced will permit me to concentrate upon the affective component of meaning. Finally, this view of meaning as an interpersonal event will permit me to draw some parallels between what is being maintained here, and Professor Tymieniecka's "Moral Sense."

I will first describe some of the work done by a group of researchers that appeared recently in *Frontiers of Infant Psychiatry* (1983). Daniel Stern and his co-workers have for a number of years carefully observed and recorded what occurs between a mother and her infant in order to determine the kinds of communication that occur in this earliest interaction. They conclude that something is communicated from the mother to the infant, which they refer to as a "resonant emotional response." The infant's response is said to correspond to an affect such as pleasure; and this affect is said to have been induced by the mothers' experience of pleasure. That is, in this form of communication, when a sender expresses a particular affect, the receiver's response is said to reflect the affect of the sender. Hence, it is claimed in this interpretation of our earliest communication that we have a direct means of sharing or communicating a meaningful experience which depends, not upon a symbolic system of communication, but upon the functioning of our nervous system. That is, our nervous system is here assumed to be innately capable of resonating to the tune of someone else's affect, much as is the case for the resonating response of our inner ear to the impinging sound waves of a tuning fork. This capacity would then stand as a fundamental aspect of human functioning; a kind of "built in" capacity for empathizing with another affectively, since affect is by this view so readily shared.

Part of the evidence Stern cites to justify this view is the familiar phenomenon of the "reflexive" crying of the newborn infant to the sound of another newborn's crying. Here, a state of distress is presumably communicated through the sheer induction of affect from one infant to another. One would have difficulty assuming prior learning, or any form of conceptual meaning to account for this demonstration of emotional contagion. To call this phenomenon "reflexive" is to explain it away, without having addressed the possibility that a primitive form of communication is indeed involved.

Another reason for describing our earliest communication as a resonant emotional response is that it helps to understand a pattern of interaction which Stern has paid particular attention to. He and his co-workers have been struck by the degree to which early mother-infant communication represents what they refer to as an affective modulation of the infant's experience by the mother. That is, the mother is repeatedly observed using gestures and speech in order to promote a particular kind of experiencing in the child. This maternal modulation is thought to represent an example of how resonant emotional responses are conveyed as our earliest form of communication. An important justification for this view, I believe, is that it permits us to see how infants overcome a problem of relating to their caregiver. Infants tend to habituate quickly to redundant stimuli; that is, they quickly lose interest in certain kinds of repetitive events. At the same time, they are disturbed by too much novelty or surprise, which tends to disrupt their relationship to another person. Thus, if mother is to sustain the infant's interest, which is to say, if she is to induce and sustain the affective involvement of the infant, then she must alter her behavior in small gradations, all the while introducing moderate degrees of novelty. Hence, Stern describes the mother as modulating the affect of her infant; and he further sees this interaction as resulting in a "theme and variation format of social behaviors," where one might further say that the mother in effect "plays" her infant, as one plays a musical instrument. For example, these researchers have found that during a particular active episode of play, the affect in question usually remains the same, unless a major change is felt by the mother to be necessary to sustain the infant's positive engagement.

The mothers in this research were also observed using what Stern describes as "prosodic" speech variations, or forms of speech which are interpreted as articulating variations of affect. Affect, in other words, is thought to be induced by the temporal rhythmic sequences of intonation and movement. These "prosodic" variations of intonation and move- ment would therefore "carry" the infants' interest past the habituation of repetition and the discordance of novelty. The affect is therefore said to be transmitted via the prosodic rhythm of mother's speech, an early precursor, it might be said, to the later significance of poetry. And, as in the meaning of poetry, the meaning of the caregiver's prosody may be further taken as the first instance of the creation of meaning via the metaphor.

III. INDUCED AFFECT AS METAPHOR

We have at this point parted company with Stern and his co-workers, for they do not describe the mother's modulation of the infant's affect as a metaphor. And in seeing the mother's modulation as a procedure she evolves for inducing a kind of experience, which will shortly be compared to the way metaphors work, I am saying that something more is involved than Stern's resonant emotional response. What more is involved, it will be claimed, is the mother's modulating procedure as metaphor. In *Language and Myth*, Ernst Cassirer states that any effort to understand the functioning of the metaphor "leads us back, once more, to the fundamental form of verbal *conceiving*." Thus, if inducing a resonant emotional response through the "prosody" of mother's speech is indeed an example of the functioning of the metaphor, then we must be on to something of fundamental importance. And yet, Cassirer's remark quoted above refers to verbal conceiving. How could this possibly relate to what goes on between the mother and her infant?

In a book entitled *Experiencing and the Creation of Meaning* (1962) Eugene Gendlin describes the metaphor in a way that would correspond to the mother's modulation of the infants' affect. He says, in effect, that this is what *all* metaphors do! Metaphors modulate our affect, or what Gendlin describes as our felt meaning, in a way that permits us to discover a new kind of experienced meaning. Stated differently, metaphors present us with an analogy that sets-off a particular kind of comparison. In the metaphor "love's sweet sorrow" for example, the terms "sweet" and "sorrow" are disparate. They do not normally belong together. Setting them off in opposition to each other in this metaphor permits us to discover a new meaning. But how could this be so, it might be asked? What is it that would get set-off that might then result in a new meaning? In "love's sweet sorrow," "sweet" and "sorrow" are set-off in opposition in a manner that attributes something sorrowful to what was heretofore experienced as only sweet; and something sweet to what was merely sorrowful. But if this transformation does indeed occur upon hearing this metaphor for the first time, what might be said to have been opposed? Clearly the words themselves, as a sequence of letters have hardly changed. Rather, it would have to be the experienced component of the words, or that aspect that is felt by the individual using the word, and which is referred to by Gendlin as a felt meaning. In other words, verbal metaphors, through the opposition of words, present us with disparate felt meanings which, when set-off in opposition to each other, mutually transform each other in a manner that results in a newly experienced meaning, as in the "sweet sorrow" which so aptly conveys the ambiguity of love.

The point I am attempting to make is a familiar one to the readers of poetry. In an essay in *The New Yorker*, for example, Helen Vendler describes the poetry of Czeslaw Milosz as giving rise to that "temperature which comes when two words that have never before lived side by side suddenly mingle — provoking what we feel ... when we read of Marvell's 'green thought,' Trahern's 'orient and immortal wheat,' Donne's 'unruly sun,' or Keat's 'sylvan historian.' This breaking down of 'natural compartments' is one of the most powerful effects of poetry, which by it's concision and free play can represent better than most prose the fluid access of a daring and unhampered mind to its own several regions."

This description of the creative heat that occurs when two words which have never lived side by side suddenly mingle is, I believe, as good a description as might be found for Gendlin's view of the creation of meaning. Stated differently, and far less evocatively, metaphors operate like procedures for inducing a novel experience, just as Stern's modulating mother is said to induce a resonant emotional response in the infant. In the metaphor, however, as so aptly described by Helen Vendler, the juxtaposition of words that have "never lived side by side" offers the opportunity for a "breaking down of natural compartments," and the "fluid access" to something which had not been meaningfully felt before.

IV. THE SELF AS FUNCTIONAL FEATURE OF INTERACTION

How, I might now ask, does this view of the metaphor pertain to the development of a self? If, as was described earlier, meaning emerges out of a reconciliation of opposite or disparate experiences, could it also be that a self emerges out of a similar kind of opposition? And that the very emergence of a self defines the self's purposefulness? The philosopher-psychologist George Herbert Mead indicated how this might be so when he described the differentiation of a self from the other as a functional feature of interaction. To describe differentiation as functional is to say that I cannot come to know and experience another without simultaneously discovering myself. Each time I touch another, I experience the resistance to my touch in my contact with the other; and with each contact, I come to know my body, as the philosopher-psychologist Maurice Merleau-Ponty also saw in relation to the body of the other. Together, in our shared interaction, we come to establish a boundary that defines each in relation to the other. However, for Mead, the essence of having a self was having the ability to view the self from the perspective of the other. Having a self means having the ability to be an object to one's experience, which for Mead, requires the development of a language. As stated by Mead, "I know of no other form of behavior than the linguistic in which the individual is an object to himself, and so far as I can see, the individual is not a self in the reflexive sense unless he is an object to himself." In other words, in order to take the self as an object of reflection, one must have the objective reference of a language system. Only then might one view the self from the perspective of the other, which is Mead's view of having a self as a self-conscious *me*. But in view of the functional differentiation of the self described above, one might still have a sense of self, in relation to another. And one might then describe how this sense of self emerges as a form of purposefulness, prior to the objective reference of language.

One would have difficulty imagining this possibility from the perspective of Stern's work, for his emphasis upon the induction of affect from the caregiver to the infant appears to make this a one way proposition. How might we be justified in viewing it differently?

V. THE GENESIS OF REFLECTION AND OF A SENSE
OF PURPOSE

As described earlier, in Mead's view children come upon the objective reality of a *me* when they can take the perspective of the other. When a child might induce a felt aspect in another that had been induced in the child, then that child is on its way to discovering a sense of self as an objective *me*. But long before the development of a language, the child has begun another form of reciprocal activity. By the third month of life, the infant demonstrates active and pervasive signs of imitation. That is, by age three months, there is an active striving to reproduce events. Among these events, smiling is a gesture that may occur spontaneously and may also occur in response to the smiling of a caregiver. Smiling may also be seen as one of the first gestures to be used purposefully by the infant, for effect. How might this be said to occur?

We might readily imagine that an infant's smile induces a similar affect in the caregiver, who in turn smiles at the infant. The experienced smiling of the infant may then be said to have induced itself, via the reciprocal response of the caregiver. That is, by smiling in the presence of a caregiver, the infant promotes the effect of a further, extrinsically induced smiling. Two features of this reciprocal induction of smiling seem noteworthy. First, it is initiated by the infant. Second, the infant's experience is reflected in its own experiencing by the induced affect conveyed by the adult. In other words, our initial experience of a reflected state is not the reflection of some independent reality; it is by this view an experienced reflection of our own experienced being. And it is this reflection of our own experiencing that permits us to grasp this experience as related to something we do.

That is, in this reciprocal induction we may be said to have an experienced smiling reproducing or reflecting itself! At this stage of development, the infant is far from capable of holding its own response in order to reflect upon it. However, a form of reflection, as described above, may be said to occur when the adult caregiver reciprocates the infant's smiling and induces a further smiling. And the ensuing juxtaposition of these two smiling responses in the infant is viewed here as critical. The juxtaposition of these responses in the infant is said to be critical, for it provides the infant with an experienced opposition (as in the metaphor) of two experiences sufficiently alike to be comparable, and yet different in one crucial dimension. The experiences are alike in that one reflects the other; they are different, in that one is initiated by the infant, and the experience that immediately follows has been induced by the caregiver. In other words, one has a sequence that would occur in any reciprocal interaction, where it is initiated by the infant and reciprocated by the caregiver; namely, one has an opposition of a self-initiated experience set-off against an identical experience induced by another. What is critical in this opposition is that the infant is afforded an opportunity to contrast two kinds of experiencing in a manner that the infant could not bring about by itself.

Stated differently, and in keeping with Mead's notion of the differentiation of a self in reciprocal engagement, the sense of a purposeful self, as discovered for example in the spontaneously generated smile, could only be discovered as set-off against a response induced by another. That is, the infant's spontaneously generated gesture must be reflected back so as to induce a similar kind of experiencing. The caregivers' loving containment of the infant, which invites and is immediately responsive to spontaneous expressions, is obviously a crucial feature of this interaction. And these reciprocal interactions must be permitted to occur repeatedly over time if the infant's active striving is to be repeatedly set-off against an induced responding as reciprocal and complementary states of being. The point here is that only the involvement of another person can serve to reflect the infant's gesture and thus permit an active striving and passive receiving, role and reciprocal to that role, to become organized in a mutually defining experienced meaning. That is, while the infant is clearly unable to conceive of itself outside of the immediate context of its experience, or to approximate at this stage anything like a willful sequence of acts, when the infant actively induces a smile, or a patty cake, or a vocalizing that reciprocates the caregiver's speech pattern, it may be said to have begun the process of initiating an experience for the sake of promoting a further experience as meaningful event.

It should be noted that in this attempt to specify how a sense of purposefulness is begun, I have avoided a number of the pitfalls that commonly present themselves in discussions of the development of a self. I have not, for example related a sense of self to a differentiation of an inner from an outer, as do psychoanalytic writers such as Gertrude Mahler. Inner-outer and self-other are clearly conceptual, and depend as concepts upon having an objectifiable self-other reference that can only appear with language. The symbiotic stage described by Mahler, where the infant is said to confuse mother's boundaries with its own, presents a problem; the infant would first have to differentiate mother's boundaries as separate, in order to then confuse them with its own. In what was outlined above, the infant achieves a functional differentiation which does not rely upon the infant's making a distinction between self and other. All that is required is that the infant experiences a self generated experience in opposition to an induced experience. This is not to say that the infant could "omnipotently" conclude, as the psychoanalyst D. W. Winnicott would have it, that the induced response is of its own making. The infant could not draw such an adultomorphic conclusion, for it has yet to develop a notion of what is its own.

VI. PURPOSEFULNESS AND THE MORAL SENSE

How, then, might this view of a purposeful self be related to the theme of this symposium, the "moral sense" as a phenomenological guideline for psychiatric therapeutics. There is little question in my mind that Professor Tymieniecka has, in her "The Moral Sense" (1978), delineated and clarified an area of critical importance, one which those of us in the clinical field of therapeutics would do well to consider with the utmost care. Of particular interest in this regard are the distinctions made concerning a moral sense which challenge our common understanding of this notion. First among these is Dr. Tymieniecka's view of the moral sense as not being a cognitive function. The consciousness of cognition, she points out, may be qualified as an "objectifying" function, which discriminates positively acknowledged, "objectified" facts. In contrast, the moral sense is viewed as an evaluative function, or a process of deliberation that proceeds on the basis of a many-sided scrutiny.

A second related distinction concerns her observation that our values might not be autonomous a priori objects, or products of "the constitutive acts" of our conscious cognition. Rather, values may be regarded as emerging in the process of valuation. A moral sense, then, might be a valuating function that is neither object oriented nor object confined. As a valuating function, a moral sense would lead one toward a self-interpretation that enables one to transform life's conditions into one's own conditions. Valuation as recognition, estimation, appreciation, and probity of judgment would then become a living aspect of our inward struggle to cultivate and preserve a sense of authenticity. In this all too brief description I must nonetheless attempt to underscore the

importance of this last statement, for it is, I believe, essential to Dr.
Tymieniecka's view, as well as to what follows in this essay, namely, that
there could not be a self-interpretation, a valuation, or even a sense of
authenticity were these procedures not grounded in intersubjective
interaction. Hence, Professor Tymieniecka's choice of the term "benevolent sentiment" to describe the "meaning giving function" of our moral
sense, which is released in intersubjective interaction. The reason for
viewing our moral sense as being released in intersubjective interaction
is not difficult to see; a self-evaluation, or any valuation of moral
significance, would be meaningless were it divorced from a consideration of the Other. Taken more broadly, a process for determining
meaning which neglected the common interest would not only lead to
evolutionary and ecological disaster; it would preclude the very process
whereby we establish a sense of self in our social realm. Hence, this
author's description of the exercise of benevolence in the moral
significance of acts, as resulting in an existential power; for it establishes
a harmony and balance which grounds our authenticity within the
broadest communal sense of the human significance of life. Hence also,
Dr. Tymieniecka's description of the benevolent sentiment of our moral
sense as a basis for how self-interpretation and objective reason become
reconciled.

Objective reason, she points out, is neutral to individual interest. Nor
might self-interest become the only means of resolving intersubjective
conflicts without a rapid deterioration of intersubjective reciprocity. A
benevolent sentiment as emotional attitude would provide a basis for
guiding self-interpretation within the framework of a common interest.
This surely must make sense if one is to account for the remarkable
degree of cooperation observed among the people of most societies.
And, yet, here as well is the rub, for it raises a conundrum that might
cause many to part company with Dr. Tymieniecka's important stance,
namely, that the minute one introduces the notion that an individual
must consider the perspective of another, then it is difficult to do so
without viewing this capacity as a cognitive function. This, I believe was
Mead's view when he described the ability to take the perspective of
another as an act of reflection that requires a language. That is, when I
can think or reflect upon the fact that the Other's impact upon me is the
same as my impact upon the Other when we alternate the same role,
then I may be said to have acquired the ability to assume the perspective
of the Other. Dr. Tymieniecka responds to this issue by stating that it is

during the individual's individualizing phase, where needs/regulation and satisfaction/relevance must be reconciled, that cognitive objectivity enters into moral experience. But the presence of cognitive objectivity in this process is not to "posit nor determine the moral quality of experience but merely its intersubjective form." One might then ask, could one have a moral quality which does not have an intersubjective form? And if objectivity is necessary to the intersubjective communication of our moral sense, how might this sense be intersubjective and hence moral in the absence of such an objective form?

Stated more generally, what form might be attributed to our moral sense in the absence of an objective form? And can we be said to have a sense of something without having a form with which to realize that sense? It is in response to these and related questions that the position I have outlined in this paper might be said to complement Dr. Tymieniecka's position in several ways.

First, I believe it may be shown that the evaluative function of a moral sense outlined by Dr. Tymieniecka is the same as what has been described in this essay as the process of reflection. Reflection for the infant is described here as occurring when a spontaneously produced experiencing is reflected via the induced experience from the caregivers's response. The opposition of a self-generated and an extrinsically induced response are said to set the one off against the other and to offer the first glimmerings of what it means to make something happen. Several other aspects of this process might now be set-off in relation to Dr. Tymieniecka's moral sense.

As is maintained for a moral sense, reflection from the start is an inescapably interpersonal process. The infant's experience could not, at the onset, be reflected back by other than a doting parent who readily reciprocates each sign of engagement by inducing a similar experiencing. Thus, by this view, what is to become the main feature of our cognitive function, the ability to reflect upon our meaning, makes its first appearance as an interpersonal event.

In addition, the process of reflection must be viewed as inherently one of valuation. As described by Dr. Tymieniecka, valuation is intrinsic to man's differentiation, for it is the means whereby we appropriate what is differentiated as having meaning. For the infant, the contrast of spontaneous and induced experiencing could not fail to involve an evaluative function, since the infant is hardly indifferent to the effect of these juxtaposed experiences, once they have occurred. The reason for this last statement may be seen in the obvious delight the infant and his or her caregiver take in the many forms of reciprocal exchanges that occur by the third month of life. And delight, by this view, is taken to be an early manifestation of our ability to discover value in our acts.

A further similarity pertains to the interpersonal nature of this early form of reflection, which introduces the Other as confirming or validating the import of the infant's experience in a manner that comes to assume considerable significance. Dr. Tymieniecka describes this effect as "the receptive interpretation of the significance of our deeds by the Other . . . [which] motivates directly our conscience toward a reflectively critical *revaluation* of this significance." This last statement, as well as the importance she attributes to *resounding* within the other's subjective evaluation, and to receiving the other's approbation, may be understood as originating in this early and crucial form of reflection.

Most importantly perhaps, the reflection described here assumes a symbolic intersubjective form in terms of the infant's gesture as purposeful act, prior to the objective-conceptual development of a self-other differentiation which comes with language. Hence reflection may indeed be seen as per Dr. Tymieniecka's moral sense, as arising within an interpersonal matrix, prior to the constitution of an objectively cognitized reality. The point here is that if the evaluating function of a moral sense is equated with the reflective process as described in this paper, then valuation is indeed intrinsic to man's differentiation, as Dr. Tymieniecka would have it, for the process of valuation becomes synonymous with the very process whereby we symbolically realize our world. By symbolic realization is meant the process in which reflection enables us to elaborate the meaning of our experience through the emergence of novel felt-symbols as was referred to earlier in this paper, and as has been described by Gendlin (1962).

In terms of what has been described, this process of symbolic realization may be said to be grounded within the caring reciprocity of a caregiver-infant relationship, where the loving administrations of the caregiver contain and sustain the infant, while reflecting back the infant's gesture in a way that permits the meaning or valuing aspect of that gesture to emerge in symbolic realization. Meaning and valuing must therefore be seen as synonymous, since the meaning that grows out of this process of reflection must be an interpersonal meaning; and as interpersonal meaning, it must have value to those who share it. For it is not the random gesture, or the somatic functions which are reflected back by the caregiver; rather, it is the all too human smile that evokes delight and is therefore reciprocated in kind as a prime and valued symbol of what human beings do. Reflection is then seen as the means whereby we discover human significance or meaning in our experience. Only in reflection might we sustain that benevolent sentiment referred to by Dr. Tymieniecka, whereby interpersonal sharing is crucial to the elaboration of meaning; and whereby the valuation of self-interpretation allows for the emergence of moral significance and authenticity.

If this view of our human condition has validity, then we will have to pay much closer attention to the process of symbolic realization, as it has been clarified in the work of Eugene Gendlin, and to how reflection as valuation might be waylaid, lost sight of, or simply ignored in our living and in our theories.

KARL-ERNST BÜHLER

THE UNFOLDING OF THE "BENEVOLENT SENTIMENT" AS THE BASIS OF PSYCHOTHERAPY

I. THE MAXIM OF "NIHIL NOCERE"

"Nihil nocere," an old maxim of medical therapeutics, seems also to hold for humanistically orientated psychotherapy. Yet scholars in this field often face unexpected problems, and not only because of conflicting psychotherapeutic methods. Not only is it difficult to decide precisely on the advantages of a particular psychotherapeutic intervention, but assistance itself may become a source of disturbance in individual cases. Do we have any guidance in this situation?

It must be the basic attitude of the "benevolent sentiment" (Tymieniecka, 1983) to have the well-being of all other living beings as its aim. But the situation is analogous to that in pedagogy: if you help the child in doing something, then he or she becomes no longer capable of doing it for him/herself. Therefore every intervention must always remain in proportion to the faculties and the potentialities of a particular person. This apparently paradoxical situation in which we disturb in the course of helping is the starting point of our investigation which begins from the intersubjective constitution of the person.

II. INTERSUBJECTIVE CONSTITUTION OF THE PERSON

According to Wyss (1976), the constitution of the person is based on the experience of a fundamental ontological 'deficiency' (*Mangel*) as the basis of communication, which latter is in my view also a process of mutual interpretation, and not merely a transmission of information. The basic features of existence are elaborated in an oscillating process of exchange and communication facilitated by this sense of fundamental deficiency. Deficiency is not to be conceived absolutely, but as a discrepancy relative to further possibilities of existence. Deficiency in this sense is not merely a motive and opportunity to communicate with the world and 'others', but also a source of basic imbalances which find

A-T. Tymieniecka (ed.), Analecta Husserliana, Vol. XX, 207–223.
© 1986 *by D. Reidel Publishing Company.*

provisional resolution in communication itself. By communication or interpretation the deficiency will be compensated for partially or temporarily until it once more recurs. The process of communication does not end in the tautologies of machine language, because the reception or sending of a message always changes the receiver or sender of the message in the communication process.

Communication is not merely a cyclic process of sending and receiving information but is also a fundamentally dialectical process. It is the intersubjectivity of this mutual dialectic in which the person is constituted. The experience of deficiency cannot be compensated for perfectly because of the difference between the actual communication and its potentialities. This intrinsic difference is one reason for the latent or even manifest violence which takes place in every intersubjective communication. This latent or even manifest transgression in every communication can only be overcome through the 'benevolent sentiment' of the communicators.

Another source of this antagonism between communicators derives from their individual freedoms; these can only be limited through mutual acknowledgment and respect. But mutual acknowledgement in turn presupposes independent and free individuals. Intersubjective encounters therefore necessarily lead to the mutual restriction of lived possibilities. Once more we meet with the psychotherapeutic paradox in this antagonism of intersubjectivity. The antagonism of intersubjectivity is a *Leitmotiv* of psychotherapeutic interaction.

In general, the aim of psychotherapy is to realize in dialectical togetherness with the patient his genuine suffering and with him to find ways of coping with it through elucidating our intersubjective relationship and its antagonism: this is the real sense of 'dialogue'. The suffering itself has to be coped with by the patient himself.

The primary aim of psychotherapy is not to inform the patient of his deviations, hidden or overt. Thus the therapist must be alive to the fundamental antagonism within every human encounter. That means, contrary to orthodox psychoanalysis, that therapy is not to be ruled by Freud's 'rule of abstinence' alone; the therapist is to lead the patient in an opening out of his potentialities and not to a solution of a particular problem. This is said in disagreement with some kinds of behavior therapy. The patient's problems and conflicts are clarified via the consistent 'working through' of the varying fundamental intersubjective antagonisms, also involving what the therapist feels, and realizes of his feeling, toward the patient. The patient's mastery has to be accomplished beyond what the help of the therapist can bring, or this will diminish the potentialities of the patient. Demonstration of such potentialities for living by the therapist remains independent of his activity in the therapeutic process, which means that both intervention and abstinence contribute to success in psychotherapy.

It is not possible here to give special recipes for concrete psycho-therapeutic interventions. But in our experience, not only "benevolent sentiment" but also the modes developed by Wyss (1982) are useful signposts for orientation in the psychotherapeutic process. Depending on the particular circumstances of the psychotherapeutic encounter the style of intervention must change. A psychotherapy in the phase of "exploring" involves different forms of interaction than one in the phase of "confronting" or even "mastering".

Here are some examples, loosely defined:

In the phase of "exploring" (*Erkunden*), the therapist encounters the patient in an open and unbiased manner, registering the other's body, his movements, gestures, posture, verbal expressions and overall appearance. The therapist refrains from giving directions, puts no accents into the interaction and gives no verbal explanations (e.g.: "Please, tell me what your problem is?"/"Everything you say is important.")

"Discovering" (*Entdecken*) signifies extension of the encounter by means of contrasts, for instance through an observing partner. The therapist takes the role of the observing person, without revealing the patient's conflicts to him. The therapist gives no interpretation, but tries to widen the experiencing of the patient through evoking an observing Thou (examples: "I notice/see/hear (. . .)"; "Is it correct if I suppose that (. . .)?").

In "opening up" (*Erschließen*) the part played by the circumstances in which the patient lives is sought for. In order to better differentiate his condition and deepen the patient's experience, new perspectives on a situation or to a conflict are suggested. Here the therapist pays more attention to the content of the message (e.g.: "What has been your main fear?"; "What forms has it taken?"; "Were you more ashamed of yourself or of what other people might think?")

The phase of "confronting" (*Auseinandersetzen*) expresses contra-position. The observations and experiences of the therapist are presented to the patient in controlled fashion; (e.g.: "I am not very convinced about what you have said."; "I have a different/opposite view about that matter"; "Perhaps there are also different views about what you mean"; "It is not easy for me to see your problem.")

Preliminary decisions are favored in "binding/loosening" (*binden/lösen*). The therapist becomes more active, taking a more supportive and carefully balanced attitude. In this way he or she seeks to loosen the therapeutic relation and to promote development of new and unique experiences and attitudes by the patient. In this phase the shift from a symbiotic to an independent form of relating takes place.

The therapy concludes in the phase of "mastering" (*bewältigen*). Preliminary changes occur for instance in the sphere of the patient's work, in his family or in his relation to a partner. The patient becomes a more consistently self-responsibly acting person, while the therapist changes from having been a more or less remote councilor to a partner who supports the newly discovered life-style of the patient.

Freud's rule of abstinence gives wide latitude to the patient in structuring his or her experience according to his or her own presuppositions in free association, within the "transference", thus making possible psychotherapeutic interventions such as interpretations especially if transference occurs in the actual encounter in a stereotyped, repetitive manner, as is generally true of transference reactions.

But to focus on the intersubjective antagonism in every act of intervention, in every speech act, interferes with the person's unfolding of his or her own potentialities. On the other hand, the latitude for the personal horizon — the human potentialities to expand — is not only the condition for manifold ways of existence but also for existential anxiety, hence can also act to constrict personal potentialities, insofar as breadth and open horizons evoke sentiments of uncertainty. I suspect the term 'shrink' for a psychiatrist is intimately linked by the general populace with this reduction of human potentialities. Here the therapist has a task of orientation and reevaluation.

Depending on the personal relations of the therapist to the patient as well as on the therapy situation, keeping silent, or giving advice or even sharp confrontation, may benefit the patient in developing his or her sense of self. Always keeping in mind the dialectic of the dialogue, therapeutic interventions should be guided by the 'benevolent sentiment' of the therapist. Yet the therapist's silence, benevolently intended to give the patient the opportunity to expand and unfold his or her potentialities, may be misunderstood as disinterest, as indifference or even rejection. Therefore, the therapist needs to pay attention to any change in the way his attitude is experienced by the patient. He should explain his particular attitude to the patient and perhaps even change it. A good therapeutic relationship, in my opinion, does not exclude confrontative interventions, which may be very helpful in advanced stages of the therapeutic process. But in the beginning of therapy, confrontations, or even the therapist's silence as far as it is misunderstood as confrontation, very often disturbs the therapeutic process. In closing however, silence may acquire the meaning of mutual understanding of the patient's success in coping. But in every instance the therapist needs to be guided by 'benevolent sentiment'. The benevolent sentiment has to be the basic felt habit of the therapist, since it connotes a basic acceptance of the other person within their dialectic antagonism.

The therapist needs to evaluate his interventions in terms of the benevolent sentiment towards his patient. Benevolent sentiment not only rules the conflicting self-interests of the partners, in our case the therapist and the patient, but also serves to balance out the antagonisms of human communication in general. The moral sense, even if a habit of experience, not a rational sense and not a sense of logos, requires unfolding; here an aesthetic sense seems to be called for. This reminds one of Schiller's treatise on the aesthetic education of mankind which I cannot discuss here. In my opinion this experienced habit, or better, this sentiment, may develop in the way described by Wyss (1973, 1976, 1982; see also: Bühler, 1979, 1980, 1981 and Bühler and Wyss, 1984), namely through structures and modes. I will elaborate after first presenting the essentials of existential and humanistic psychotherapies.

III. THE ESSENTIALS OF EXISTENTIAL AND HUMANISTIC PSYCHOTHERAPIES

The development of 'humanistic psychology' in the United States at the end of the '50s and the early '60s may be regarded as a kind of reflection of the European traditions of phenomenology and existential philosophy. At this time psychology departments of European universities, especially in Germany, were mainly concerned with scientific method. Those interested in philosophical issues, or in psychology of a phenomenological or hermeneutic origin, found themselves relatively isolated.

As of now no renaissance of phenomenological method or of the ideas of existentialism is apparent, save perhaps as a reflection of the U.S. movement, particularly the modified client-centered psychotherapy of Carl Rogers.

A basic difference between the European tradition of existential psychotherapies (see: v. Gebsattel, 1954) and American humanistic psychology appears in the idea of "failure", one intrinsic in the human situation, and one which has slight importance for the more optimistic American tradition.

The basic allegiances of all schools of "humanistic psychology" are to a holistic and personalistic perspective of human existence which — as an open system — lives in a varying manifold exchange with its surroundings. Capable of setting standards and goals, the human being assumes the risk of self-responsibility in being creator of his own life. The aim of man as a creative and intentional being is self-realization. Defection from the ideal of self-realization is experienced as "existential guilt".

The idea of indetermination and of the freedom of the will, in the sense of choice or free decision, derives from European humanism. Responsibility must be recognized as the necessary consequence of the freedom of the will, choice, and free decision. It is the precondition of conscience and accounts for the experience of guilt.

In the tradition of humanism from Plato to modern existential philosophy the person is self-constitutive, regarding conscience, intentions and freedom of choice as fundamental for existence. Therefore humanistic psychology views neurosis as failure in the attainment of self-realization. This unidimensional conception of neurosis was criticized by Kunz (1956) as overly presumptive. He contrasted this overestimation with the equally unidimensional anthropology of Freud and his concept of the person as a passive object of unknown powers and drives.

These two very different conceptions of neuroses were integrated in the "existential-integrative psychotherapy" of Dieter Wyss.

"Existential-integrative psychotherapy" has been influenced in its theoretical foundations by three philosophical positions:

— phenomenology and phenomenological method;
— existential philosophy (especially that of Kierkegaard);
— the humanistic of early Greek philosophers (pre-Socratics, Aeschylus).

The core of "existential-integrative psychotherapy", as of all existential philosophies, is the individual person in his or her actual situation, which is also one that is social. The person is not to be formed according to an abstract conception but is to create his or her own personality.

The humanism of the early Greek philosophers is a humanism tragically conceived. This allows for all human capabilities being in potential conflict with one another. The tragic is given by an antagonism or an antithesis which is insurmountable. Only with compromise can the tragic situation disappear.

The existence of the tragic human being remains ambiguous. Perfect harmony with oneself, with others or with one's surroundings means stagnation and finally decay. The contradiction within the "tragic situation" is not limited to the individual alone but also exists between the individual and society. Greek philosophers called this condition "antipalous harmonia", i.e. "rebellious harmony".

The fact of "failure" in human existence, the fact of unresolvable guilt, marks the difference between the "existential-integrative psychotherapy" and the optimistic humanism of "humanistic psychology". Thereby it differs also from Marx's idea of an "absolute" humanitarianism and a belief in the determination and ability of mankind to bring itself and its world into harmony. It is uncontroversial that it is each person's task to realize his life as self-responsible. But without certain norms for his existence, it will remain ambiguous, and in risk of guilt and failure.

In this regard, "existential-integrative psychotherapy" with its specific conception of humanism is much closer to the psychoanalysis of Freud — despite great and significant differences — than to humanistic psychology. In Freud's view a person exists not only through biological instinct (as homo natura) but also in an unresolvable conflict situation (e.g. eros/thanatos). The reductionist conception of "homo natura" in Freud's theories is opposed to "existential-integrative psychotherapy", in which the person's relationship to conflict is ontologically understood.

The existential psychotherapies have understood themselves as a countermovement to Freud's mechanistic—reductionistic thought. The indebtedness of Freud's theories to the concepts of 19th century physics has been dealt with by Wyss (1977). For example the term "psychodynamics" is analogous to "thermodynamics" in physics. Furthermore, consider the numerous references to "psychic apparatus". But the development of the person is overdetermined with respect to causes (causa efficiens) not understandable by scientific methods alone; on the other hand it is imperfectly determined by purpose (causa finalis). Both are presuppositions of choice, responsibility, freedom, and even guilt.

The conception of "existential-integrative psychotherapy" can only be outlined here. Wyss has elaborated basic "structures" and "modes" which represent fundamental ways of existence, comparable in some ways to the existentialia of Heidegger. Wyss proceeds from the core of the human paradox of existence — namely being at once the subject and the object in the act of self-understanding. The division into subject and object eventually leads to a negation both of subjectivity and of the experiencing person. This fragmentation must be overcome. A theoretical idea for accomplishing this program lies in Heidegger's concept of "being-in-the-world".

IV. EXISTENTIAL STRUCTURES AND MODES

The four structures "space" (*Raum*), "time" (*Zeit*), "body" (*Leib*), and "achievement" (*Leistung*) can be considered as "crystallization nuclei" for communication. Here we find remarkable analogies in humanistic psychology especially with Charlotte Bühler's (1959) four basic tendencies of human existence, namely "satisfaction of needs and drives", "self-limiting adaptation", "creative expansion", and "maintainance of inner order".

The structure of "space" (*Raum*) according to Wyss means the differentiation of the person in space including the surroundings such as the geoclimatic and social-historical "milieu", as well as the "inner space", i.e. orientation to and coordination of customs, habits, traditions, rules, and standards. Once more we find similarities with Bühler's tendency of "maintaining the inner order". But Bühler has not made explicit — as has Wyss — the structure of "time" (*Zeit*), referring to the person's temporal orientation, his development in relation to past, present and future. Our relation to time and temporal development are the preconditions of reflection and human responsibility alike, because it is within time that the person experiences him- or herself as a responsible actor. Such responsibility is an essential of humanistic psychology. The "personal aspect" of the subject is realized within the perspectives of "space" (*Raum*) and "time" (*Zeit*) together.

But a human being also experiences himself or herself in his or her work, his or her achievements, in self-actualization, in creativity. Here there are links with Hegel and Marx. For Hegel, work is but one of several relations between person and world, while for Marx they are related by work alone. Wyss' structure of "achievement" (*Leistung*) includes the subject's relation to his productivity and his products, both in their pragmatic and intellectual aspect.

"Body" (*Leib*) finally refers not only to lived bodilyness and "organic functions", but also to emotions, feelings, and moods. These last two structures present some analogy to Bühler's tendencies, i.e. to her "creative expansion" and "satisfaction of needs and drives".

The interrelations among these structures are pervaded by a virtual antagonism, one which entails a continuing risk for the person, and the possibility of disintegration. Here we see the core of Wyss' conception of mental illness. But before continuing any further let us first summarize what he has written concerning "modes of existence" in existential-integrative psychotherapy. Deficiency as the pro-motor of personal development gives rise to successive modes of existence: "exploring" (*erkunden*), "discovering" (*entdecken*), "opening-up" (*erschließen*), "confronting" (*auseinandersetzen*), "binding/dissociating" (*binden/lösen*), and "mastering" (*bewältigen*).

These modes of existence, in their reciprocal determinations, depend on the constitution of the person, on his or her specific knowledge and self-expression in actual communicative situations. Nor are these modes realized only in active performance; they are also contemplative.

The "exploring" person begins to deliberately understand himself or herself and his/her surroundings. "Exploring" means more than just perceiving and acting, it has a creative character. But as yet there is little visible dialectic. Despite its creative moment, "exploring" is principally registering, while important relationships and interconnections remain obscure.

"Discovering" (*entdecken*) announces the reception of new and important themes within an opened-up curiosity. The discovery of new themes in the surroundings of the person is accompanied by the discovery of new modes of experience.

The emotional qualities of the environment and the person himself as well as the opening of new horizons are developed in "opening-up" (*erschließen*). The network of correlations among the emergent topics is grasped. "Confronting" involves an experience in which doubt, breakdown and destruction of the person are implicit. Conflict is realized, even actively sought.

"Binding/loosening" is represented by decisions through dialogue with others as well as with oneself. These modes of relationship between the person and the world, and even between the person and the self reach a uniquely individual end in "mastering" (*bewältigen*).

This ideal sequence of modes from "exploring" to "mastering" implies the correlative ideal development of the person in relation to the world and/or the self. Thus Wyss' schema of structures and modes is to be understood as an outline for enabling one's existential development as a person in the situation of "existential-integrative psychotherapy.

This is only a summary of the structures and modes developed by Wyss (1973, 1976, 1977, 1982). Some of these modes are reminiscent of Cusanus' differentiation of the stages of understanding in terms of degrees of dialectic power. His initial stage, "sensus", is one of confused icons. "Ratio", the second stage, sustains the opposites in the form of the principle of contradiction. "Intellectus" realizes the opposites as reconcilable.

V. MENTAL ILLNESS AS FAILURE OF COMMUNICATION

The dynamically interrelated structures and modes may be mutually restrictive depending on the situation, the subject's condition, his or her life history and current emotional state. The constriction of one structure or one mode may result from the expansion of another, and thus a certain balance occurs. Yet it is imbalance between structures and modes which makes illness possible. The more a person is limited to one realm of existence, the greater the danger of illness; the more various a person's communicative possibilities, the greater are the possibilities and opportunities of resolving a conflict or of mastering an existential discrepancy. Wyss views illness as an inability to communicate with the world on diverse levels of "being-in-the-world", and as evidence finally of the latent ontological fault which he calls deficiency (*Mangel*). This conception of illness as dysregulation and malfunction of exchange and of communication implies "loss of norm" and "loss of balance" in physical behavior and in experience of the body and of the person. On the organismic level malfunction means the imbalance in exchanges with the natural environment, and on the personal level disturbances of communication. In this understanding of illness the overall structure of communication is of central importance, along with the part played by a specific and limited conflict. Therefore, this kind of existential psycho-therapy implies the ideal possibility of communicating in terms of all structures and modes. With this an ideal norm is acknowledged. Because of the ontologically prior imbalance of structures and modes this ideal norm is always to be sought anew. Existential-integrative psychotherapy is not confined to the definition of norms or of empirical standards; it seeks to do justice to "subjective standards" in its therapeutic practice. The therapeutic process in "existential-integrative psychotherapy" is a continuing movement between the extremes of "exploring" and "mastering", facilitating an increase in the communicative possibilities of the structures and modes.

VI. THE SIGNIFICANCE OF THE BIOGRAPHY FOR EXISTENTIAL PSYCHOTHERAPY

Biography is a delineation of personal "becoming". It is expressive of the course of development of the individual personality. The life-span is not an aggregate of individual data but a *Gestalt*. For Dilthey (1927) a "plan of life" (*Lebensplan*) provides integration for decisions, actions, desires and hopes. Such integration was spoken of as "life-style" by Adler (1929). The biography is a "via regia" to the understanding of the person in his uniqueness.

Central for self-understanding of the person is an understanding of his or her growth, for that deepening understanding of oneself which can become therapeutic.

The structure of biography is narratively ordered, i.e. its order is not determined once and for all. It is a dynamic order in which life events are being continually reinterpreted; the realization of biography in psychotherapy is always a joint work of patient and therapist together.

All schools of depth-psychology are biographically oriented, and it is this which provides the link between existential and psychoanalytic traditions. Psychoanalysis, however, with its orientation to the past, generally neglects the future dimension of existence. A human being is neither a punching ball of his drives nor the simple product of his past. On the contrary, exclusive focus on the past and to factual events of the past is a constriction as well of the possibilities of the future, and may be exactly characteristic of the patient's illness. Personality is not a static pattern of properties but a dynamic process of becoming, which results in illness if sufficiently disturbed.

A necessary condition for healthy personal becoming is an acknowledgement of the facticity of the past events; such an acknowledgement brings new possibilities into relief, if only through reinterpretation of past events. A fixation on the past, with inability to reinterpret the meanings of its events, is basic in mental illness, as von Gebsattel (1954) has shown. He referred to this as the "inhibition of becoming" (*Werdenschemmung*). Failed reinterpretation of past events, together with the need to deal with them (*das Sich-nicht-Abfinden-Können*), are characteristic of the neuroses. Past experiences fail to be productive for the present and future. The neurotic person is unable to transcend his or her temporality and temporal process, because of the inability to reinterpret the past on the one hand and to accept the past events on the other. He or she is therefore unable to mould the present and future. Fixation in the temporal dimension, with the temporal conflict, is the basic conflict in the neuroses. The consequence, according to von Gebsattel (1954), is faulty experiencing of the present, with repetition of the unresolved past, which he called "presentification-of-the-past" (*Präsentisch-Werden-der-Vergangenheit*). Study with the patient of his or her biography also serves to disclose to both the possibilities of coping with conflicts and problems.

To be sure, neither historiography nor biography creates an overall and formalized theory of their objects with which a structure or a pattern could be imposed on the data. Consequently the data can only be ordered according to an a priori, selective perspective, but one which is not arbitrary.

A preliminary structure of biography in psychotherapy consists of the clinical symptoms, psychic or somatic, and their interrelations. But this categorization yields but a rough grid. Consequently they require to be completed by an anthropological frame of reference. The structures and the modes of Wyss are appropriate for such a frame of reference because they are comprehensive enough, without requiring rigid application.

Beyond the delineation of lawful relations and the general anthropological frame of reference which enable only a global structuration, the use of paradigmatic biographies may help make a therapeutically beneficial selection from the clinical data.

Paradigmatic biographies are ideal type ("idealtypische," Max Weber) biographies constructed from the abstraction and generalization of many individual ones. Empirical regularities make possible such constructions. Paradigmatic biographies function like general theories, not on the basis of general laws, but on the basis of the likeness of patterns to the "paradigma". Thus they provide a context for plausible interpretations. The understanding of the biographies through the use of paradigmatic biography takes place through the structure of "family-resemblances", a concept of Wittgenstein's (1971).

What is the importance of the conception of paradigmatic biographies for psychotherapy?

The biography of a person consists of a variety of narrations told by, or about, the person. Their coherence is given through the concept of family resemblance. This coherence exists not only for synchronically, but also for diachronically reported biographies. The paradigmatic biography of a person may be thought of as a kind of lifestyle, not one that is rigid, but dynamic. A person can only understand his or her actions or experiences − such as psychotherapy is concerned with − in the context of his or her own paradigmatic biography.

VII. THE PERSPECTIVE OF PSYCHOTHERAPEUTIC PRACTICE

Any kind of intervention that enhances the subject's communicative possibilities can be considered psychotherapeutic. The theoretical conception of "existential-integrative psychotherapy" may be employed as a guide for a variety of techniques with an existential stance.

Successful psychotherapy demands openness to the possibilities of living not only from the patient but also from the therapist. The therapist's task is to show the patient his or her own possibilities of living, even while remaining aware of the latent antagonism of his or her interventions, even in the ideal communication or benevolent encounter.

The therapist, like the patient, may gain openness for the possibilities of living in self-experience, whether concentrated in a dyadic encounter, or a many-sided one in a therapy group. The experience of rarely occurring, so-called abnormal, events is possible in Balint-groups, where usually nonspecific forms of the relationship between therapist and patient are made the focus.

Such self-experience is, along with the original personality of the therapist, the source of the so-called nonspecific influences of therapy which are important for psychotherapeutic success. These nonspecific factors override the theories of psychotherapeutic schools and are usually inappropriately categorized with the qualities of the therapist. Self-experience is not a quality of the therapist but an intersubjective process, important in all psychotherapy. This crucial point must guide the education of psychotherapists.

The fundamental element of all existential psychotherapies is the intersubjectivity of the encounter, not only that of verbal exchange, but also of shared silence. But, as we said, the intersubjective process is ab initio antagonistic, because a mutual acknowledgement is presupposed, one which is possible only for independent subjects. The aim of existential psychotherapy is mutual acknowledgement between free persons, acknowledgement which is very often disturbed in neurotic patients. A symmetry of personal encounter is sought in contrast to the asymmetrical relationship of therapist to patient in orthodox psychoanalysis.

Intersubjective encounter is necessarily based on mutual control of the intersubjective antagonisms. For example, the latent contradictions of dialogue or of the dialogical principle are even manifested in maeutic discourse, a subtle form of confrontation. The maeutic discourse is a form of conversation used by the Platonic Socrates in his dialogues. This socratic maeutic discourse is characterized by a logical and an aporetic moment by a sensitive orientation to the partner on the one hand but which introduces an aporia on the other. Although this crisis of an aporia may promote psychotherapy it does so at the cost of doubt and despair on the part of the patient. The roles in a maeutic discourse are clearly asymmetrical between the leader of the dialogue and his partner. The relationship of the partners, although benevolent, is nevertheless a strategic one since the partner is to be led to insight or understanding through an aporetic crisis. The partners are not equally privileged; their relationship is not symmetrical nor reciprocal. The lesser partner is not only constrained by the aporia but also by the arguments, both of which may interfere with the psychotherapeutic process. This kind of discourse is not very suitable in psychotherapeutic dialogue, save for the phase of "confronting".

The confrontation in psychotherapy proceeds not only from the therapist, but also from the patient, who is free to refuse assistance. The therapist can do nothing but accept this "annihilation" by the patient. Here we see an important difference between psychotherapeutic encounter and that of everyday life. But even in seemingly open and straightforward encounters, tendencies to disguise and to withhold may persist. V. E. von Gebsattel (1954) has pointed out the presence of "masochistic elements", for example, in a seemingly open dialogue. The issue of truth in psychotherapy generally and of psychotherapeutic interpretations in particular reaches to the core of our concern. This joint search for truth has at least two requirements, the fundamental equality of the persons engaged in finding out the truth and their honesty and sincerity. Both must be presupposed by the psychotherapist even in the face of the facts. This is necessary for the psychotherapeutic dialogue; such symmetry must be conceived "as if" it were the case. This supposition of the partners' equality, of mutual acknowledgement and respect, of sincerity, truthfulness, honesty and confidentiality are the conditions of a viable psychotherapeutic interpretation and a vital psychotherapeutic encounter.

The aim of the initial phase of the psychotherapeutic process is to reduce the asymmetry in the relationship of the therapist with patient. This is promoted by the earlier mentioned basic attitudes. At the outset the therapist should refrain from explanations in order to avoid suggestion. But explanation becomes common to the extent that intersubjective asymmetry decreases.

The dynamic process of mutual interaction is crucial for interpretation in psychotherapy, based as it is on the judgement of a state of affairs. To the extent the judgement is fixed and unrevisable, it becomes a condemnation of the person even if it is not negatively stated. Every rigid judgement fixes the person assessed, attributes a specific past to him or her, and thus may prevent further possibilities of living, concrete possibilities of his or her personal development. Careful attention lest these possibilities for development be curtailed represent the sense of "nihil nocere" for psychotherapy.

Therefore a therapeutic attitude of "Not-wanting-to-know-but-being-allowed-to-know" (Nicht-wissen-wollen-aber-erfahren-dürfen) is required, since intersubjective encounter is possible only if objectification and alienation of the other person can be avoided. Such disturbances are avoided in a meditative encounter (cf. Bühler, 1983) because intersubjective contradictions and antagonisms are temporarily suspended. Developing the meditative attitude may be a way of coping with the antagonisms of intersubjectivity and existence.

In closing, let me review those features of the psychotherapeutic process in which benevolent sentiment plays an important part:

(1) the antagonisms of communication are temporarily overcome by the benevolent sentiment of the participants; (2) the mutual acknowledgement of free and individual partners calls for benevolent sentiment, otherwise interaction would be an armistice only; (3) the therapist induces orientation and valuation on the part of the patient, with the guidance of benevolent sentiment; (4) therapeutic interventions, such as remaining silent, giving an interpretation, giving advice, or even confronting, should be controlled by the benevolent sentiment of the therapist.

Benevolent sentiment remains for the therapist the basic guide, but with knowledge of the fundamental dialectics and antagonisms of communication and of the human condition in general.

Institute of Psychotherapy and Medical Psychology,
The University of Würzburg

ACKNOWLEDGEMENT

I am indebted to Erling Eng for his thorough revision of my translation into English.

REFERENCES

Adler, A., *The Science of Living* (New York: Greenberg, 1929).
Binswanger, L., 'Lebensfunktion und innere Lebensgeschichte,' in: Ausgew, *Vorträge und Aufsätze*, Bd. I (Bern: Francke, 1947).
Bühler, C., *Der menschliche Lebenslauf als psychologisches Problem* (Göttingen: Hogrefe, 1959).
Bühler, C., *Wenn das Leben gelingen soll* (München: Droemer u. Knaur, 1969a).
Bühler, C., 'Die allgemeine Struktur des menschlichen Lebenslaufes,' in: (1969b).
Bühler, C. u. Massarik, F., *Lebenslauf und Lebensziele* (Stuttgart: Fischer, 1969).
Bühler, K.-E., 'Existential-integrative Psychotherapie,' in: *Mensch, Medizin, Gesellschaft* **4** (1979):236–240.
Bühler, K.-E., 'Der Krankheitsbegriff der existential-integrativen Psychotherapie,' in: *Psychother., Psychosom., Med. Psychol.* **30** (1980):121–126.
Bühler, K.-E., 'Neurosenkonzeption der anthropologisch-integrativen Psychotherapie,' in: *Fortschritte Neurol. Psychiat.* **49** (1981):366–370.
Bühler, K.-E., 'Meditation in der Anthropologischen Psychotherapie,' in: *Integrative Therapie*, Heft 4 (1983).

Bühler, K.-E., 'Über die biographische Methode in der Psychotherapie,' in: *Jb. Psychoanal.* (i. Druck).

Bühler, K.-E. u. Wyss, D., 'Die existentiellen Psychotherapierichtungen,' in: Toman, W. u. Egg, R. (Hg.), *Handbuch der Psychotherapie*, Kohlhammer-Verlag (Stuttgart: Kohlhammer, 1984).

Dilthey, W., 'Der Aufbau der geschichtlichen Welt in den Geisteswissenschaften,' in: *Ges. Werke*, Bd. VII (Leipzig: Teubner, 1927).

Gebsattel, V. E. v., *Prolegomena einer medizinischen Anthropologie* (Heidelberg: Springer, 1954).

Kunz, H., 'Die latente Anthropologie der Psychoanalyse,' in: *Schweizer Zeitschrift für Psychologie* **15** (1956):84.

Tymieniecka, A-T., 'The Moral Sense,' in: Tymieniecka, A-T. & Schrag, C. O. (eds.), *Foundations of Morality, Human Rights and the Human Sciences, Analecta Husserliana*, Vol. XV (Dordrecht: D. Reidel, 1983).

Wittgenstein, L., *Philosophische Untersuchungen* (Frankfurt/M.: Suhrkamp, 1971).

Wyss, D., *Beziehung und Gestalt* (Göttingen: Vandenhoeck u. Rupprecht, 1973).

Wyss, D., *Mitteilung und Antwort* (Göttingen: Vandenhoeck u. Rupprecht, 1976).

Wyss, D., *Der Kranke als Partner* (Göttingen: Vandenhoeck u. Rupprecht, 1982).

BRUNO CALLIERI

CLINICAL PHENOMENOLOGY AS THE "DEMYTHOLOGISING" OF PSYCHIATRY

The Movement toward the Other

Objectifying thought and knowledge characterise the sciences of nature, that is the world where as a consequence of its being subject to determinism neither liberty nor future exists. Both the Greek and Western cultures generally maintain this way of thinking, which results in considering each person as an "object" of knowledge.

This objectifying type of knowledge, an expression which can already be found in myth (Bultmann), is — as a rule — incapable of understanding what is the meaning of will, decision, hope, liberty, of "I and Thou", of interpersonal relationships; that is, it fails to grasp existence in its nature of temporality and historicity. And yet it is here that the significance of expectation, the Blochian "not-yet," the Utopic ontology of man — which is something that must be discovered in the "non-place" — all this can be found. Here is the exodus. Here the expectation as *memory of the future* as well as the genuine source of subjectivity is given as possible. Indeed, following Husserl who, analyzing the structure of the transcendental Ego, discovers that its being is temporal (and therefore the subject is given as temporal), we can well accept the centrality of duration and expectation in our being-in-the-world. Temporality is what makes us *open, ec-static*, and projects the future of the encounter with the Other; it is disclosure of oneself and reception of the presence of someone else. Therefore it is *transition.*

Instead, objectifying knowledge tends by its nature to dispose of the world, of itself, of the others, and even of God, mastering them by means proper to knowledge and reason. In this fashion objectifying knowledge precludes and rejects any true alterity, that is the non-objectifiable reality, which should approach and investigate the subject, defining it in terms of appeal (*Anrede*), event (*Ereignis*), encounter (*Begegnung*), existence.

No-one may or has the right to violate by means of reason the liberty of the Thou while approaching him, to attempt to deduce or understand a priori his behaviour.

I am, in fact, in no position to manipulate the Other either by empirical or transcendental means just because of his being what he is:

225

A-T. Tymieniecka (ed.), Analecta Husserliana, Vol. XX, 225—229.

appeal, event, encounter; not even knowing the common character of human nature, the Thou remains forever the alterity in opposition to me whether he is in his full-fledged capacity or whether, immobilized, he lies hopelessly in bed.

This alterity pertains to *historicity*; it results from the fact that man exists not in a naturalistic but in a historical way. Man's historicity signifies, in point of fact, that he finds fulfilment in encounter. And encounter, which consists in the original unity of reality and knowledge, is given only and always through and in *language*.

It is the so-understood subject — in his existential encounter with me, in his existential relationship with his history — that constitutes the fundamental preliminary datum toward his "clinical" understanding. This latter is not a subjective comprehension resting upon the individual ideas of the clinician. Instead one has to assume that the subject, especially as regards his objective content, can only be understood by a subject who is existentially oriented.

Here the ineffable, the region of what cannot be expressed by words, but can be identified as it tends to be transferred and removed towards a body language, comes into play and is essential. The lexicon of the body acts as the clue to the clinical insight and — let us bear it in mind — the look, the tone of voice, the gesture, the motility, in one word that which allows the understanding of the patient's personality in his irreducible ecceity, all of them are "body", lived body (*Leib*), incarnated consciousness.

The so-evoked ineffable "speaks" so as to reveal the distance between the hence emerging significance and the category crystallized in the "text" of the linguistic expression. The inexpressible, the singularity of the "clinic", is made visible to the mind's eye by the omnipresent counterpoint of the body. The clinical bodiliness (*Leiblichkeit*) remains only tangential to the expressed forms of the text. Here phenomenological anthropology regains the clinical sense of the body, the very concentration of its reinsertion in the inner biography (*Lebensgeschichte*), where structure and theme converge, where symptom and communication-link refer back to the presence of the patient, to the relationship. Clinical-phenomenological knowledge, therefore, can never be considered as either conclusive or definite. The phenomena, which in the single encounter are revealed to us, cannot undergo neutral observation. Their meaning is made clear only to the investigator who is intensely touched by them. For this reason they become understandable only when they refer to the present-moment situation which is a historical one.

No doubt a great deal of insecurity disappears with rational knowledge and technical control, but the essential insecurity remains and moreover penetrates into the very security of science and technique (R. Guardini). Only the clinical phenomenology is a true experience, a message rising from the innermost structure of existence. That is one reason why clinical research is endless and must constantly be carried on. Founded as it is on a "vital relationship" it cannot be free from presuppositions. However, the presupposition (Bultmann's *Vorverständnis*) does not establish a prejudice, neither does the choice of a certain perspective. The important thing is that the presupposition is not considered as a definite comprehension. This is true only when, in a clinical sphere, it is justly said: *tua res agitur*, that is, I am personally concerned with the case with which I am dealing.

There lies the risk any true psychiatrist has to take in encountering the appeal (*Anrede*) of an another-inviting self: not to reduce the latter to a simple contemplation object, but to comprehend him within the actual experience of passage and transit, the thisness of the present — Heidegger's horizon of *nunc* — where the movement of pro-ject and proposition undergoes the superimposing movement of re-tortion and reflexion.

Time is therefore the subject which emerges in reality; not empirical time (resulting in a succession of countable *nows*) but the time that sets the subject in reality. It is the time of the clinic, which is itself history, enlargement of the field of presence, established by the *look* (the "clinical eye," perhaps?).

However this is not enough. Something which will disturb every Pharisee (according to Heine) has still to be said.

Since the text refers to existence, it is never understood in a definite manner. Clinical knowledge, from which comment, organization, arrangement derive, cannot be conceptually transmitted, but must always be taken up anew. The greatest danger for the psychiatric practice consists in self-abandonment to the collectivized and codified power of the transmitted body of knowledge; it results in the ignorance of the symptom and in the desertion of the individual.

Obviously this does not imply that in knowledge conceptualized in a systematized text continuity cannot be preserved. The results of psychopathological research can be transmitted in an objectified form, even though they can only be appropriately taken into account under the condition of a constant critical re-examination (as pointed out by Gadamer).

But also as regards the text clinically established, there is continuity, insofar as it is an indication of direction for the future (e.g. the comprehension of DSM-III concerning the symptoms of schizophrenia proposed by Kurt Schneider). Considering that this comprehension and arrangement of a text must always be renewed through discussion and confrontation with the classical text, any new authentic arrangement is always an indication of direction and a requirement to confront it seeking for a correspondence each time it is taken into consideration in an independent, that is clinical, manner.

The clinician must always listen to the word of the text, as it is addressed to him with-in the context of the specific situation of the clinical encounter.[1] In this fashion he will understand the ancient word in an always new manner. He will then have to express his re-affirmed insights with constantly renewed ideas. This need for an ever-to-be-renewed interpretation is essential in order to thoroughly understand the lesson of anthropological psychiatry in which the interest in either general or universal laws — let us say for the invariant aspects of concrete insights — is always linked to the awareness of the relative autonomy of the systems which fall each time under a more general legality.

One of the essential points is the great leading principle that marks complementarity between autonomy and interdependence, invariance and relativity. The very notion of *structure*, understood in a strictly Husserlian sense, represents an evident specification of that principle. The fundamentally directional nature of the anthropological program has to be emphasized. The latter is not a proper theory of psychiatry, nor does it aim at producing true descriptions of systematic inter-relationships; on the contrary, it appears as a framework of reference with respect to which the theoretical hypotheses can be measured each time and above all as a powerful integration device towards the generation of more comprehensive hypotheses (for example of an interdisciplinary nature). The only valid indication that psychiatry in a strict sense has so far produced to enable us to understand the functioning of a schizophrenic mind is of an anthropological nature; that is anthropology offers the only criterion which explains the inner splitting of his discourse: ambiguous, polysemic, always reinterpretable, and at the same time, withdrawn, compact, self-reclining.

The clinical encounter with the objectified data fixated in the text could lead either to the *yes* or to the *no*. The clinician stands before the text as before a decision. And the comprehension (of the case) is always a true answer to the request of the text; answer which cannot be confuted because of its being an existential decision, not an objectifying knowledge.

For this reason there is in the present day — peremptory and unarrestable — the recall of clinical phenomenology in psychopathology. I greatly fear the psychiatrist well fed on text and little on clinic. This I fear because he runs the risk of understanding exegesis as fixed once for all. The clinic is indeed the true means of demythologising psychiatry.

(Translated by Paul G. Weston)

NOTE

[1] In the editorial of the *Rivista di psicologia analitica*, Dec. 1981, it is stated that "the clinician must face novelty not as if it were something to reject on account of not being in agreement with his own conception of the world, but as something worthy of understanding which could possibly change the above-mentioned conception. The profound teaching inherent in the clinical diagnosis is perhaps never fully appreciated".

JOHN DOLIS

THEORETICAL FOUNDATIONS OF PSYCHIATRY: THE (K)NOT OF BEING AS A (W)HOLE

I

In her essay entitled "The Moral Sense: A Discourse on the Phenomenological Foundation of the Social World and of Ethics," Anna-Teresa Tymieniecka attempts "to free moral experience as *valuation* from its direct submission to objective factors."[1] As she observes, the history of value inquiry — whether an ethics of reason (ancient and medieval) or feeling (Brentano, Husserl, Scheler) — traditionally assumes that values are "the *a priori* determining factors of morality"; hence, "this formulation of the issue presupposes the role of objective cognition in moral valuation" (*AH*, 15:11). Thus, while Husserl and, even more so, Scheler emphasize the role of emotion in value experience, both nevertheless assume "that objectivity belongs to the constitutive-intentional nature of consciousness, which is directed toward positing objects through the noematic content of intentional acts" (*AH*, 15:13). Over and against its cognitive ground in (transcendental) subjectivity, Tymieniecka would shift the locus of value to its origin in the *human condition* and, if I understand it correctly, thereby recuperate intersubjectivity as ontologically prior. The very nature of subjectivity itself depends upon this formulation.

It is commonly acknowledged that Husserl's thinking does not sufficiently accommodate this problem. The transcendental reduction inevitably points to the monadic essence of consciousness, for as the ground of the "world," it is the very whole upon which all the parts (objects) depend. As Husserl remarks, "Each part that is independent relatively to a whole *W* we call a piece (portion), and each part that is nonindependent relatively to *W* we call a moment (an abstract part) of this same whole.[2] Here, the notion of foundation provides the noematic basis of eidetic insight, for, as Timothy Stapleton suggests, "it points to the presence of an essential relationship between elements or objects such that the condition for the possibility of the objects' existence lies in their

231

A-T. Tymieniecka (ed.), Analecta Husserliana, Vol. XX, 231–245.
© 1986 *by D. Reidel Publishing Company.*

necessary correlation with other objects."[3] Transcendental subjectivity is this absolute "concretum," the founding (and thus not founded) whole that needs no other to exist.[4] Once the object-horizon structure of the transcendental turn has been effected, this applies to consciousness, and pure consciousness alone.[5] In the *Cartesian Meditations*, Husserl thus concludes:

> Every imaginable sense, every imaginable being, whether the latter is called immanent or transcendent, falls within the domain of transcendental subjectivity, as the subjectivity that constitutes sense and being. The attempt to conceive the universe of true being as something lying outside the universe of possible consciousness, possible knowledge, possible evidence, the two being related to one another merely externally by a rigid law, is nonsensical. They belong together essentially; and, as belonging together essentially, they are also concretely one, one in the only absolute concretion: transcendental subjectivity.[6]

If we recognize transcendental consciousness as such, there need be no confusion between the "psychological" ego and its "phenomenological" counterpart. The two are one and the same:

> What distinguishes the "two" egos is the context or horizon within which they are seen and interpreted, or the whole of which they are parts. When the ego is seen *as* worldly, then the horizon for its interpretation, the whole to which it belongs, is the world. When seen *as* transcendental, there is no external horizon, for there is no whole or more comprehensive unity upon which it depends. It has only the *infinite, inner* horizon of temporality. It is in this sense that the transcendental ego is a pure "in itself," as related only to itself. This enclosure of subjectivity upon itself, however, does not preclude relation to the world; rather, it *constitutes* it. The world first comes to be as the necessary correlate of such consciousness.[7]

In other words, it is assumed from the outset that, if there is a part, there must be a founding whole which confers upon the object its sense. For Husserl, this can be nothing other than subjectivity — that is, transcendental subjectivity (ego-cogito-cogitatum): subjectivity as a whole.

In contrast to the cognitive essence of moral value inherent in Husserl's notion of the world, and its foundation in consciousness as object-oriented, Tymieniecka proposes "that values have their origin not in constitutive acts, but in the process of valuation" (*AH*, 15:13). Despite Husserl's assertion that affectivity functions as the *sui generis* independent source of emotion, feeling, desire, and so forth, Tymieniecka suggests that the experience of values, to which we are lead, "is nevertheless object-oriented and thereby is submitted to the objectifying laws of the intellect" (*AH*, 15:9). Hence, she seeks to alter the focus from the "objectivity" of value to the "subjectivity" of moral valuation — that is, to its process, to the interrogative, self-interpretive experience of the "natural" subject in its existential significance. Here, the act of "making valuable" calls into being a language of its own, a "moral language." The axiological experience is not a private event; rather, its very structure always already inosculates the Other. It is essentially intersubjective. This sense and language of morals — in opposition to objectifying consciousness — Tymieniecka calls the "Moral Sense."

Apart from, and "prior" to, the noetic-noematic structure of cognition, the moral sense bestows upon existence a meaningful interpretive paradigm — one whose inner direction takes the following form: "Everyone ought to valuate and act 'benevolently'" (*AH*, 15:23). The precognitive presence of this "ought" always already locates the subject in its primordial, intersubjective dimension; it is, at all times, a unique "operative" or "subliminal" spontaneity precisely in so far as it is "virtually present in the *Human Condition*" (*AH*, 15:23—24). As such, the moral sense founds any and all "social" *significance*. In essence, then, a subject is "for" the Other, "for the sake of" the Other. The referential center of the moral sense remains always and everywhere the same: the "good" (of the Other). Its negative moment, on the other hand, seeks to obstruct the Other "for the sake of" the Self, whence there obtains a negative sense of "aesthetic" (Kant) enjoyment. Now to the extent that the reflective nature of interpretation returns the subject to itself, this experience — in both its positive and negative moment — articulates that Self. As positive, the Self comes to be fulfilled or whole, and can be characterized as such (self-contentment, self-satisfaction). As negative, conversely, the moral sense confers upon the subject its very absence (to itself). In turn, its mode of existence comes to be characterized by discontent, dissatisfaction — at best, regret or remorse. This (self)-fulfillment is paradoxically to be found only in the good, in the Other as a "whole," whereas the subject as a "(w)hole" discovers itself as discontent, empty, or meaningless. Of course, by way of repression, the subject has at its disposal a means of avoiding the void of itself, its lack of authenticity — a mode for which we commonly employ the term "hypocrisy."

By situating moral valuation in advance of the noematic "object," as ontologically prior, Tymieniecka simultaneously establishes the ontological priority of intersubjectivity whence there emerges to consciousness (the individual subject as ego) the possibility of "fulfilling" its own significance, the meaning of its own existence. From out of the Other, there arises the question of Being itself: what does it mean "to be?" Thus intersubjectivity inaugurates the subject to value: the moral sense here constitutes a unique mode of the "primogenital modality of the *logos*." Yet if the being of a subject discovers its meaning in the Other, there nevertheless remains the question of the meaning of being itself — that is, the question of the being of meaning itself.

II

While Tymieniecka clearly locates the being-question in the logos, she does so in relation to a "whole." Like Husserl's whole of transcendental subjectivity, this whole confers upon a subject the full significance of itself. If not the whole of consciousness, the "fullness" of significance is now to be found in the Other, in the relation of a part (subject) to the whole (Other). In one sense, then, the subject is a "moment," to use Husserl's term, of this whole, this (self) fulfilling context. Though given over to the logos, at least the subject can be whole. Tymieniecka has already secured this possibility by laying down the origin of man's "constructive interpretation of his existence" as that which comes to him from out of the context of the Other, the self (Same) in its very other-ness: "man's self-explication proceeds at all the stages with reference to the radical 'otherness,' the 'other self,' or the irrespective other."[8] In this regard, the subject composes itself "by *inventing* and *organizing* the 'ciphers' themselves" (*AH*, 6:155). Thus while the subject no longer merely "deciphers" itself in terms of a "*preestablished meaningfulness* proper to the *vital progress of the objective givenness*" (*AH*, 6:154), it nonetheless composes itself with reference to the Other. Only thus can it establish itself (fulfillment) as a whole. The notion of "textuality," then, takes us to the very heart of the logos. Here thinking is offered a direction from Heidegger.

Leg (Indo-European) — to collect, to choose, to speak; *legein* (Greek) — to gather, to speak; *legere* (Latin) — to read, to gather: language (*logos*) "gathers." It plays across the abyss of (inter)subjectivity; it draws subjects together as one to another. "To discuss language, to place it, means to bring to its place of being not so much language as ourselves: our own gathering into the appropriation. ... Language itself is — language and nothing else besides."[9] Language "bespeaks" the corre-spondence between self and other; in this sense, Heidegger sets forth his subtle dictum: "*Language speaks.*"[10] Here one belongs to another at the very origin of alterity, at the threshold of an event which is itself the advent of what Emmanuel Levinas has called the "interval of discourse." Whenever one stands out (ek-sists) and into this graft of displacement, there being "comes to pass," and reverberates the interval of silence wherein abides the alterity of the Other. As A. F. Lingis has suggested, this is the very locus of language: "Language arises in the interval of discourse. . . . we are concerned . . . with the structure of the situation in which a discourse becomes possible, with the kind of distance, of interval, in which a discourse plays. This distance . . . is essential to the event of the ingathering Logos. *There is* the Other, and because there is the instance of the Other, there is discourse. This alterity is irreducible, primordial; it is essential to the "dis-stance" of discourse."[11] It is precisely in this sense that we are to understand Heidegger's remark: language speaks. Language speaks as the peal of stillness.[12] Within the structure of this event, only that being who hears the call can reply. Herein occurs the moment of correspondence itself. He who hears the call speaks back. This correspondence therefore sets the self into its own: it "stays" the self as something other than the One (*das Man*). And though it cannot be explained, this correspondence is itself expressed whenever being is provoked to secure a place for itself, whenever it hearkens to the invocation to inscribe a text of its own: to be its own (authority). Being called to speak for itself, it thereby expresses the inaugural problematic of authorization in general.

What, then, is the meaning of being itself? As Jacques Lacan cautions, "Any statement of authority has no other guarantee than its very enunciation, and it is pointless for it to seek it in another signifier, which could not appear outside this locus in any way."[13] Setting out from the conception of the Other as the locus of the subject, Lacan defines a signifier as that which represents a subject for another signifier. The subject is this very "cut" or "wound" (*blessure*) in discourse itself, the eradication (erasure) of the phallus which allows it to (re)emerge as the signifier of desire. The subject cannot accede to itself by means of being designated as a whole "in(−)itself," but is the constant remainder of the discourse as a (w)hole. Herein, also, there appears the very "trace" of the signifier as the desire of another (signifier). We detect in this bind a structure not unlike that which Derrida calls "differance," and which Gayatri Spivak transcribes as "*pre*monition" and "*post*ponement."[14] If the sign might both be said to defer itself and differ from itself, the same can be said of the subject itself, an ego whose "text" excludes a privileged authorial position − one whose beginning, in fact, originates elsewhere and otherwise. The signifier considered in and of itself is nothing but the isolated subject "taken by itself" (narcissism): the subject objectified. Thus would reason — "ratio" or "correctness" — reduce (castrate) meaning to the "integer" of self-"identity." A fact is similarly meaningless precisely to the extent that it coincides with itself; identity does not give rise to meaning, but rather — like meaning itself — arises from discrepancy: division within.

Similarly, the question of "self"-identity returns every subject to another. In this regard, identity is always a "fiction," a text whose context, whose very pretext, needs to be delineated: a story whose knot demands to be (un)tied. Likewise, historical authority "lies" in the subjective register, the domain of an author — though for its institutionalization (that is, in order to become a fact) it clearly requires the public consent. As Foucault repeatedly observes, authority belongs to discourse itself, a discursivity, moreover, whose origins remain forever closed off to any and all forms of the absolute. Discourse displaces the insignificant certitude ("correctness") of fact with the "sign" of fiction: in language, there is (Dasein) inscribed the na(rra)tivity of being itself. Regarding this discursive fabric of all authority, it is from the beginning — indeed, in the beginning (was the Word, the logos) — devoid of propositional logic. Insofar as no text (much less one which claims authority for itself) can ever be said to supply its whole field — or even its intention — in advance of itself, as Edward Said points out, it can properly be said to begin, therefore, only with a large supposition; if meaning is to be produced in writing, this beginning intention remains always and everywhere a fiction.[15]

Concerning its claim to authority, moreover, the text refers us purely and simply to the domain of (good) faith and interpretation. Authority displaces the logic of the "line" with the discursive knot of textuality. The signifier inaugurates this dislocation; it opens being to meaning — a series of substitutions; language constitutes the beginning of another enterprise, an intentional structure signifying a series of displacements: language replaces genesis with paragenesis, origins with beginnings, continuity (the line) with contiguity (the knot), genetics with textuality. In other words, textuality transforms an original object whose (in)significance is fixed (factual) into a beginning intention (consciousness) whose significance is open and multiple (fictive).[16] In this beginning (fabrication), there emerges the intention to mean — an intention which, because it is bequeathed to language, allows the meaning to "stray" toward multiplicity, permits the possible forms of discourse to merge one into the other. Authority is nothing less than this inaugural intention of the text.

Because meaning is grounded in intentionality, authority primordially refers us back to (self)-expression — the fictive or fabricated correspondence between subject/object, signifier/signified — and its reversible gestalt against the ground of repression. To begin, as Freud observes of the Hexateuch, there is a "violence" in texts:

almost everywhere noticeable gaps, disturbing repetitions and obvious contradictions have come about — indications which reveal things to us which it was not intended to communicate. In its implications the distortions of a text resemble a murder: the difficulty is not in perpetrating the deed, but in getting rid of its traces. We might well lend the word "Entstellung" [distortion] the double meaning to which it has a claim but of which today it makes no use. It should mean not only "to change the appearance of something" but also "to put something in another place, to displace." Accordingly, in many instances of textual distortion, we may nevertheless count upon finding what has been suppressed and disavowed hidden away somewhere else, though changed and torn from its context.[17]

This textual disinterruption is traceable to the discrepancy and substitutability of the sign itself. Meaning emerges from this (violent) gap in signification: the self-reflexive subject reveals that its "identity" in no way coincides with itself, but rather is constituted by another in its origin. Otherness, then, represents the boundary or limit of the subject as signifier (of desire). If origins, however, remain forever inaccessible, beginnings do not: beginnings inaugurate the signifier to meaning. Here we stand witness to the Freudian subject and its (de)termination (in language): to write itself off. As signifier, the subject experiences the double-cross (the chiasm, self-other) of its discourse, the duplicity of a demand to "eye" its "I," to pen itself what has been hitherto inscribed within the other and therefore express that authority which is its own — though yet it "lies" repressed. What is absent (the phallus) cannot hide — what is present can.

From the structure of repression, there emerges to being its lack (of meaning), its need to be expressed: the signifier transforms subjectivity to discourse. In being given over to dialogue (contextuality), the subject translates itself toward its (un)final truth: writing (existence) is a dead end. This too suggests that language does not simply play at "childbirth" across the abyss of paternity (the line) alone, but rather exposes the "androgynous" abyss across which signs signify, that continually problematic status of being as a (w)hole which combines the absence of the (dead) father with the presence of the (living) mother: "Writing leaps back and forth across this impossible interval, doubling, multiplying, with no escape save annihilation."[18] Yet, in as much as the subject refuses to authorize itself, it gives itself entirely over to the other. As Geoffrey Hartman remarks: "the word that is given up is not given up: it must inscribe itself somewhere else, as a psychosomatic or mental symptom."[19] To be means to "write" the story of one's self — the "maternal" cathexis (narrativity) in (the) place of the (missing) phallus (nativity).[20]

In all of this there is implied the ontological status of language, the "textuality" of human existence — "the book of life" — and being itself. Narration occupies both "story" and "history" as their beginning, their point of embarkation — that locus of authorization which makes existence (a) work. Meaning is born through exposition, the temporal fabrication in which existence is made to stand out (eksist): *there* (must) *be* a narrator. Expression enables existence to stand "out" as, and "in" for, itself. Adverse to expression, to singing the song of oneself (one's being), repression circumscribes the subject as an "impostor" (as not its own), and therefore articulates the very impossibility of being. If at its origin the Self is always already "written" by another, it nevertheless is called upon to inaugurate the manner and meaning of its significance. In its intention to secure a beginning, the self thus authorizes its emergence into the world as an authentic "work" — one of its own making. The world (of the work) becomes the work (of the world) exposed. With this expressive exposition, being is provoked to itself. Repression, however, revokes the self to the failure of being, its being spoken for and written by the other — upon the "line" of resistance to the self at the insistence of another, existence is repressed. Authentic existence would yet undo or (un)tie this original (paternal) line with a significant k(not) of its own expression, its story: the subject itself is always a fiction.

Here the subject inscribes itself upon the scene at the very point where "something" is missing — a lacuna which echoes Freud's brilliant observation of the "fort/da": gone (away)/there (here) it is! To the extent that existence is given over to language, to the abysmal structure of the sign — the irreducible discrepancy between signifier and signified — the subject enacts both presence and absence as separate aspects of the same phenomenon, the otherness of being itself. A subject both defers and differs from itself. Provoked to "sign" the Other with a discourse which encircles the (w)hole of being, its (in)significance, the subject exposes a circle of denegations which recover the self by means of expression. Expression redeems the subject from the abyss of nothingness across which the signs signify and for significance itself; it articulates the "heartfelt" meaning of being in its exposition — the subject's correspondence to the heart of things (*res*): a matter for discourse. This logos or "logic" of the heart (Pascal), secures the self to a "significance" which precedes it and thus occasions the very possibility of (self) expression. Discourse interrupts the ratio of reason in as much as it returns the subject to an irreducible locus of signification whence any and all meaning (sense) originates; it bestows upon the subject the possibility of expressing itself and thereby exposing what at the very center of its existence — as a repressed signifier — inheres as its irreducible kernel or *kern* (to use Freud's term): a heart of nonsense.[20] As Lacan suggests, all discourse harbors within it this locus of the "imaginary" by means of which the subject constructs the object (image) of both the (real) world and itself: "The *I* is not a being, it is a presupposition with respect to that which speaks."[21] Regarding the textuality of existence, this "lack" returns the subject to the heart of all significance, its center of nonsense (nonbeing) around which all signification (existence) would tie the k(not) of being. It exposes what in the register of the symbolic must remain forever unspeakable: existence is a dead end. Around this (k)not, the subject inscribes itself. The presence of desire but marks the point of origin of this absence, this (missing) scene, that most simple discontinuous space (0, 1) of the Freudian *fort-da* where the presence of the father (to be) "playing" with the absence of the mother engenders discourse itself: "Writing oscillates between a name that cannot be inscribed and the dead body, a corpse-effect whose intrusion into the real is the sign and signature of this impasse. . . . the tomb is the point at which name and body are wed in their common impasse."[22] Existence encircles this (black) (w)hole which would draw everything into it. It should come as no surprise, then, that in the failure of adequation, in this very lack, in the trace or mark of the signifier as remainder of another — that which represents a subject for another signifier — death has already taken place, has taken up its place "in" the end, and done so from the beginning. Death has secured the end at the front. This (double) "crossing" of signifiers constitutes, in effect, the very locus of significance as that which comes between: the gap or "fold" of being (subjectivity) itself.

Merleau-Ponty explores this "crossing" of signifiers, the chiasm of self and other, with his felicitous analogy of the glove:

Reversibility: the finger of the glove that is turned inside out — There is no need of a spectator who would be *on each side*. It suffices that from one side I see the wrong side of the glove that is applied to the right side, that I touch the one *through* the other (double "representation" of a point or plane of the field) the chiasm is that: the reversibility —

It is through it alone that there is a passage from the "For Itself" to the For the Other — In reality there is neither me nor the other as positive . . . subjectivities. There are two caverns, two opennesses, two stages where something will take place — and which both belong to the same world, to the stage of Being

. . . They are each the other side of the other. This is why they incorporate one another: projection-introjection — There is that line, that frontier surface at some distance before me, where occurs the veering I-Other Other-I —

The axis alone is given — the end of the finger of the glove is nothingness — but a nothingness one can turn over, and where then one sees *things* — The only "place" where the negative would really be is the fold, the application of the inside and the outside to one another, the turning point . . . the things, realized by the doubling up of my body into inside and outside — and the doubling up of the things (their inside and their outside).[23]

Regarding being and its "lack" (not-being), "to be" means to be missing from one's self in such a way that it matters. Language speaks this fold in being, its nullity; discourse discloses this (w)hole. In one sense, discourse covers itself; it hides; it obscures the Other to whom the discourse is addressed: the alterity of the Other must be forgotten or repressed by the discourse. And yet, the exposition of the Other simultaneously "opens the interval in which language is possible, the interval across which the signs signify."[24] From out of this abyss, the Other calls. It is precisely against the contour of this ontological phenomenon, this "interval," that Lacan defines the unconscious as the discourse of the Other. For in the register of the psychoanalytic, being shelters this absence, this nullity, within the structure of language itself.

Within this "house," the subject is provoked "to be," to dwell, to inscribe the text of itself: the meaning of its being. As Heidegger observes, language is the temple of being. This metaphor, moreover, solicits the ontological structure of existence at its core, its kern, its heart: for being itself is a figure, an expression — one whose meaning but figures forth against the ground of not-being. The figure already resonates this silence, this gap or void, as a reversible gestalt around which being, in turn, inscribes or encircles ("entombs") itself. To use Hegel's figure, the subject is a hole in being. Meaning (truth) here constitutes the play of presence and absence, disclosure and concealment, whereby the "ghost" of substitution (desire) assumes its place — its "in" (the place of). Language expresses this (silent) correspondence between signifier and signified, subject and object, self and other, being and meaning: the insertion of the subject in (the place of) the (w)hole of being.

The structure of signification here plays across this silent abyss in which the subject is both its cutting edge and knot. Regarding all the forms of being, there is the gap, the chiasm which yawns — waiting for the subject to insert itself, to uncover its meaning, its truth: to inscribe the text of existence with its (k)not. There is (Dasein) the interval of discourse: amid this interval, the Other is exposed. If in the Other there arises to being its exposition, the possibility of being-expressed, discourse sounds the silent correspondence between this being and its meaning: the very heart of nonbeing (nonsense) at its center. Discourse implies an ontology wherein both presence and absence equiprimordially obtain: the construction and repetition of a text in which the subject is at all times missing from itself, and thereby stands out (eksists) in (need of) interpretation. Interpretation, in turn, returns us to the (inter)face or "persona" of textuality — disclosure in concealment: a form of existence as being (exposed or unmasked) in (the very face of) the (masked) text. Over and against the irresponsible imposition of being as repression (concealment), responsibility expresses (discloses) the signature of being one's own. In re-sponse-ability, there is given to being its exposition. And in the roundness of this correspondence, the subject is bequeathed to itself.

III

Tymieniecka's notion of the moral sense obtains, then, only insofar as language opens the subject onto responsibility. Language gathers. The moral sense of intersubjectivity always already implies this correspondence, a correspondence whose whence and whither refer the subject to the question of being itself: the possibility of not being one's (own) self is the possibility of one's own not being. Whenever being is called to itself, it ought to be its own. In this event (*das Ereignis*), the value of existence (ought), its moral sense, arises from its aught — the nullity already in (the place of) the signifying structure itself. Psychoanalysis, for example, attempts to reinscribe the "text" of the subject within the context of this (missing) "scene" — that locus of what Lacan refers to as the "real," that which always comes back to the same place, and before which all symbolic discourse falters: what is at all times held in abeyance. It is "another locality, another space, another scene": the encounter in as much as it is missed, "in so far as it is essentially the missed encounter."[25] That is to say, psychoanalysis repeats the very trauma of interpretation, that "primal scene," which inaugurates the subject to meaning, as Barbara Johnson suggests: it is "the traumatic deferred interpreptation not *of* an event, but *as* an event which never took place as such. The 'primal scene' is not a scene but an *interpretive infelicity*. . . . Psychoanalysis has content only insofar as it repeats the dis-content of what never took place."[26] By provoking a subject to interpret the meaning of its being, psychoanalysis invites the subject to repeat in "forward recollection" (Kierkegaard) that very scene which never (yet) occurred but will: one day, I (ego[cogito]) will not be. One day, the subject will be irrevocably missing from itself in such a way that it *cannot* mean — though now it *can* not-mean. Indeed, if psychotherapy — in any and all its forms — is to effect (affect) a "cure," it must in-"cite" the subject to interpret its absence thus; it must repeat the whence and whither of the way in which the subject essentially "departs" from itself, the way in which the subject must insert itself into the w(hole) of being — the only way in which the subject can ever be itself: as a (w)hole. Only in departure, on the other hand, can being be the whole. The essential nullity of the subject therefore gives the self over to the possibility of being its own, of Being itself the very exception to the whole, the One (*das Man*). In this parting of the ways, there is inserted, then, the possibility of being (exceptional): the possibility of being the exception to being — of not-being. Amid the roundness of this correspondence, being is expressed.

Signification already carries with it this normative sense: the valuation of being-for-itself. Being in language means always already being with the Other (*mit-sein*). As Wittgenstein so aptly argues in the *Philosophical Investigations*, there is no such thing as a "private" event. To be (in language) means to be already in the Other. Language gathers. And yet it provokes the subject to appropriate for itself that which can only ever be called its own: its own not-being. While in the register of the psychoanalytic the subject can deny itself (the discourse of the Other), and in the register of the existential analytic being can avoid itself (the discourse of the One, *das Man* — idle talk and gossip), yet neither One nor the Other can (be)speak the subject's own not-being. In death, the subject itself cannot be spoken (for); only the subject can die (for) itself. If anything can be said "to be" only the subject's own, it can be only its own not-being. Adverse to the repression of being — that is, the forgetting of Being itself — language speaks as the peal of stillness; it calls the subject to repeat itself in forward recollection, to express the meaning of its being: the yet-to-be is not-being.

Repression here constitutes the very realm in which Heidegger translates the well-known Epicurean admonition, "Live in hiding"; it is that ontological dimension in which Heraclitus discovers Being itself: how can one hide himself before that which never sets?[27] In the discrepancy between being and meaning, moreover, the one stands in relation to the other as signifier to signifier. Each does so as but a different "aspect" of the same phenomenon: a structural gestalt, as it were, which takes the form of a reversible figure-ground relation — a structure, furthermore, which incorporates the psycho-logos and onto-logos simultaneously. For at the level of the ontological as well, expression/repression articulates the structure of both the meaning (of being) and being (of meaning). On the hither side of this event, each is grounded in temporality. While language thus bespeaks the correspondence between being and meaning, this correspondence itself neither represents a coincidence of the one with the other, nor even a plenitude (whole); for here the irreducible abyss between the two implies that neither can be fully present to the other. A consciousness of the facts can never account for the fact of consciousness: One speaks for the Other; the Other speaks for itself. The structure expression/repression, then, implies an ontology in which both presence and absence equiprimordially obtain. Though language speaks in the mode of calling, the place of arrival to which the subject is called is therefore a presence already sheltered in absence.[28] Upon the threshold of this event, the place of arrival is always already also understood as the place of departure as well. Language provokes the subject to assume its position here: in (the) place of the departed One. In contrast to the irresponsible imposition of being as repression, responsibility expresses the signature of being itself. In this event, there is given to being its (own) exposition; in this perpetual inadequation, the subject is bequeathed to itself. Herein dwells the genuine work of existence, the tangle whence emerges its narrativity and toward which psychotherapy must ceaselessly direct its solicitous dialogue. For in this "fabrication," there is (Dasein) inscribed, in (the) place of the missing (phallus), the magic "limen" — the thread of a story designed to be, from the outset, fated to (dead) end.[29] The neurotic subject — and, to a near-absolute degree, the psychotic — would rather escape its own time, the k(not) of its being; it would be timeless and thus undo its knot by passing time, spending time, killing time. To psychotherapy is entrusted the task of once again "making" the subject timely, of restoring it to textuality, of recuperating this (k)not of being as a (w)hole.[30]

Pennsylvania State University, Scranton Campus

NOTES

[1] Anna-Teresa Tymieniecka, "The Moral Sense: A Discourse on the Phenomenological Foundation of the Social World and of Ethics," in *Foundations of Morality, Human Rights, and the Human Sciences: Phenomenology in a Foundational Dialogue with the Human Science, Analecta Husserliana*, vol. 15, ed. Anna-Teresa Tymieniecka and Calvin O. Schrag (Dordrecht: D. Reidel, 1983), p. 13; hereafter cited in the text as *AH*, 15.
[2] Edmund Husserl, *Logical Investigations*, trans. J. N. Findlay, 2 vols. (New York: Humanities Press, 1970), p. 467.
[3] Timothy J. Stapleton, *Husserl and Heidegger: The Question of a Phenomenological Beginning* (Albany: State University of New York Press, 1983), p. 59.
[4] See Edmund Husserl, *Ideas: General Introduction to Pure Phenomenology*, trans. W. R. Boyce Gibson (New York: Humanities Press, 1967), p. 152.
[5] See Stapleton, *Husserl and Heidegger*, p. 65.
[6] Edmund Husserl, *Cartesian Meditations: An Introduction to Phenomenology*, trans. Dorion Cairns (The Hague: Martinus Nijhoff, 1960), p. 84.
[7] Stapleton, *Husserl and Heidegger*, p. 75.
[8] Anna-Teresa Tymieniecka, "The Creative Self and the Other in Man's Self-Interpretation," in *The Self and the Other: The Irreducible Element in Man, Analecta Husserliana*, vol. 6, ed. Anna-Teresa Tymieniecka (Dordrecht: D. Reidel, 1977), p. 154; hereafter cited in the text as *AH*, 6.
[9] Martin Heidegger, "Language," *Poetry, Language, Thought*, trans. Albert Hofstadter (New York: Harper & Row, 1971), p. 190.
[10] Heidegger, "Language," p. 190.
[11] A. F. Lingis, "On the Essence of Technique," in *Heidegger and the Quest for Truth*, ed. Manfred S. Frings (Chicago: Quadrangle Books, 1968), p. 135.
[12] Heidegger, "Language," p. 207.
[13] Jacques Lacan, "The Subversion of the Subject and the Dialectic of Desire in the Freudian Unconscious," in *Écrits*, trans. Alan Sheridan (New York: W. W. Norton, 1977), pp. 310–11: "which is what I mean when I say that no metalanguage can be spoken, or, more aphoristically, that there is no Other of the Other."
[14] Jacques Derrida, "Differance," *Speech and Phenomena*, trans. David B. Allison (Evanston: Northwestern University Press, 1973), pp. 129–30; Gayatri Spivak, "The Letter as Cutting Edge," in *Literature and Psychoanalysis; The Question of Reading: Otherwise*, ed. Shoshana Felman (Baltimore: Johns Hopkins University Press, 1982), pp.

209—10. If postponement recollects forward there being a future, premonition recollects backward the future as that which has already been. Postponement "knows" behind itself from out of the future; premonition "knows" ahead of itself from out of the past.

[15] Edward W. Said, *Beginnings: Intention and Method* (Baltimore: Johns Hopkins University Press, 1975), pp. 59—60.

[16] Said, *Beginnings: Intention and Method*, pp. 65—67.

[17] Sigmund Freud, *Moses and Monotheism, Complete Psychological Works*, vol. 23, trans. James Strachey (London: Hogarth Press, 1964), p. 43. Cf. also Said, *Beginnings: Intention and Method*, p. 59.

[18] Daniel Sibony, "*Hamlet:* A Writing-Effect," in *Literature and Psychoanalysis*, p. 74.

[19] Geoffrey H. Hartman, Preface to *Psychoanalysis and the Question of the Text*, ed. Geoffrey H. Hartman (Baltimore: Johns Hopkins University Press, 1978), p. xviii.

[20] See Philippe Sollers, "Freud's Hand," in *Literature and Psychoanalysis*, p. 337, n. 1: "Freud makes the following suggestion: that writing was invented by women through the weaving and braiding of their public hairs."

[21] Jacques Lacan, "From Interpretation to the Transference," in *The Four Fundamental Concepts of Psycho-Analysis*, trans. Alan Sheridan (New York: W. W. Norton, 1978), p. 250; cf. also Spivak, "The Letter as Cutting Edge," p. 223. We might read into this a correlation between an "original" signifier or *kern* of irreducible nonsense and the religious sense of an original "sin." If signification originates in a si(g)n or (fortunate) "fall" of the signifier into the "de-nominator" as zero, it constitutes the very possibility of value and signification, and "kills" all meaning. The "void" of the subject (as represented by a signifier for another signifier) is therefore an infinity (of possibilities: $s/o = \infty$) against the finitude of desire. This, in turn, constitutes the subject in its freedom. See Lacan, "From Interpretation to the Transference," pp. 250ff.

[22] Spivak, "The Letter as Cutting Edge," pp. 219—20.

[23] Sibony, "*Hamlet:* A Writing-Effect," pp. 82, 75.

[24] Maurice Merleau-Ponty, "Working Notes," in *The Visible and the Invisible*, trans. Alphonso Lingis (Evanston: Northwestern University Press, 1968), pp. 263—64.

[25] Lingis, "On the Essence of Technique," p. 136.

[26] Jacques Lacan, "Tuché and Automaton," in *The Four Fundamental Concepts of Psycho-Analysis*, pp. 55—56.

[27] Barbara Johnson, "The Frame of Reference: Poe, Lacan, Derrida," in *Literature and Psychoanalysis*, p. 499.

[28] Martin Heidegger, "Aletheia (Heraclitus, Fragment B 16)," in *Early Greek Thinking*, trans. David Farrell Krell and Frank A. Capuzzi (New York: Harper & Row, 1975), pp. 106—7.

[29] Heidegger, "Language," p. 199.

[30] See Sollers, "Freud's Hand," and Sibony, "*Hamlet:* A Writing-Effect," pp. 337, 53. See also my article, "Expression and Silence: The Responsibility of Language/The Language of Responsibility," *Journal of Phenomenological Psychology* (forthcoming 1986).

PART III

CIRCUITS OF COMMUNICATION

AARON L. MISHARA

A PHENOMENOLOGICAL APPROACH TO LANGUAGE ACQUISITION AND AUTISM IN TERMS OF A MOTOR UNCONSCIOUS

In recent years there has been an increasing effort to relate the study of "developmental kinesics" to a psychology of the acquisition of language (most notably, Walburga von Raffler-Engel in, for example, "Developmental Kinesics: The Acquisition of Conversational Nonverbal Behavior"). Whatever empirical results may be obtained by such an approach, the assumption that kinesics itself has its own structure or "grammar" shared by Ekman, Birdwhistell, and others, whether this structure be biologically or culturally conditioned or both, can at present be explained most persuasively by a "motor metatheory of mind" which Weimer, the philosopher of science and cognitive psychologist, offers as the only model "adequate" to the present research problems facing cognitive psychology on all fronts. The motor theory is not only relevant to the "physiological basis of action," but to all "higher mental processes," including language use and comprehension, which can be conceived as "constructive motor skills." Nevertheless, Wiemer writes:

> We have no theories, even on the farthest horizon, that account for the problems of the unconscious represented in tacit awareness and skill. But we need not be saddled with a conceptual framework that renders no solutions possible. From the motion theory framework one can at least attempt to assay what is involved without running into contradiction and inherent confusion. ... The problem for the future psychology of knowledge and action appears somewhat paradoxical: we require a theory of tacit knowing and doing that will also account for that "little bit" of conscious or explicit phenomena that we sometimes exhibit.[2]

By conceptualizing the "unconscious" in terms of a theory of mind based on bodily movement, our understanding of the nature of language acquisition will be enriched. I hope to demonstrate in the present paper how a phenomenological approach not only clarifies a motor theory of the unconscious, but also how the acquisition of speech is already embedded in the child's experience of his own body in relation to a world inhabited by others. In this way a phenomenological approach may offer advances on both ends: a motor theory of mind, and how a

249

A-T. Tymieniecka (ed.), Analecta Husserliana, Vol. XX, 249–264.
© 1986 by D. Reidel Publishing Company.

motor theory may contribute to an understanding of language acquisition in terms of the recent studies that attempt to relate the latter to a "developmental kinesics." Once this has been developed, some means for testing and further understanding these suggestions will be offered.

Phenomenology begins with the problem of existence. To isolate any aspect of it through "abstractive reduction" is to take a perspective that organizes itself according to an internal necessity with regard to an emergent group of phenomena which similarly organize themselves into a coherent unity along with the theory and are preselected by the perspective to which they correspond. It is not in a dissimilar manner that Weimer writes about metatheories as the shared (unconscious) context of a perspective and the object it studies:

> Metatheories are like perspectives or vantage points: they provide a point of view from which a domain may be scrutinized. In providing such conceptual underpinning to a domain of inquiry a metatheory is in itself all but invisible: one "sees" the domain through the conceptual glasses that constitute the metatheory, but one does not see the metatheory itself. Thus, a metatheory is characterized only indirectly, by pointing out those aspects that structure and constrain the domain addressed. . . .[3]

If Merleau-Ponty chooses the phenomenological method in his 1949—50 lecture course at the University of Paris, published as *Consciousness and the Acquisition of Language*, over the "reflexive" and "inductive" methods, as the most fruitful approach toward conceptualizing the acquisition of language, it should not be too surprising that the central problem for understanding language acquisition becomes the perception of other people by the child. For this reason, as we shall see later, imagination and affectivity play a central role in the development of language comprehension and expression. This is not to say that more formal approaches such as the Jakobson approach to structural linguistics or even Chomsky's approach to deep structures of syntactical operations do not have their place in such an approach, it is rather that they have their place as metatheoretical perspectives which take abstractive systematic relations to only certain aspects of the data the total context of which is to be laid out in a phenomenology of human relations. Merleau-Ponty writes:

> Jakobson's interpretation would be acceptable if language had only a representative function. But we have stated along with Karl Bühler that language is indissolubly: (a) representation; (b) self-expression; (c) appeal to others.
>
> The child's movement towards speech is a constant appeal to others. The child recognizes in the other one of himself. Language is the means of effecting reciprocity

with the other. This is a question of a vital operation and not only an intellectual act. The representative function is an aspect of the total act by which we enter into communication with others. . . . The phonemic system is a style of language . . . .[4]

Because language is acquired in a context of other people toward which the child appeals in his emergent sense of self, Merleau-Ponty turns in his course to the analysis of "imitation" as it is discussed by Guillaume. "To imitate is not to act like others, but to obtain the same result as others," Merleau-Ponty writes.[5] We are not first conscious of our body and then the things or goals towards which the body acts: "Before making a movement, we do not represent this movement to ourselves; we do not envision the muscular contractions necessary for effecting it."[6] The attention rather is to the outcome and whatever bodily skill one has so far acquired is rather brought to the task on the level of a "tacit" know-how which if it requires further thought only distracts one from absorption in the object of the task: "Guillaume says that the child first imitates the result of the action by using his own means and thereby finding himself producing the same movements as those of the model." Therefore, "the imitation is immanent; it aims at the global result and not the detail of the gesture. . . . In short, one makes use of his own body not as a mass of sensations, doubled by a kinesthetic image, but as a way of systematically going toward objects."[7]

Although Merleau-Ponty cites examples such as the child who at thirty-two months is asked to imitate an adult shifting his eyes from side to side, and who responds by turning his entire head, in order to show that the child begins with the result rather than the means, he has nevertheless made the problem, perhaps deliberately, artificially simple. The relation between perception, bodily movement and speech is far more complex, such that transitions between these are best described by Viktor von Weizsäcker, the philosopher and physician, who is largely responsible for introducing the psychosomatic approach in medicine to Germany in the early 1930s, and whose work was well known to Merleau-Ponty, in his concept of the *Gestaltkreis*. Dieter Wyss, the student of von Weizsäcker and who today further develops von Weizsäcker's thought and research, writes:

The fusing of movement and perception means that the activity which prompts a perception, i.e. which prompts a look, is not itself perceived but takes place unconsciously. Such activity is negative achievement and can only become an object of perception after it has taken place. In other words, the spontaneous movement involved in the act of looking, the movement of the head in a certain direction, does not determine (cause) the perception, rather the perception itself is a spontaneous movement. In this case the unity of perception and movement does indeed operate on the principle of a revolving door: the inside of the house is only perceived when you enter the house, once you go out it is no longer visible. Or to put it another way: perception and movement are mutually concealed from one another. Although the act of perception is itself a spontaneous movement, the movement, as it were, "knows" nothing about the perception and the perception knows nothing about the movement: they are concealed from one another just as the inside of the house is concealed until you enter the house. . . . The *Gestaltkreis* is now seen to consist of the unity of the subject with the environment, which unity is constantly established by the subject by means of movement and perception.[8]

We have in the notion of the *Gestaltkreis* a "model" that shows how transitions from perception to movement in imitative efforts are able to occur unobtrusively in a constant shifting of attention from perception of the other in movement toward activating one's own body, and toward the object of the action. At each moment all the previous phases fall into the background in a kind of tacit grasping of the situation Any one of these phases may become for a moment again the focus of awareness during the accomplishment of the task by means of a Gestaltcircle which has already connected the various phases: perception of the other, movement and perception of the object toward which the movement is gauged, as they pass out of the shifting awareness. This ability to shift awareness toward an aspect or object of our experience away from some other object or aspect, or from action to perception or from perception to action, without fear of being unable to shift back to what was the theme just a moment before in terms of oblique or lateral connections in an "operative" field of consciousness, I would like to call a "tacit" operation of the "lived-body" (*Leib* as opposed to *Körper* in German). Moreover, these tacit operations of the lived-body that are engaged in everyday consciousness as the conditions of coherent experience, but nevertheless remain invisible or in the background of our experience, show up on their own in the imagination, especially in the dream.

But what has this to do with the child's acquisition of language in his or her early efforts at speech? Merleau-Ponty writes:

Vocal imitation is a particular case of imitation in general. But it has the advantage of being precisely controllable by hearing (*l'ouïe*): one is always witness to one's own speech. . . . We can note that the child reproduces new sounds by assimilating them to those that he has already spoken. Here, too, imitation signifies carrying oneself by one's own means towards a goal (heard speech). The child imitates as he goes along (*dessine*), not by following the model point by point, but by carrying himself towards a global result.[9]

Speech and comprehension then involve their own circle: first passing from perception of the other, to imitation of the result of his or her behavior, spoken sounds which have their own order and rhythm, to perceiving oneself speaking which is fundamentally a motor operation. All this then becomes the background to what it is one is referring to or trying to say in a further circle whereby the means or know-how of speaking is "operative" in the background of what it is the speaker intends to say. This occurs in such a way, however, that the way one expresses oneself, the intonations, rhythms, words selected, *can* become the focus or "theme" of one's attention whereas the subject matter which, for example, could be a trivial conversation at a party, becomes background to what one believes oneself to be expressing in the very manner of one's speaking, gesturing, etc. Moreover, one can emphasize it is in such a way that one can also direct the listener's attention to what is expressed "between the words." This observation led to Stanislavski's well-known distinction for actors between a "subtext" of motivations: what one really wants in a situation and the means one choses to express or disguise it in terms of one's entire bearing, and the "manifest-text": what one actually says. It is part of the subtext to determine which of these one wants to be focal for the others experiencing one's expressive actions.

Merleau-Ponty believes that during the babbling phase, "even before speaking, the child appropriates the rhythm and stress (*accentuation*) of his own language." This is because the acquisition of language works in circles of anticipation, where determinate sense emerges from the spoken sounds around one and from the experiments of one's babbling through a "selection" and "impoverishment" of range to imitate the surrounding sounds, intonations, and rhythms, against a background which is at first indeterminate.

The moment a phonemic opposition is introduced, however, replacing the freeplay of babbling but already anticipated by its rhythms, rhythms that melodically echoed spoken patterns in the environment, an "interior" of systematic opposition occurs that serves as the background of all further structuring of phonemic oppositions into the emergent sense of "linguistic Gestalten, general structures". These are appropriated "neither by an intellectual effort nor by an immediate imitation." Merleau-Ponty writes:

Phonologists have succeeded in extending their analysis beyond words to forms, to syntax, and even to stylistic differences because the language in its entirety as a style of expression and a unique manner of handling words is anticipated by the child in the first phonemic oppositions. The whole of the spoken language surrounding the child snaps him up like a whirlwind, tempts him by its internal articulations, and brings him *almost* up to the moment when all this noise begins to mean something. The untiring way in which the train of words crosses and recrosses itself, and the emergence one day of a certain phonemic scale according to which discourse is visibly composed, finally sways the child over to the side of those who speak. Only language as a whole enables one to understand how language draws the child to himself and how he comes to enter that domain whose doors, it is believed, open only from within. It is because the sign is diacritical at the outset, because it is composed and organized in terms of itself, that it has an interior and ends up laying claim to meaning.[10]

It is to be noted that this last text, written some ten years after the lectures on language acquisition, appears to be more under the influence of the "structuralism" of Saussure than the earlier text, which ascribed to language acquisition the existential-phenomenological themes of Gestalt-perception and the expressiveness of the motoric lived-body always in a context of being-with-others (*Mitsein*, Heidegger, Scheler, Binswanger) to the point where some scholars have suggested a "structuralist" turn in the later Merleau-Ponty. I think that this assumption is fundamentally a misunderstanding of his deepest philosophical convictions, which were humanistic, and the apparent shift, as I will hope to demonstrate, is one of emphasis. At any rate the patterning that the child first experiences in the rhythms of spoken sounds that it discovers in its voice and the voices around it, and which anticipates the way linguistic *Gestalten* of significative sense emerge from the opposition between signs, seems to emerge from motoric sources in the nervous system. Here I will allude to my argument given in detail elsewhere. If one goes to the very beginnings of the ontogenesis of mammalian organisms, to the neurogenesis of the embryo, the motor system is developed prior to the sensory system not only in anatomical structure but also in function. Very specific behaviors emerge in sequence according to a schedule and in fairly regular periodicities of outbursts.[11] This suggests that along with the relatively early anatomical innervation of the muscles, particular around the head and neck and working their way downward, one has complex, scheduled movements that predate anything of equal sophistication in the perceptual system. It would seem, then, that the nervous system would supply the structure and periodicity of its own motoric patterns prior to anything that learning theory would be able to call reinforcing input. Similarly, sense or meaning, i.e., the tendency of the nervous system to construe motoric and sensory patterns into well-formed shapes would seem to antedate any capacity of language to signify or refer to identifiable objects in a so-called "object recognition" phase. The nervous system seems able to construe patterns of meaning as early as the first two weeks as Goldstein has noted in the smiling response.[12] It is in light of these considerations that I wish to show that Merleau-Ponty's account of language acquisition, even if it appropriates certain features of Saussurian linguistics, remains "humanistic" insofar as he finds the experience of human encounter, of perceiving the other motorically and affectively in terms of one's own postural schema, to be fundamental in the acquisition of language.

Even if we grant that the nervous system has its own properties to spontaneously construe sensation and motility patterns meaningfully in terms of a Gestalt-circle, Merleau-Ponty is firm in stating the social origin of language acquisition. He writes about the babbling stage:

In conclusion, the child receives the "sense" of language from his environment. Imitation plays absolutely no part at this stage. However, one must emphasize the importance of the child's involvement in the mode of speech of his environment (i.e., rhythm, pitch, etc.) the effect of which is a general attraction to language. (Again, remember Delacroix's statement: "The child bathes in language.") Wundt says that the development of language is always a "premature" development. In fact, it is impossible to deny a kind of spontaneity; but the child's relationship with his environment is what points him toward language. It is a development toward an end defined by the environment and not pre-established in the organism.[13]

Since this was written, results in empirical research in the fields of psycholinguistics, neurology, and even the newly coined discipline "human ethology," may have caused Merleau-Ponty to allow the maturing nervous system to play a greater role in the acquisition of language. Nevertheless, Merleau-Ponty's main argument remains untouched. In treating psychology as a human science, he always tried to avoid the extreme opposition between "intellectualist" and "materialist" metatheories. Despite their apparent opposition, they often repeat the same fundamental errors on account of shared, implicit metatheoretical assumptions. (Thus, Weimer finds that cognitive psychology stands in danger of repeating many of the errors of behaviorism by not replacing the sensory metatheory with a motor theory of the mind that employs constructive skills in its tacit understanding of events.)

For Merleau-Ponty, "at decisive periods of development, the child appropriates linguistic *Gestalten*, general structures, neither by an intellectual effort nor by immediate imitation."[14] Because we cannot reduce the acquisition of language as a human phenomenon to either certain properties of the nervous system which produce meaningful or expressive motor patterns spontaneously (an intellectualist *a priori*), or to certain properties of the sign-system of the mother tongue as a diachronic field of differences — that could be incorporated in terms of the material aspects of the sign (spoken, written, etc.) through imitations with regard to background totality of signification which increases in complexity with each new learned opposition (a materialist explanation) — we ought to examine the "in between" area where the properties of the nervous system encounter those properties exhibited by the environment, the area of our own experience. It is this area that phenomenology as a method of systematic reflection seems particularly well-suited, and in so doing we find that the so-called "intellectualist" and "materialist" perspectives constantly flip-flop over into their opposite, showing their interdependence, by taking up a perspective that embraces both.

As was suggested above, Merleau-Ponty finds that the child in the babbling stage first senses and responds to a certain rhythm in the language even before he is able to transform his babbling into the appropriate opposition of phonemes:

In the eight month, the child can begin to repeat words when they are spoken to him with the expectation that he will repeat them. He introduces these words into kind of a sentence, that is, a kind of imitation of the sentence according to its rhythmic aspect. This is the pseudo-language.[15]

In this sense, we may say that the nervous system is mysteriously an accomplice in providing for anticipations of development, that is, a rhythm sense, which anticipates articulate patterns at a later phase. Here, there is a movement from the indefinite to the more definite where each phase involves a "prematuration" that anticipates "in outline" the accomplishments of the phase to follow it. In the same way that we are able to shift our attention from what is perceived to our bodies actively in motion in terms of a Gestalt-circle, so it seems that we first acquire language in terms of a similar circle between what is in focus and a tacit background of operations that may be more or less defined. Merleau-Ponty writes that there is a

circle, according to which language, in the presence of those who are learning it, precedes itself, teaches itself, and suggests its own diciphering, (which) is perhaps the marvel which defines language .... Language is learned, and in this sense one is certainly obliged to go from part to whole. The prior whole which Saussure is talking about cannot be the explicit and articulated whole of complete language as it is recorded in grammars and dictionaries. Nor does he have in mind the logical totality like that of a philosophical system, all of whose elements can (in principle) be deduced from a single idea. Since what he is doing is rejecting anything other than a "diacritical" meaning of signs, he cannot base language on a system of positive ideas. The unity he is talking about is a unity of coexistence, like that of the sections of an arch which shoulder each other.[16]

What Merleau-Ponty does not say here is that this unity is already antic-ipated by the rhythm sense, by the nervous system's own contribution, even before there is phonemic opposition and the field of the signified which receives further differentiation with each new opposition. There is to begin with a field of organization that emerges from the Gestalt-circle between perceptual and motile patterns whereby the organism comes into contact with the surrounding world and in which the encounter with others takes place. Merleau-Ponty writes about the phenomenon of imitation in language acquisition:

This initial imitation presupposes that the child grasps directly the body of others as the carrier of structured behavior (*conduites*). It also presupposes that he experience his own body as a permanent and global power capable of realizing gestures that are endowed with a certain meaning. This means that imitation presupposes the apprehen-sion of a behavior in other people and, on the side of the self, a noncontemplative, but motor, subject, an "I can" (Husserl). The perception of behavior in other people and the perception of the body itself by a global *corporeal schema* are two aspects of a single organization that realizes the identification of the self with others.[17]

It is this identification which first makes imitation and, thus, language acquisition possible. In fact, the child seems far more predisposed to identification than the adult, an operation which itself turns on an origi-nal "unconscious egocentrism" of the child. Yet, this egocentrism of perspective beyond which the child sees nothing except perhaps pro-visional "ultra things, i.e. entities of which the child has no direct experi-ence, which are at the horizon of his perception, like the sun, the moon, etc . . . [and] remain for the child in a state of relative indetermination," is based precisely on the child not yet arriving at any definite perception of himself through which he could transcend his situation.[18] Merleau-Ponty quotes Guillaume, "the self is ignorant of itself in that it is the center of the world."[19] The child only comes to himself through others, through taking up their behaviors, their language, their goals into him-self. He sees himself first in others which he can only recover, or rather first appropriate, through taking on their perspective. At first, "the child considers himself as 'another other'." It is this vulnerability of self, this spontaneous tendency to identify the self with the surrounding world, "transitivism", "i.e., the absence of a division between myself and others that is the foundation of syncretic sociability,"[20] which first enables language acquisition. So far as the autistic child never surpasses this vulnerability as an initial phase, however, it paradoxically prohibits further development. In fact, this first susceptibility is not itself imitation, which is an active appropriation:

When one adopts an aspect of other people's behavior, the totality of consciousness takes on the "style" of the person being imitated. In other words, true imitation permeates beyond conscious limits and becomes global: once it has been *accommodated*, imitation supercedes itself. It is this kind of superceding (*depassement*) that permits the appropriation of new structures and, for example, the acquisition of language.[21]

This operation "presupposes a quasi-magical relationship with our own body and the acts of others which are perceived by us as melodic totalities (to the extent that we have the same capacities)."

Imitation is a spontaneous operation of the motor subject as "lived-body." It is the "act by which identification with others is produced."[22] In this regard, it is interesting to note a study which Walburga von Raffler-Engel considers to have very significant implications. She writes:

Waterhouse and Fein (1978, "Patterns of kinesic Synchrony in autistic and Schizophrenic Children") compared eight autistic children, aged six to fifteen years old, with normal children of the same chronological age in reference to kinesic synchrony. They matched these children with children of the same language with reference to kinesic synchrony. The disturbed children had normal but delayed verbal language and the researchers matched them with normal children of the same language age, one-and-a-half to four years old. Waterhouse and Fein divided kinesic synchrony in full match and partial match and found that within basically the same kinesic behavior the disturbed children show greater variation across individuals but less variation within individuals. They conclude that "Clearly, each individual disturbed child is sticking to some kinesic synchrony more closely than any individual normal child."[23]

Autism seems to involve a heightened susceptibility to "kinesic synchrony" without the normally ensuing phase of active imitation that curtails or tempers the susceptibility in terms of the goal appropriated from the other's behavior, an attitude which is "incorporated" into one's own postural schema. Merleau-Ponty defines the "corporeal scheme" or "body image" as the "scheme of all possible activities, rather than a scheme of the actual body state."[24] In autism, then, something prevents the susceptibility to others (which comes with the "unconscious ego-centrism" of the child) from becoming an active imitation in terms of a bodily schema of a motor subject. With this schema there is naturally an incentive to spontaneously move toward others and the objects which affect the interest of others (*Bewegungsdrang,* Buytendijk). This incentive is expressed in the child's imagination and affectivity. I propose that what is missing is an ability acquired on the level of constructive motor skill, or, as I prefer to call it, a tacit operation of the lived-body in terms of an emergent corporeal schema of possible actions in situation, that is, in interaction with others. This may or may not require specific precon-ditions of the nervous system at certain critical stages of maturation. What is missing, then, is a "surpassing" of one's own bodily state in terms of a more "global" response. Here, the tacit operation of taking on the goal of the other's interest affectively is allowed to "inhabit" one's own body in a way that escapes conscious effort and control. In other words, if I may put it in Sullivanian terms, there is a fixation of the attention *away* from experiences or contexts that are tacitly held to be devastating to the self, in whatever undefined form the self remains. There is so to speak a lack of cooperation between what is consciously held in attention and its tacit motor background, thereby rendering the Gestalt-circle which normally allows unobtrusive transitions between perception and action, susceptibility and imitation, always incomplete. For lack of more precise language, there is a disconnection or disen-gagement between the conscious awareness and the tacit background operations that orient the awareness toward taking affective interest and imitative responses with regard to others' activities. If the autistic child avoids eye contact, it is because the consequences are too great: total submersion, absorption in the Other in a moment of susceptibility to kinesic synchrony. This devestating moment does not resolve itself, then, in a further alleviating moment of active imitation and appropria-tion, as the bodily schema is simply not "lent" in that direction. It is as if the function of the imagination, which enables transition of the Gestalt-circle in motor projected achievements and which at each moment is an expression of the tacit underlying operations, became too crystallized, or was taken too literally. In a similar way, Melanie Klein describes childhood phantasies about part-objects to be an expression of a too precocious imagination, which by virtue of their extreme implications prohibit further reality-testing. A rigidity of alternatives is established whereby the survival of the self is interpreted in terms of power, a struggle to the death that is undertaken and lived from day to day, in an autistic "logic" which can only circularly confirm itself by not allowing further information to enter into its inference chain: "either I close off contact with others or I will be destroyed." (Cf. Binswanger on Strindberg's *Schicksalslogik*.) Merleau-Ponty writes:

[T]he surpassing of childlike egocentrism will be characterized, not by a "departure outside of oneself" (the child does not know the individual self), but by a modification of the relationship between self and others.[25]

It is not the place here to discuss the manner in which the phenomenological approach to psychopathology demonstrates how a process of "surpassing" or "transcending" in situation (*übersteigen*) can be essentially interconnected with tacit operations of the body in terms of an ongoing "temporalizing" of experience. This temporalizing has its foundations in a continual opposition and interplay between care for the practical world of things and love for others beyond their situational circumstances (as in Binswanger's *Grundformen und Erkenntnis mennschlichen Daseins*).

Scheflen states at a conference on interactional rhythms (held at Teacher's College in 1980) that the discovery that people move in shared rhythms, which serve as a basis of their being able to communicate with each other, "forces us to a new methodological, personal and theoretical perspective." He does not at the same time acknowleldge, however, that the phenomenology of Merleau-Ponty some thirty years earlier was able to show that "the perception of behavior in other people and the perception of the body itself by a global corporeal schema are two aspects of a single organization that realizes the identification of the self with others," and thus enables language acauisition.[26] Scheflen sees the need for a new paradigm that would replace the "prestigious" paradigms of "organismic biology" and "individual psychology" which are based on an "Aristotelian" epistemology of the subject. Such an epistemology prohibits the observer from looking at more than one person at a time and thereby forecloses in advance the study of the interactional synchrony which is present in the rhythmic patterns essential to the movement of any interaction:

Biology, psychology, and sociology adopted and held on to various versions of Aristotelianism, and in the main, they still do. That is why expression and interaction theories are still popular in both academic and clinical circles. But no matter how much you elaborate on a thing-action model, it is still subject centered, and it still depends on the postulation of heuristic foroces. It still depicts the world as boxes representing things or people and lines that depict abstracted actions. An Aristotelian epistemology still ignores the observables, that is, the patterns of action and change, and keeps us in a conceptual universe of imaginary forces. ... Once we recognize that participants regularly, continually and generally act in synchrony, we need no longer entertain an action-reaction model or any simpler Aristotelianism as a basis for our theory. We were forced as were Einstein and Wiener and others some three generations earlier, to adopt a field epistemology.[27]

Scheflen believes that in order to accommodate the paradigm shift our tacit knowledge, or our scientific know-how, must similarly adjust:

We can no longer point our foveal vision at one participant *and then* at a next one. We must learn, consciously and deliberately, to look at more than one person at a time. ... We must learn to watch our own bodies in relation to the bodies of others, instead of looking (consciously) from the face of one speaker to the face of a next one as we have done in the past. ... And I think we will be able to help a group of people we now call schizophrenic in a much more purposeful way. At present we give these people insight therapy or drugs with doubtful results, or we do try to engage them in shared activity, such as dance therapy, hoping that they will somehow achieve an ability for co-action in some automatic way. ... We can learn to watch *and* listen and lay to rest the useless dichotomy between "verbal" and non-verbal communication. We can learn to study the simultaneous relationship of all modalities of the communication code. And, when we have achieved a more holistic and less reductionistic ability to observe, we must turn this ability to the study of interactional rhythms themselves.[28]

In terms of empirical research, we can engender something that has been called the "prometheus principle" (Connie Fischer): in the assessment or experimental situation, we do not try to isolate a feature of pathology as a dependent variable, but rather attempt to help the subject toward recovery and at each moment see what features of that person's style in the situation prohibit recovery. This is, then, somehow shown or discussed with the person and the whole process is begun again: engaging certain prohibitive features of style, reflecting them back again to the subject, etc. What emerges from this kind of encounter situation are general features of a certain "syndrome." In the case of autism, I would try to help the child trust the more tacit operations of the lived-body which would allow a surpassing of synchronous susceptibility to active imitation. No doubt, typical prohibited responses would occur. I have hypothesized that the role of the imagination and the underlying affective life with regard to others is obviated in favor of a circular self-fulfilling, autistic "logic" that keeps attention disconnected from more tacit operations that would allow a surpassing of the susceptibility, i.e., toward a natural movement and interest in others. What is inserted in its place is a fear of others as devastating to the self, particularly as the self is reached, or transformed, in the imagination. I would stress activities, i.e., behaviorally reinforce activities, that involve the imagination: drawing, spontaneous movement, speech. By mimicking their gesture in form, I would at the same time gradually introduce imaginative transitions (in the same way that an actor may suddenly alter his voice or gate slightly and thereby introduces a totally new shift in character portrayal which nevertheless is perceived as a probable variation of what went on a moment before). Thus, if the autistic child insists on shrieking on one pitch, I would immediately copy the tone, only to alter it slightly in the next breath. In this way, I introduce gradual change through imaginative variations at the same time suggesting on an implicit level that impending destruction does not occur with such gradual shifts in our motoric relation to others.

It is to be noted that autistic children become focally absorbed in whatever activity is at hand so that they seem to block out any peripheral notice of the surrounding environment, specifically, the responses of other people to their gestures. I suspect, however, for brief moments, there is a monitoring of the perceptual field which alternates with the focal absorption, i.e., the rudiments of a suppressed Gestalt-circle. This would have to be confirmed by careful scoring of videotapes with each individual child. At the moment of this monitoring where the focus may periodically shift to the field only to withdraw again in perhaps milliseconds, the kinesic posture of the child is to be noted. It is at these moments that there would be any residue of the natural transition of the imitation response from mere synchronous susceptibility. Whatever patterns can be obtained by scoring the slowed down video, whether these be according to a temporal periodicity of, for example, "every forty-five seconds," or in terms of a characteristic posture of the child which would denote the momentary monitoring response and thus partial openness to whatever is going on in the surrounding field. It is at these moments of perceptual susceptibility, which would become observable in terms of the scoring of the videotapes, that the experimenter would introduce his gradual shifts of imaginative variation in the otherwise constant mimicking of the child's behavior. Insofar as this enters the field of awareness during a monitoring response, we have introduced some information that may disconfirm the autistic logic in its otherwise impenetrable circularity. The child's behavior would then be recorded to see if eventually any variations were introduced in the direction of the experimenter's deliberate shift during the mimicking of the child. The experiment is then to be repeated with the further refinements of information supplied from previous trials. If significant results followed, it would show that we have found a method for "awakening" the imitative response in autism by finding the weak spot in their defensive system during perceptual-motor coordination.

In conclusion, I can only make reference to Merleau-Ponty's very suggestive but difficult final notes in *The Visible and Invisible* before his untimely death. In the following, we see a phenomenological description of how some overheard bits of language collect in a tacit field before they organize into significative sense. Perhaps this can suggest some further direction of how we are to think through the problem of the relationship of a motor subject, tacit operations of the body, and speech in relation to others:

The taxi driver at Manchester, saying to me (I understood only a few seconds later, so briskly were the words "struck off"): I will ask the police where *Brixton Avenue* is. — Likewise, in the tobacco shop, the woman's phrase: *Shall I wrap them together?* which I understood only after a few seconds and *all at once* —. . . . This means: there is a *germination* of what *will have been* understood. (*Insight* and *Aha Erlebnis*) — And that means the perception (the first one) is of *itself* an openness upon a field of *Gestaltungen* — And that means: perception is unconsciousness. What is the unconscious? What functions as a pivot, an existential, and is in this sense, is and is not perceived. For one only perceived figures on levels — and one only perceives them by relation to the level, which therefore is unperceived. . . . The occult in psychoanalysis (the unconscious) is of this sort. Cf. a woman in the street feeling that they are looking at her breast, and checking her clothing. Her corporeal scheme is for itself-for the other — It is the *hinge* of the for itself and the for the other — to have a body is to be looked at (it is not only that), it is to be visible. . . .[29]

It is this "hinge" that interconnects the motor subject with the appearance of his body for others that enables the Gestalt-circle between motility and perception of a world and which is a first indication of tacit operations of a motor unconscious.

The Pennsylvania State University / Würzburg

NOTES

[1] An article which appears in *Aspects of Nonverbal Communication*, ed. W. von Raffler-Engel (Bath: The Pitman Press, 1980).

² Walter B. Weimer, "A Conceptual Framework for Cognitive Psychology: Motor Theories of the Mind," *Perceiving, Acting, Knowing,* ed. Shaw and Bransford (Hillsdale: Lawrence Erlbaum, 1977), p. 281.
³ Ibid., p. 269.
⁴ Maurice Merleau-Ponty, *Consciousness and the Acquisition of Language,* trans. Hugh J. Silverman (Evanston: Northwestern University Press, 1973), p. 31.
⁵ Ibid., p. 33.
⁶ Ibid.
⁷ Ibid., pp. 34—35.
⁸ Dieter Wyss, *Depth Psychology: A Critical History,* trans. Gerald Onn (New York: Norton, 1966), p. 418.
⁹ Merleau-Ponty, *Consciousness,* p. 36.
¹⁰ Merleau-Ponty, *Signs,* trans. Richard C. McCleary (Evanston: Northwestern University Press, 1964).
¹¹ The scheduled sequence of behavior was demonstrated in G. E. Coghill's research with salamander embryos in the early 1930s. See, for example, "The Neuro-embryology study of Behavior," *Science* 78 (1933), 131—38, and F. J. J. Buytendijk's *Allgemeine Theorie der mennschlichen Haltung und Bewegung* (Berlin: Springer-Verlag, 1956), pp. 258—62.
¹² Kurt Goldstein, "The Smiling Response of the Infant and the Problem of Understanding the Other," in *Selected Papers* (The Hague: Hijhoff, 1971), pp. 466—84.
¹³ Merleau-Ponty, *Consciousness,* p. 14.
¹⁴ Ibid., p. 21.
¹⁵ Ibid., p. 15.
¹⁶ Merleau-Ponty, *Signs,* p. 39.
¹⁷ Merleau-Ponty, *Consciousness,* p. 36.
¹⁸ See his "The Child's Relation with Others," in *The Primacy of Perception,* ed. James M. Edie (Evanston: Northwestern University Press, 1964), especially p. 99.
¹⁹ Merleau-Ponty, *Consciousness,* p. 37.
²⁰ Merleau-Ponty, "Child's Relation with Others," p. 135.
²¹ Merleau-Ponty, *Consciousness,* p. 40.
²² Ibid.
²³ Walburga von Raffler-Engel, pp. 153—54.
²⁴ Merleau-Ponty, *Consciousness,* p. 68.
²⁵ Ibid., p. 54.
²⁶ Ibid., p. 36.
²⁷ Scheflen, *Interaction Rhythms, Periodicity in Communicative Behavior,* ed. Martha Davis (1982), p. 19.
²⁸ Ibid., pp. 15—16.
²⁹ Merleau-Ponty, *The Visible and Invisible,* trans. Alphonso Lingis (Evanston: Northwestern University Press, 1968), p. 189.

EUGENE T. GENDLIN

PROCESS ETHICS AND THE POLITICAL QUESTION

Two questions will be discussed in this paper: (1) can ethics be found on a certain manner of process, the kind of decision making, rather than the content or conclusions?; and (2) does our personal decision-making process merely reflect social and political control? Or can more arise from the individual than what society has built into the body?

I

Ethics is often said to be lost by an emphasis on authenticity. Heidegger and Sartre were more successful on the negative side, in showing the breakdown of any code, content, or formulated ethics. What they put in its place seems less: the call of conscience, responsibility, an authentic way of deciding moral issues. This is often taken as mere caprice, as if authenticity requires only that I decide, rather than following conventions. Such an interpretation of authenticity would result in Ivan Karamazov's dictum: "Anything is possible." It might be authentic for someone. I want to show, on the contrary, that "authenticity" can name a distinguishable kind of process.

Ethics is best cared for as distinctions between kinds of processes. After all, it is the process which determines the contents. Thoughts, feelings, desires, and other experiences are not just given things. They are generated by processes. A certain kind of process creates the ancient virtues. It is not the case that just anything at all can be the content of just any process. Far from it.

Suppose your good friend has decided to marry someone, and you like the person. Marrying that intended spouse seems (in general) a good thing. Is that enough for you to call the decision right? Would you not need to know more about *how* your friend decided? What if the decision was made on a drunken afternoon to get married that very day? Suppose your friend badly wants money and the intended spouse has some? Suppose the wedding was announced and your friend wants to back out but is scared of disappointing the relatives? What if your friend talks mostly of not wanting to live alone?

A.-T. Tymieniecka (ed.), Analecta Husserliana, Vol. XX, 265–275.
© 1986 *by D. Reidel Publishing Company.*

We commonly call these "wrong reason." Why? They indicate some-thing about the process of arriving at the decision. The trouble is not exactly these reasons themselves. After all, one rarely marries for the "reasons" one gives oneself. Many people spend the rest of their lives trying to discover why they married as they did. In my examples, these clearly wrong reasons indicate something more: the lack of the kind of decision-making process we respect.

It is difficult to delineate "the right kind of process" even though it is well known. Therefore, people describe it indirectly. For example, they describe the process in terms of time. You might ask how long your friend has thought about it. Or, it can be described in terms of a spatial analogy called "depth": how far down inside has your friend examined his decision? Or, we describe the process in factual terms: how much is known about the person, the family, where will you live, and so on.

The questions of length, depth, and knowledge show that we know of a "right" process, but time, space and facts are accidental parameters. We do not mean mere length of time. We hope the time was not wasted going round in circles. We mean the process which is opposite to going round in circles. We mean a process of steps which can correct what one thought, felt, or was before. I will say more about such "steps." Similarly, depth does not help, if nothing changes while digging deep. Sheer depth does not make something right. If the friend comes upon a deep wrong motive, we trust that the decision will change, somehow. Nor are the most relevant facts those that now exist as facts. Rather, the process should raise new questions, newly needed facts.

We depend on this kind of process, but we need to speak from it more precisely.

So far, I said that the "right" kind of process has (1) steps; (2) in which what is found can also change; (3) and in which new facts can appear. My three involve more coming out of the process than went in. I spoke of change. Into what? More than the person plain is. I do not ask what fits how the person is. Instead, I assert that human nature is a different sort of "is," an is-for developing into what (we later say) the person really "was." That development cannot be decided or directed by what a person now is, thinks, or wants, nor by anyone else. Purposes and motives develop. If one's present purposes were to determine the development process, one would be permanently stuck. Persons and situations are a single intelocking system. In our kind of process both person and facts turn out to "have been" more than they seemed.

Let me say that if this kind of process were an assumption, I would reject it myself. But we observe it, we do it, we must face it — but as the rejection of another assumption: this kind of process forces us to reject the assumption that events occur only as previously existing forms and units. More order, forms, and units arise from this kind of process than existed at the start.

We must reject the usual science model which "explains" every event by constructing it out of the forms and pieces of earlier events. But in practice even science does not work that way. Like our process, science first studies the events, then redefines what the earlier one "was" so it can explain the later. New forms and units can come with each discovery. The results are put in linear order retroactively, from the discovery process. In ordinary life we do not assume fixed forms and units in advance. For example, an odd "situation" requires to be "met" by some action that has never as yet existed. After our actions change the situation, hindsight can say what it "really was." But even now, when it demands to be met, we may sense that familiar actions do not meet it. We stay stuck, rather than acting in a way we know.

That a situation can demand a new action is familiar, but odd to say. The new action is neither determined nor not determined by the facts of the situation. If they determined it, we could derive the action from the facts. All situations would be easy. But if the action were indeterminate, any action would do. Again all situations would be easy.

Both "determined" and "indeterminate" assume the same kind of order: fixed form and logical derivation. But situations have more than that kind of order. A situation is not only its formed "is." Rather, it is-for action to meet it and change it. It is-for a change in itself. Therefore, the change cannot be derived from the existing forms. But the change is very finely required. The requiring is not unordered, or only half determined, with half leeway. Rather, it is more ordered than the given forms and facts. That is why we may fail to devise an action to meet it. That is very common, and no assumption at all. But to speak of it involves rejecting the theoretical assumption that fixed form is the only type of order there is.

An unfinished poem also shows the other kind of order. It implies and requires an ending that cannot be derived from what has been written so far. Poems would be easy to finish if their endings were determined by what is already written. But finishing them would also be easy if the ending were not determined, or only partly determined. Then

most any nice ending might do. When the ending does come, it brings a
change in the poem. Therefore, the unfinished part cannot determine
the ending. And yet, what is written does very finely and demandingly
imply and require — what has never been said in the history of the
world. One prefers to leave a poem unfinished, rather than violating
what the unfinished poem "needs."

I can now add a fourth and fifth to my three earlier characteristics of
authentic decision-making process. (4) There is often a "sense" of something not known but needed, required, more finely ordered than the
existing forms. For some moments or months, there is a direct sense of
some. . . . When that moves, opens, releases (these words work newly
here) steps of the process bring new actions and new ways of using
words. (5) Now one reads these new ones back. Now we say we know
what the needed action "was," or what the poem's needed ending "was."
But the word "was" works in this way, here. There may be many steps,
each further forming what the previous "really was." We say that the
previous step "was in the right direction," using that word in an odd way
— since the new step has just changed what "direction" usually means.
Elsewhere I have delineated many more signposts of this type of
process.[1]

We could argue that such a process cannot possibly be a mere
reflection of social forms, since it has more intricacy and order than the
existing forms. But one should neither dismiss nor agree with this too
quickly. Let me develop the problem further. What we can be sure of is
only that we can differentiate this kind of process from other kinds. It is
also a social process. Anything human is also social. We have to
examine more carefully in what ways this kind of process can change or
exceed the social forms and in what respects perhaps it cannot. There
are two questions here. In Freudian terms, can there be a morality other
than the superego's? Again, can there be a social and political "reality"
other than the ego's?

If you come to recognize the kind of steps I described, you will prefer
them to the imposition of forms upon yourself. Society does impose
existing forms — bodily felt collective forms of how people ought to be,
how women, men, sons, teachers, students, should be. We feel role
models of what is strong, independent, productive, creative, good, and
right, and what is not.

Freud called it "the superego" — that inner agency which threatens
and punishes us if we do not fit the forms we ought. And Freud was

right that every person has that inner agency, usually a nasty, destructive, primitive, ill-willed voice inside which says: "Anything you try won't work"; "If it's your desire it's probably bad, selfish, immature, unrealistic"; "If it's your own perception it's probably wrong"; "You're wrong even if it isn't yet clear how. You must have done something wrong." Freud said: "The superego dips deeply into the id." In Freud's jargon this says that the superego is crazy, a primitive channel for destructiveness. And he was right. But is that the only kind of morality?

Freud identified the superego with the moral conscience, and there he was wrong. Conscience is traditionally "a still small voice." You must become very quiet to hear it. In contrast, the superego is by far the loudest voice inside. There is another difference: Superego guilt makes you self-concerned. If you have injured some person, the superego attacks you, depreciates you. You shall be cast into the outer darkness. During guilt attacks you lose sight of the injured person altogether. You become constricted, smaller than usual. Genuine morality comes rather in the sort of process I am defining here. There is a movement outward from within. You feel care and concern for the injured person. Those feelings extend and expand you.

We can tell the difference between these two kinds of process. But now the second question: Freud held that only the ego provides order. The "id" consists of chaotic drives. Ways of action (ways of "discharging" drive energy) are given only by society. The ego is the "reality principle" and reality is the social forms.

For Freud, any experience that does not fit social forms must be a throwback to pre-ego infancy. It cannot bring apprehensions of reality. Training along these lines makes each person feel crazy, unrealistic, inappropriate, since we all have such experience. But, actually the sense which does not fit the forms may apprehend reality more accurately than the forms we try to impose on it.

Language is not a system of static forms. It has the body's moving type of order I call "is-for." Words evolve; they often work newly in ways that can not be derived from extant forms. All human situations are patterned with language, but language is not alone their order. On the contrary, language is never alone, it is implied by the body in situations. When new and odd situations leave us at a loss for words, we can feel that the usual words will not do. Old and new language (and other actions) are implied by the body. It senses new actions and phrasings which do not yet exist, but can come in the steps I described.

The social patterns are not imposed on mere chaotic drive energy. They are imposed on a more intricate texture, a greater order — but this shows itself only in process. There is not a second, natural person under the socially formed person. But all order is not from society. Animals are already very complex. Society develops that further, but the body also develops these forms still further. It is not just a copy of society.

Very complex inherited behavior patterns have been discovered in every animal species. Today the *tabula rasa* hypothesis can no longer be held. Animals without language have complex nesting behvior, intricate mating dances, many sequences that are inherited via the body, not learned.

Every major therapeutic theory since Freud attributes its own order to the body. They rename Freud's "id," usually as the "organism." This is because in therapeutic steps one observes more order, more intricacy, and an inwardly arising "direction" which is very different from imposed form. But these theories do simplify the issue. I say that the organism does implicitly include the social forms. It is not just separable from them. We cannot say what is from the organism, and what is imposed. We can differentiate only the kinds of process, the kinds of further steps. Let us see to what extent that solves the problem.

Just because a situation is called "psychotherapy" does not insure the kind of process I described. But when it does, one notices a far more powerful ethics than the conceptual arguments. From its process-characteristics one can derive the ancient virtues.[2] The process is a deeper honesty than the usual kind. One soon prefers the sincerity of living from that process. One senses one's care and need for other people. But the process functions more intricately than the abstractions in which these virtues have been conceptualized. The human being lives with others, and body-life implies them. Isolation, withdrawal, missing the fullness of other humans feels bad, stifling, thin, dull, weak, and avoidant. Exploitative patterns can feel like that too; there is no company from the other person, only a stand-in for the patterns of one's autism. One may sense one's fear behind the macho poses. One also comes upon denials of oneself. Hiding feels false. One senses the cowering that avoids confronting the other. It is lonely. The other is cheated as well.

Distinctions in this process are often new and finer than the common ones. Here is an example of such steps. Note that the old training is certainly built into the body, but we can distinguish it from steps in which the body feeds back with new form. In this example the conceptual ethics of equality is never questioned or changed. But, at first it functions to block a more intricate mesh of would-be experience which only develops in the steps themselves. The isolating "superior" feeling changes in these steps. What she then says it "was," was not there at the start.

Patient: And I'm mostly alone, and then when I get with people I feel strange. Either I criticise everything or I keep quiet. I feel superior, they were playing cards all night, and I just looked down on them, and I was mad at myself. . . . (Silence) I hate to say anything because it might come out then, that I feel superior, and I *know* I shouldn't. Nobody's superior to anybody. I don't want to feel that way, it's wrong.

Therapist: Your values are that humans are inherently equal and to feel superior is wrong, and can't be true. But let it come for a minute, so we can sense what it's like. Is it like "ech" (sweep-away hand-motion).

Patient: (Silence) No. Feels high and mighty. Kind of good.

Therapist: Sit forward a little . . . yeah, like that . . . Loosen your body so you can let it come in more.

Patient: (Giggles.) I'm the queen . . . but that's stupid and selfish and, um, it's wrong. But it feels good. It's uhm . . . it's sort of self-confirming. Actually it's not even superior so much as . . . um. . . . (Silence) It's like "Make room for me!"

It turns out to "have been" very different from feeling superior. She has not discarded her equality values to permit feeling superior. Rather, what was there changed in the steps, and turned out to be more ethical than it seemed. Her conceptual evaluation played a role, but not the only role.

Here is another example. (The therapist is omitted here.) Note the training built into the body — but also that the body has its own order. At first there is no way to say the steps that come.

I've been looking forward to coming, much more than I did the other times. I've had a crummy week. My job is really bad . . . and everything seems flat like I'm just watching. . . . (Long silence) I have lots of energy there, but it's tied up. . . . (Silence) It's like a heavy wall in front of it. It's behind that. . . . (Silence) It's a whole part of me that I keep in. Like when I say it's OK when it's not. The way I hold everything in. . . . (Long silence) There's a part of me that's dead, a part that isn't. . . . One is dead, one survived. . . . (Silence) It wants to scream. . . . To live. . . . (Silence) And there's also something vague. I can't get what that is. . . . (Silence) It's like I want to run. Someone will be mad at me if I let that part live, and that's very uncomfortable. . . . (Silence) I want to run and never look back and just be free. . . . (Silence) Then that's sad. Yes. Running from the vague thing is sad. . . . (Silence) Some of me wants to find out what the sad thing is, some of me doesn't. . . . (Silence) I'm very angry. It's a big loss, something missing, That's what the vague thing was. . . . (Silence) And my energy is right there, too. Yes, I feel lighter!

At each step the bodily sense is implicitly meaningful in a way that then turns out to be speakable. Note the effect on body-energy at the end. These steps are from the body's own order. All therapists since Freud have noticed that order.

II

Unlike the therapists, philosophical and political writers still side with Freud on this issue. Adorno and Foucault[3] find it simply impossible that a freeing process could arise within the organism and go counter to the socially imposed forms. Worse, to assert such a thing is to sustain the political status quo; one seems to promise individual freedom under present political conditions. In not even thinking about politics, one does reactionary politics.

Social change has its own supra-individual laws and developments, like the evolution from agriculture to industry. Individual bodies live and are programmed by these social developments. But the theories since Marx also assume that there can be no feedback from the organisms except perhaps disorder and resistance. These theories deny the body an order of its own and imply that social change can come only from engineering on the social level. It must be imposed on individuals, since it cannot come from them. But recent history shows that such social change is anything but freeing. Therefore, Adorno and Foucault see no freeing possible on the political level, either. Social patterns are imposed by people in some positions on people in other positions. Change can only be different imposed forms, or different people in power. Change in that fact seems impossible.

For example, Foucault rightly points out that all social functions involve control. Medical people and institutions have aquired a lot of control, which does not always help them cure. Psychiatry is an agency of social control and privilege. Education shapes people to want to obey and fit in, rather than think. Churches control attitudes more surely than they offer spiritual experience. But we cannot help deploring this fact. We see that the social functions are something other than control. It is not the control that cures, or develops thinking. At least in principle there is also health, help with personal problems, the discovery of thinking, and spiritual experience.

We know the difference, even when there is no therapeutic help, only socialization imposed by professionals who know little more than their certificates and roles. We know the difference, even when schools do not develop thinking and prepare only for obedience and repetition.

I do not argue that the genuine social functions can be found without being largely defeated by the inevitable control side. But the import of Foucault's work is not discouragement. We can certainly lessen the control aspect and devise social forms that provide more and more of the actual functions. We must always again see and struggle with the inevitable control aspect of each new form. But the control is not the functions.

But can education for thinking be distinguished from teaching what to think? Can genuine help be distinguished from socialization? We need my distinction between the two kinds of processes again here, on the social level. We know this difference here too.

The theories make a conflict between the individual and the social level of analysis, in order to assert that the social level is independent and determinative. But it is wrong to identify individual with genuine process, and society with control. Then it seems too bad that individuals are impossible without society.

But the genuine individual process is also social. Thinking is inherently social, too! But it is a different kind of process than obedience. The intricate texture of personal feelings is social too. It is from and about living with others in social patterns of love and work. But these can be newly elaborated and more complex than the imposed forms.

Foucault for one assumes that this is impossible. For example, he says: "In the California cult of the self one is supposed to discover one's true self . . . thanks to psychological and psychoanalytic science which is supposed to tell you what your true self is."[4] Foucault assumes it can only be some kind of science, socially imposed forms, that "tell you what your true self is."

Most political theories hold that the direction can only be one way: social control provides order for individual experience. There can be no orderly feedback, certainly not the more intricate feedback we actually find. Foucault thinks that creative feedback assumes a nonsocial individual, and rightly denies that possibility. But "subjectivity" is not the separated unsocial source he denies. Its own order is also "social," but we cannot let that word mean only imposed form. Imposed form is not the only kind of order. And we can distinguish the other kind.

I report publicly verifiable measures of authentic psychotherapy steps distinct from other manners of process.[5] I do not deny that much of what goes under the name of psychotherapy is mere social control. I do not rebut Foucault's critique of psychiatric power to control people. Coming from Carl Rogers' work, many of us have been saying something like that for many decades. We welcome Foucault's excellent critique of the medical and psychiatric profession. The question is only whether it must be so. Or is there a distinction between socially imposed form and another kind of social process?

But do not even now agree too quickly. We can distinguish this difference. But can we be sure that our process will undo every imposed form we would wish to overthrow if we could be aware of it? We cannot say that. Political analysis can show us what we might otherwise never question. Conceptual thinking alone does not usually change us, but it has an essential role in the process I described. But this role of concepts differs from the usual. In out excerpt, her step ("Make room for me!") might not have come without her conceptual ethics of equality which made a conflict with her superiority feeling.

Theories, concepts and values do not merely float on an independent conceptual level. They may also enable a step of process. But when they do, what comes is more intricate than the concepts were. We have seen that concepts and old forms do not "determine" such a step, they do not impose their form on it. The process can lead to rejection or modification of the very concept that helped the step to come.

The process I describe does not eliminate controversy and pluralism. The varying concepts remain. But, despite their conflict, they can point up something experienced. If we make the process central, conceptual pluralism does not destroy ethics. Let me give a self-illustrating example. Aristotle concludes that we are not ethically good until we enjoy doing good acts. Until then we are merely practicing at it, acting only from knowing what is right. Kant flatly contradicts this. He says that we are ethical only when we act out of duty. He says that even then we are "in danger" of doing it for the wrong reason, if we also enjoy it. Both views are right and needed. But we cannot simply merge them. That would only dull the clarity of thought. Each is systematically connected to other concepts that must not be lost.

If only conceptual form is considered, ethics disintegrates into competing dogmatisms. This is largely the current state of ethics. But the conflicting conceptual systems must be retained; there is no way to "resolve" them on their own level. Nor would we want only one system!

Of course we would rather be (and deal with) someone whose inclinations are good. But, we must also be able to challenge what is sensed as good at a given moment. People with good inclinations become accustomed to doing what "feels right," which is often superior to thinking. Then the day comes when they mistreat us. They can not see what is wrong; it "feels right" to them, as usual. We must be able to reason with them, appealing to something other than their good inclinations. But neither can we expect mere argument to determine ethics. Our appeal must get the process moving again.

That is why ethics must involve both inclination and concepts. Neither can simply impose itself on the other. A finer cognition feeds back from the body's implicit order in process steps. But a vital role is played by conceptual cognition (as the ethics of equality did in our excerpt). The body is not chaos with merely imposed form. Neither is it all-wise so that we would not need to think. Both are needed to see and change unconscious oppressive forms. That is one reason we cannot be sure that this process will overthrow every unconscious oppressive form, or arrangement of life.

All we can say — but it is a lot — is that this kind of process reveals a more intricate order which can exceed and reorder existing forms. Imposed form is not the only kind of order.

University of Chicago

NOTES

[1] See my *Experiencing and the Creation of Meaning* (New York: Macmillan, 1962, 1970); "Experiential Phenomenoloy," in *Phenomenology and the Social Sciences,* ed. M. Natanson (Evanston: Northwestern University Press, 1973); *Focusing* (New York: Bantam Books, 1981); "Two Phenomenologists Do Not Disagree," in *Phenomenology, Dialogues and Bridges,* ed. Bruzina and Wilshire (Albany: State University of New York Press, 1982); "Dagenais' Direction Beyond Presupposition," *Journal of Religious Studies* 11, 1—2 (1984).
[2] See my "Values and the Process of Experiencing," in *The Goals of Psychotherapy,* ed. A. Mahrer (New York: Appleton-Century-Crofts, 1967); "Neurosis and Human Nature in the Experiential Method," *Humanitas* 3, 2 (Fall 1967).
[3] M. Foucault, *Power/Knowledge* (New York: Pantheon, 1980).
[4] Foucault, interview with Rabinow, 1984.
[5] M. H. Klein, P. Mathieu-Coughlan, and D. J. Kiesler, "The Experiencing Scales," in *The Psychotherapeutic Process: A Research Handbook,* ed. W. P. Pinsof and L. Greenberg (New York: Guilford Press, 1985).

PART IV

PSYCHIC CIRCUITS OF SENSIBILITY
AND MORALLY SIGNIFICANT
SPONTANEITIES

CHUNG-YING CHENG

NATURAL SPONTANEITIES AND MORALITY IN CONFUCIAN PHILOSOPHY

The concept of spontaneity is as rich as it is important. It is rich in meaning and ambiguity, and yet no thorough explanation has been given it in philosophy, although the word has been used in various contexts of common language. It is also important because it arises from some fundamental experiences of life and reality in their process aspect and seems to capture an essential characteristic of life and reality, thereby contributing to their understanding. Western philosophers in general have not particularly focused on this concept in order to explore its reference and structure, but seem instead to have attempted to explain it away. The mechanistic model of classical physics is a good example. Even in contemporary philosophy of mind and history, systematic explanation is based upon considerations of workings of mechanic laws or law-like regularities. Spontaneity is not considered as a principle of its own; it is not analyzed or made precise. There is an escape from spontaneity. On the contrary, Eastern philosophy, particularly Confucianism and Taoism, from the very beginning have focused on spontaneity as an ultimate principle: a principle of explanation and a principle of justification as well as a principle which is identified with the most general and most profound experience of man and his relation to the world. Vested in the concept of spontaneity is the whole corpus of our understanding of nature, life, reality, and the ultimate destiny and fulfillment of man. It can be shown that all the major concepts found in Confucianism and Taoism are based upon or derived from our understanding and experience of spontaneity.

What then is spontaneity? Although Taoism and Confucianism did not give an explicit analysis of the notion of spontaneity as a form of experience in the fashion of modern analytic philosophy, nevertheless they have described or illuminated certain ways of understanding life and reality which upon reflection make the concept of spontaneity highly structured and significantly clarified. In this paper, I will first try to describe and illuminate, in an analytic manner, this concept of spontaneity in the light of my understanding of Confucianism and Taoism; I will then relate this concept to some of the important paradigms of

279

A-T. Tymieniecka (ed.), Analecta Husserliana, Vol. XX, 279–287.

Confucian philosophy; finally, I will draw out some important conclusions regarding spontaneity as an ultimate principle of metaphysics.

When we talk about spontaneity in such contexts as "the wood burns spontaneously," "a child awakens spontaneously," "he understands spontaneously," and "the artist paints a painting spontaneously," it is clear that spontaneity applies to a natural event, to a human activity, to an action of mind, and to a creative act of cultivated human talent. Although the context may vary and the conditions for the application of the term may differ, there is a unifying factor underlying all these contexts which can be connoted by other terms such as "naturalness," "immediacy," "effortlessness," and "unpromptedness." Phenomenologically, we may consider that the subtlety of meaning conveyed and the variety of labels applied indicate how real the experience of spontaneity is. I would describe the reality of spontaneity in terms of internality and self-sufficiency versus externality and conditonal dependence. An event or phenomena is spontaneous if it arises from an internal, self-sufficient source that needs no external causality nor depends upon conditions outside the source. It is apparent here that the internal and external, the self-sufficient and dependent must be recognized as categories which we generally understand in talking about things and ourselves. With regard to things, we can make a distinction between what is external and what is internal to a thing. We also recognize that things are either self-contained in their existence and transformation or dependent upon external causes for their workings. With regard to ourselves, it is even more clear that we distinguish what is internal to me and what is external to me ("me" being a self-contained and unifying energy which will transform or act either by itself or under the influence or impact of external causes). Here I do not want to go into detail about how an individual person comes into being, or how we classify acts as internal or external. I merely want to call attention to a basic condition that we do recognize: we do act as self-agents; the term "self" is not an empty term, but contrasts in meaning with entities external to self, at least on a phenomenological basis.

I want to maintain that the sense of distinction between internal and external and that between self-sufficient and external dependence represent a genuine and meaningful experience which carry metaphysical import, giving structure to the concept of spontaneity. As has been noted, spontaneity applies to natural events, human behavior, and mental and creative acts of man. This means that spontaneity can be stratified on different levels. When spontaneity occurs on the natural level, we call it "natural spontaneity" or "spontaneity of nature"; when spontaneity occurs on the human level as far as human existence is concerned, we call it "spontaneity of human person" or the "spontaneity of human nature"; when spontaneity occurs in relation to mind and creative human activity, it may be called "spontaneity of human mind and human talent." Mind and talent are extensions of human nature, so spontaneity of human mind and talent can be regarded as forms of spontaneity of human nature. As human activity is not to be confined to human mind or human talent in an individual alone, so spontaneity of human nature may be also extended to collective activities of man and society. We may apply spontaneity to community, society, and the government of man. On each level a sense of internality and self-sufficiency are pertinent. In this sense of spontaneity, a society which functions spontaneously requires the society to be well-integrated and harmoniously interrelated so that it may become an organic unity and function with self-sufficiency. A government which functions spontaneously will be one which is well-ordered and maintained without inner conflict or tension, but harmoniously executing and administering its rule. The Taoistic concept of government in the *Tao Te Ching* and the Confucian concept of "sagely rule" in the *Analects* present ideal models for government in spontaneity.

A question arises here concerning the origin of the principle of spontaneity. To resolve this, we must introduce and recognize the metaphysics of spontaneity. The ultimate reality is *sui generis* (self-caused) in the same sense as Spinoza's God. This reference to Spinoza's God is important, for it is not only self-caused, *sui generis*, but forms the very basis of the activities of all things at any time. But Spinoza's account suffers from two drawbacks which prevent him from developing his concept of God into the ultimate principle of metaphysical spontaneity. (1) His God is a universal substance, which does not allow plurality of individual, self-sufficient entities. (2) His God does not transmit the same degree of spontaneity to all modes of individual things in the world. If we remove these two drawbacks and conceive of spontaneity as a fundamental characteristic of the ultimate reality which manifests itself in a multiplicity of things which retain the same degree of spontaneity in particularity as in universality, we have the ultimate principle of spontaneity which should explain spontaneity in all contexts of our experience. We are saying, then, that the ultimate principle of meta-

physical spontaneity should allow spontaneously a multitude of spontaneities to come into being. Each individual of the multiplicity of
spontaneities manifests the ultimate principle of spontaneity. This
amounts to the recognition of the organic unity of the pluralities in both
qualitative and spatio-temporal dimensions.

It is clear that the ultimate principle of spontaneity is the principle of
the *Tao* as expounded by the Taoist Lao Tzu in the *Tao Te Ching* and
Chuang Tzu in the book by the same name. I want also to argue that this
principle of the *Tao* as the ultimate principle is evident in the metaphysics and the epistemology and morality of Confucianism. First let me
explain the ultimate principle of spontaneity in relation to the *Tao*.

The *Tao Te Ching* has described *Tao* as many things, but in particular
Lao Tzu presents *Tao* as doing-nothing-and-yet-everything-is-accom-
plished. This concept has not been properly understood metaphysically.
It is not as if the *Tao* does something which is nothing so that everything
can be done. *Tao* is doing-nothing; it is the essence, the principle, and the
reality of doing-nothing, and simultaneously there is doing-everything.
This means that everything is creatively developed on its own. Everything
contains the *Tao*, namely, doing-nothing, as its basis of being. This means
Being and Non-Being, Form and Void are co-existent and mutually
transforming. This also means that reality is a ceaseless creative process;
ceaselessly, inexhaustibly, and effortlessly present things are being transformed. It is in this sense that Lao Tzu speaks of "ten thousand things
self-transformed"; also he speaks of *Tao* as "producing things, but not
possessing them, making things happen without claiming credit, making
things grow and prosper but not dominating them." In the same spirit,
Chuang Tzu further develops the metaphysics of the *Tao* in terms of
self-transformation and mutual transformation. *Chuang Tzu* particularly
stressed the individuality of things in the multitude, and emphasized also
the natural process of the spontaneous transformation of things. In light
of *Chuang Tzu* we can speak of innumerable natural spontaneities on
different levels and at different stages of the transformation of things. In
essence, transformation becomes the core of spontaneity. *Tao*, in light of
this understanding of spontaneity, is considered to transcend common
language by both *Tao Te Ching* and *Chuang Tzu*. It is a concept which
we cannot adequately express, though we can experience it. We can only
say that *Tao* is *chih-tao* (self-speaking), *Chih-hua* (self-transforming),
chih-ch'eng (self-completing), and *chih-chen* (self-evidencing, self-completing).

Although *Tao Te Ching* and *Chuang Tzu* have developed the principle of spontaneity from a metaphysical point of view, they have been unable to apply this principle to the activities of man, at least, without conceiving them beyond the context of natural spontaneities. They refused to recognize the automony of human activities and an independent category of human spontaneities, namely, spontaneity that occurs uniquely in relation to human behavior, human mind, and human talent, on both the individual and societal planes. In fact, in light of the corruptions and obsessions found in human civilization and society, they reject the meaningfulness of an autonomous human spontaneity. If human at all, they regard human activities as a degeneration and degradation of natural spontaneity. For this reason they denounce the Confucian virtues and the Confucian form of government by sagehood. But this, of course, can be regarded as a form of Taoistic reductionism which need not to be consistent with the ultimate principle of spontaneity, namely, the principle of the *Tao* in Taoism. Human nature and human mind, human society and human government, arise without contrivance in the course of time. There is no premeditation or planning on the part of the *Tao* for the evolution of human existence. On this basis we cannot, therefore, simply reduce to, or identify human activities with, the activities of nature.

Human activities and human beings are part of nature and yet more than simply nature on a nonhuman level. We must allow human development on both individual and social levels as a part of the spontaneous activity of the *Tao* and yet with a richer content and with a specific quality of its own. Hence we can distinguish between natural spontaneity and human spontaneity. The latter, though embodying the former, has an independence and uniqueness of its own. We may indeed consider human spontaneity as a rising above and a refining of the natural spontaneities, but not a degeneration or degradation of nature. In other words, human nature and human culture are both extensions and refinements, both embodiments and consummations of nature.

To know morally all forms of human spontaneity is regarded by the Confucianists as the central mode. Morality is central in the sense that it is essential for developing a human person as a full individual and for integrating human society into a life-enriching harmony. In fact, morality can be conceived as embodying all forms of human spontaneity because it leads into all forms of human spontaneity. In stressing the importance and autonomy of morality, the Confucianists, however, do not forget that human spontaneity, and hence morality, is not separate or separable from the natural spontaneity and the ultimate principle of the *Tao*. In the *Chung Yung* this point is made emphatically. "The *Tao* cannot be separated [from things] for a second. What can be so separated is not the *Tao*." The *Chung Yung* indeed makes it explicit that human nature (*hsing*) in the sense of universal humanity and in being human is endowed from heaven (*t'ien*), heaven being an aspect of the *Tao* or ultimate spontaneity in the form of internal will. Can will be spontaneous? The answer is yes, if will is freely exercised without determinacy or the influence of external powers. In this statement human nature as a *natural* extension/consummation of the *Tao/T'ien* is recognized. This use of human nature is a spontaneity of natural spontaneities resulting from the process of transformation. Hence human nature represents a starting point of human development. It is also the center and norm of being human in the individual/social activities of man. The *Chung Yung* indicates how man should be cultivated into a full individual, an individual who fulfills himself, authenticates himself, and liberates himself as he creates his own true identity by acting out his potentiality. The full theory of self-cultivation which is a form of human spontaneity and the basis of morality is discussed in the *Ta Hsueh* (*Great Learning*).

I will not give here details or analyze such a theory. It suffices to point out that the *Ta Hsueh* together with the *Analects* and *Mencius* have developed the concept of the self-cultivating individual as an individual fulfilling the spontaneity in him. It is important to point out that the process of self-cultivation is a process internal to the human self. It is also development of something internal to the human self. The *Mencius* argument for the goodness of human nature is relevant here. Human nature is good, not because it learns to be good, but because it can directly, effortlessly, and spontaneously bring out something which is good. The immediate and unpremeditated response for helping a helpless child (the child about to fall into a well) is both a manifestation of human nature and an example of human spontaneity. What is even more important is that goodness is spontaneity of nature. If nature manifests its internal stirrings from self-sufficiency of self (not a divided or impoverished self), then what nature manifests is goodness. In this sense goodness can be identified with spontaneity of human nature. Morality is just the extension, elaboration, and extensive application of spontaneity to all human contexts and occasions.

In this light we can understand the *Analects* of Confucius much better: *jen, yi, li, chih, hsing,* and all the other virtues are simply forms of human goodness as they come from the human spontaneities of human nature. *Jen* in particular is the centralizing and universal form of human spontaneity because it is the most integral and most immediately experienced form of spontaneity. It is also the most fertile form of spontaneity, because it focuses on man as man and enlarges the circle of human self-fulfillment in terms of human reciprocity and togetherness, exemplifying also spontaneities of nature. Confucius is able to show that man can become fully himself if he can fully develop *jen* and also that man can fully develop spontaneity within himself if he can fully develop *jen.* Hence *jen* is the central key to fulfilling human spontaneity in man. A sage can be said to be a man of full spontaneity in nature and in action. Thus Confucius says, "When I reach the age of seventy I can follow my heart's wish without transgressing against the rules of morality" (*Wei-cheng*). This indicates that the human mind/heart of the sage is fully spontaneous as it is fully developed in *jen*: the subjective meets the objective, the internal meets with the external in total harmony and unison. The rules of morality are themselves results of the development of human spontaneity in accordance with human nature. Confucius also says, "Is *jen* far away? If I desire *jen,* lo! *Jen* is here" (*Shu-erh*). *Jen* is spontaneity presented because it is the content of human nature as human nature spontaneously flows from Heaven and the Way.

It is not difficult for us to point out that morality, as the central form of spontaneity in human nature, leads to the refinement of government so that government becomes another form of human spontaneity as family and community. As each form of human spontaneity has its uniqueness and autonomy, so government must have its own uniqueness and autonomy in spontaneously functioning. The *Analects, Ta Hsueh,* and the *Mencius* have discussed this issue in great depth. It suffices to say the Confucianists' view is that government centers around man so that the men in government (ruler and ministers) must themselves exemplify virtues and therefore spontaneities of creativity and unselfishness. Only when this fundamental prerequisite is satisfied can those governing be able to fulfill other requisites for governing such as rectifying names and the development of economic welfare planning. The ultimate ideal for government again is the ideal of the spontaneity of activity and positioning, as indicated by the following statement of the *Analects*: "If one governs by means of virtue, [then one's government] is like the Pole

Star: it simply positions itself and all the stars will center around it [in order]" (*Wei-cheng*).

After having explained human spontaneity in the form of morality on the levels of human nature and human cultivation, and on the levels of human society and government, it is important to point out that the Confucian philosophy still inquires deeply and reflectively into the source of human spontaneities. In this it does not fail to bring the metaphysical principle of cultivated spontaneity into the picture since it is the concentration of both natural spontaneities and human spontaneities. In fact, Confucius himself has spoken of both *Tao* and *T'ien* in the *Analects*; he seldom mentions them, but he does not totally ignore them. The mention of "What does Heaven say? And yet the seasons rotate, and a hundred things are generated. Yet what does Heaven say?" (*Yang-ho*) is a clear indication of how the ultimate spontaneity manifests itself. It is an ideal Confucius clearly wishes to embody and emulate. In the *Chung Yung* the distinction between *chung* (centrality), *yung* (ordinary and constant), and *ho* (harmony) is important as a distinction of different aspects of human spontaneity. It is also important as an indication of some universal base of human spontaneity. Together with other passages of the *Chung Yung*, the metaphysical principle of ultimate spontaneity (which can be identified with self-realization [*chih-ch'eng*] and the fulfillment of all natures [*ching-hsing*]) can be seen as the very foundation and constant source of human activity. It is also true of the great commentaries of the *I Ching* : the principle of *cheng cheng* (creative creativity) is precisely such a metaphysical principle of ultimate spontaneity, for it leads to all forms of human spontaneities such as human thinking, human knowledge, human action, and human civilization, as explained in the great commentaries of the *I Ching*.

Confucianism would lapse into artificiality if it failed to link human spontaneity with the ultimate spontaneity. In the *I Ching* both natural spontaneities and human spontaneities (including morality) are woven into a unity of the symbolic system capable of onto-hermeneutical interpretation on different levels for different stages of development. The Taoists complement the Confucianists by reminding them that natural spontaneities arise separately from human spontaneities, so that the latter will not fall into indifference, evil, and poverty of life. The Confucianists complement Taoists by pointing out that the *Tao* creatively furthers itself into more rich forms of self-realization in the form of human spontaneities. Here we recognize that both natural and human spontaneities manifest the principle of ultimate spontaneity at the same time in a whole organic unity without which neither spontaneities can be spontaneities in the fullest sense of the term.

University of Hawaii at Manoa

ERLING ENG

PATHEI MATHOS – THE KNOWLEDGE
OF SUFFERING

There is only a transcendental psychology, which is
identical with transcendental philosophy

(*Crisis*, 257)

I

I am a psychologist in a veterans' hospital, and it is not uncommon for
some of the men with whom I talk to be afflicted with nightmares of
wartime experience. It is often many years afterwards that these
memories return at night with particular virulence. It is quite possible
that with advancing age that youthful insouciance with which they were
earlier brushed aside becomes attenuated, allowing the banished
demons to return in greater force than ever.

But the veteran whose one particular story (for there were others
too) I am about to tell you was a 32 year old Marine infantryman in
Vietnam who suffered as he told me from "having been turned into a
killing-machine." He came to the hospital after having thrown up a
responsible position, and left his family without any reason he could
give, and with partial amnesia for what had happened during that
period. He was afraid of exploding, and as it turned out had been
suffering from suicidal phantasies going back to active duty when he
kept hoping that he would be killed. As time went on it also became
clear that he had unaccountably run off from job and family many times
since leaving service, and that he had thus broken up his first marriage,
and was about to possibly lose his second for the same reason. Little by
little over a two month period, during which time I saw him five times a
week, a nuclear scene of his disturbance emerged, one which formed the
centerpiece of repeated nightmares.

He recalled the night of his first Christmas day in Vietnam when he
and four other men were setting an ambush for the Vietcong. They
heard some indistinct sounds ahead of them, and to illumine the area
ahead of them, radioed for flares to be dropped. Then, in the lit up
landscape ahead he saw a figure, shot at it, and saw it fall. When he later

289

A-T. Tymieniecka (ed.), Analecta Husserliana, Vol. XX, 289–296.
© 1986 *by D. Reidel Publishing Company.*

went to it, and turned it over, he discovered that it was an unarmed girl, 13 or 14 years old. She was beyond help and he held her while she died. He said it was the first person he had killed in combat, and that after that he only wanted to die. On her person she had a little pouch with identification papers and a picture, one which he had been keeping for many years and had only recently destroyed. I said to him, "You loved that girl." He replied, "She was more to me than even a sister, I would compare her with a wife, it was as much as if my wife died." He went on: "She's helped me a lot, and she's hurt me too. She was illumined by a flare. People couldn't figure out my depression, and it angered me that people didn't understand. I was never depressed before that happened. Not depressed about others. But with her it was different. Now my daughter is almost her age. For a long time I want to go back and hunt her people up, but I never could. It was definitely wrong. I took her life. It would have been different if she had had a weapon, but she didn't. I killed her on Christmas night. It was on Thanksgiving that I ran away before coming to the hospital. It's always been close to holidays that I've had the habit of running away."

I said to him, "It has been said that each man kills things he loves. It is also true that each man loves the one he kills."

I do not wish to leave you with the notion that it was simply the crystallization out of this otherwise obscured incident that made it possible for this man to leave the hospital with a very different attitude, looking altogether different; we talked about many other matters that were also of great consequence. Nevertheless I have chosen this paticular event to consider more closely with regard to the way in which moral sense can emerge in psychotherapy.

But before I do that I would like to add a remarkable parallel, in a way a kind of confirmation of my belief that this scene of the discovery of hidden love through accidental killing was crucial for this man's new understanding of himself. After I had presented the story of this man to a small group of colleagues with whom I meet regularly, I received through the mail a case vignette which had just appeared in the *Journal of the American Medical Association* (2/17/84) from which I have taken the following extracts:

The patient stated that since the end of World War II he had rarely had a full night of sleep. At least three nights each week he dreamed in exact and unaltered detail an event that had occurred just shortly before the war ended.

He had landed in Europe on D-Day and was a veteran of many hardfought battles. In a cleanup operation in a small village, he and his squad had come upon a group of fleeing enemy solidiers. One soldier was especially visible as he ran and the patient had taken careful aim, fired, and the soldier had fallen dead immediately. When he turned the body over he saw that he had killed a soldier who was "a beautiful blond-haired blue-eyed boy not more than 14 or 15 years old."

Shortly thereafter, the patient's recurrent dreams began in which the exact event was fully re-created, and although more than 35 years had passed, the dreams were still vivid and emotionally intense re-creations of the original evvent. They ocurred at least three nights every week and at times in nightly sequence. The patient always awoke from them in an extreme state of agitation, horrified at what he had done. He was then overcome with feelings of anxiety, sadness, and remorse. He had never, not even once, been able to resume sleep during the remainder of the night. Throughout each following day, he regularly felt overcome with guilt and moodiness and was preoccupied and totally withdrawn from his family.

This man was seen for nine sessions over a two year period. After six sessions,

As I listened to him talk and reflected on the many years of agony and the still intense aliveness of his suffering and pondered how I might help him, I remembered and remarked to him that in the first session he had told me of his three fine sons and had not mentioned them again. I then said that I must presume that all three were beautiful blond-haired blue-eyed boys. Amidst near-overwhelming choking sobs he said, 'And, Doc, would you believe that every day of their lives when I looked in their faces I saw the face of the boy I killed.' He wept at length and finally left quietly.

On the occasion of the last and final meeting, he said he didn't need to talk to me. He said he hadn't any more dreams, had been sleeping well, and felt good. He then gave me a large basket of fresh vegetables from his garden. He said he and his grandsons had grown them. From the way he said it, I knew he could look them in the face. (Fulmer, Thos. E., M.D., "Nightmares and the Blue-Eyed Boys", *JAMA* **251**(7), 1984, p. 897. © 1984 American Medical Association.)

The correspondences with my experience are remarkable. They don't seem to me to require spelling out. What is of interest to us is the question of what took place in both cases.

Let us see now what these two reports together have to tell us. First of all there is the suddenness of discovery of the unexpected, in the place of death, life, out of darkness light, as the body is turned over, revealing a sublime beauty by which the proud warrior is humbled, recognizing not merely his own life in his victim, his prey, but discovering it there now, having lost it from the own self. Hence the inexplicable attachment to that scene, blessing and affliction alike commingled in the encounter of lives. This is the moment that Cesare

Pavese has described, in which "A man becomes the thing he kills." It is undoubtedly the moment which is employed to evoke the presence of the divine in and through the plant and animal sacrifice, from the Eleusinian ear of wheat through the Minoan and Mediterranean running of the bulls. Gratitude for the epiphany crosses with grief for its price, a moment which binds the slayer and the slain as irrevocably as the mother and her newly born infant. The self that is both and neither is discovered within what now appears to have been a delusion of the reality of the enemy. Everything is altered. Nor is there any afterward, the indelible image, with its bane and its blessing, continues to obsess its bearer day and night until its moments of freedom and necessity can be reconciled. But until then this metanoic instant is fulgurative. It is deadly in its intensity, its utterly fatal *anagnorisis*. Its bitter truth, that of Wilde's "Each man kills the thing he loves" introduces us to that quality of love" that in Jouve's words "is hard and unpitying as hell." This is the love from which one may die while another is being born. It is the sacrificial character of flesh, flesh of the *Urstiftung*, primal constitution of meaning, material substance. It is the time of the *mater dolorosa* with the limp body of her son draped across her white marble lap. It is a revelation of that flesh which Tertullian referred to as "the hinge of salvation."

But why then should it be so cruel, so remorseless? Because it carries the message we all bear within us, but have not yet deciphered, that not only will this same thing, one way or another happen to us, that we will deliver up our life as well to another, and others, but that in having become bodily, in having been born, this has always already happened. It is a discovery, as it were of our natal condition, in which our susceptibility to loving is indistinguishable from that of dying. Or, to put it in yet more staple terms, sexuality here is discovered through one's power both to inflict mortality on another, correcting the customary reverse, the identity-bestowing emphasis on power of sexuality, to bestow, as it were, immortality. So the emphasis of the early Freud in his middle age undergoes a surprising reversal. Mourning, in which loss becomes transparent to the gift of everlasting presence, here is proleptically accomplished, in an excruciating instant like that of *The Tempest*, when Prospero says of the meeting of Ferdinand and Miranda: "At first sight they have exchanged eyes." But the mourning has only taken place by half; the other's life has vanished, without its reappearance in and as one's own. Only when the event could be acknowledged to another who represented at once both the powers of love and death in the flesh could the hangman's knot of guilt be loosened. Now it became possible to discover the return of the other's life in one's own, in the one case that of his daughter, in the other that of his grandchildren.

As long as the sense of Anaximander's *life-wordly* words, "But the source of genesis for existing things is also that into which they pass to make payment, for they must make restitution for their injustice in the sequence of time," as long as these words have not been understood, the harm we do another remains accursed, in our infantile or youthful pride of innocence, refusing to hear the message of our indebtedness, the knowledge in the nightmares.

The doctor can become the place and passageway through which an opening can occur, by which what was known only in the natural attitude can pass over into a contrary sense which, together with what was first experienced, provide evidence of the *Lebenswelt* which takes place through both. But to become this does not lie altogether in the doctor's power. More often than not its secret seems to lie in his capacity for realizing his own helplessness, when the patient is unable to discover where *his* helplessness lies, when what the patient has learned to deny has become a barrier reef for both.

Each of the two men from whom we have heard were young when they encountered the face of beauty, in the one they believed to have slain as their enemy, and their nemesis in each case was even younger, like a sister or a brother. We are in the land of Cain and Abel. Both men moreover rediscover in the visage of their descendants the features of the one who died, as it were that their children might live. With this, the otherwise fatal debt is discovered to have been absolved. This is to say that they had not been in a position to realize the life-giving moment of sacrifice until the moment each began to discover, acknowledge, the life of which we are, in a very real sense, the often obscure conducting bodies. Here a momentary digression: it is not without reason that Freud discovered the *Todestrieb* when he began to dream of the death of his son who was a soldier in the Great War, that he was preoccupied with the strange, apparently oxymoronic notion of "death instinct," which our two veterans have helped us to understand.

For Freud this experience of "moral sentiment" in which one half remains hidden required the myth of its origin in a murder of the primal father by his sexually jealous sons, a murder followed by remorse, and

the founding of the moral order as a commemoration of their in-
humanity. It is also the myth to which Darwin has recourse in the
Descent of Man. Here we skirt a topic which I would like to reserve for
another occasion, namely, that of the idea of evolution as Christian
gnostic doctrine of cosmic salvation, one which in some sense came to
be understood by both the patients whose story we have heard. The
offering of first fruits by the older veteran from the garden of his
grandsons tells more than appears on the surface.

Brought down to Husserlian terms, *Bewusstsein,* i.e. conscience, is
always *Ge-wissen* as well, this "Ge-", having both the meaning of
collective as well as of past experience. But this collective that we only
know as of the past, is also one of living origin, so that, "What's past is"
always "prologue." If from naive consciousness we discover conscience,
it is also the case that from conscience there is possible a return to
amplified consciousness. Husserl's final word on consciousness is that of
the Greek idea of syneidesis that comes of first deepening the sense of
the eidetic through emptying it of noematic reference to the point of a
transcendental solipsism, within which there begins to dawn, as a
movement of consciousness itself, that of the world itself of which we
are members, as syneidos. The Lebenswelt as syneidesis, its telos also
that of its richer accomplishement: this appears to be the sense of
Husserl's vision of the earth as oikumene.

In his brief paper "On Transience" written just after the outbreak of
the Great War, Freud touched on the matter of the learning of suffering,
pathei mathos. He wrote: "to psychologists mourning is a great riddle,
one of those phenomena which cannot themselves be explained, but to
which other obscurities can be traced back" (S.E. 14, 306). It is also a
paper in which he enlarges on the way in which the moral lesson of
beauty is inseparable from the learning of our mortality, directly or
indirectly.

II

Love, learning of the death it unwittingly seeks in the beloved, wishes
to delay it, and draws back. By the same token the death inflicted
on the other discovers too late that love which alone gives meaning to
death as death. That is reason of the "repression", the latency of death
in all mortal loving (Marcel's "death" is loss of the beloved), and

apocalyptically, the latency of love in all dying. Death and love are the two faces of mortality. When the face of the one who has fallen is turned upward to light the defenseless face of love is revealed, streaming with a different light than that of day, Plato's light of the eyes, now still luminous from the face around their closure.

Transience discloses being. In killing, one way or another, I inflict my own transience on another. In having slain I behold a manifestation of that being in and of which I am constituted. I suffer the fatality of being, the being of fatality. The fatality of mortality is discovery of the own being. But we betray the first fruits of our knowledge if we believe they can be eaten by all.

Both of these patients whose experience has made them our teachers assumed the enemy as collectively identified. But they lived to discover that a far more intimate enemy was an ignorance of their being as manifestation, as phenomenon, living appearance and self-appearance. Being in manifestation is always already directed. Its discovery can take place through the recognition of one's own injury in having injured another, in having failed to realize the *Wisstrieb* (Freud), the urge to know, as the deepest of all urges, since that it manifests our incomplete being in time, that desire in and through which we were instituted, and are re-instituted. It is this shock of self-recognition in the fading of the mirroring eyes of the other we have hurt, even slain, which reveals the deadliness of our being in the insufficiency of self-knowledge. But to know all were equally impossible, since this were to deny the difference of the other as known from ourselves as knower. Moral sense is realized in the discovery of the impossibility, the shared aporia of our situation as human beings, i.e. to retain the innocence and guiltlessness of infancy, oblivious to one's indebtedness to others for the discovery of one's being, or on the other hand to ever come to know enough to dispense with the necessity for these occasions of fatal or all but fatal knowledge. The inevitability of this vital aporia is also its lesson, the lesson of commemoration, one which continues to teach even the one who is learned and teaches, namely the doctor. The doctor is one who has embarked on this course of learning from suffering, even with the desire to learn from suffering (the *pathei mathos*), without on that account escaping from the additional suffering entailed by such knowledge. If suffering teaches, it is also the case that teaching itself continues to learn suffering. The learning of suffering is never exempt from suffering in applying its lessons to other sufferers. Once more, it is like Husserl's incessant beginnership, only to realize the suffering more and more intimately, and as intimacy, one of our very being.

The commemorations of those fatal occasions in which we have glimpsed the good, the true, and the beautiful through the ill not only suffered, but through the suffering of the ill we have done to an other is what I take to be the discovery of that moral sense in which the situation of psychotherapy, and above all that of psychoanalysis, is constituted. Now it becomes clear that suffering teaches only in the first instance the possibilities of alleviating suffering, of forestalling subsequent recurrences. But it also teaches suffering as a doctrine of being. As long as, and as much as, I resist that fatal knowledge, I must continue to suffer it in a more shameful manner, continuing to inculpate myself through inflicting it on others. It is just here that we touch on what might be termed a gnostic moment of Christianity, and which as philosophy, or as psychoanalysis, makes a necessarily figurative language attractive, even perhaps obligatory.

Now what is said itself comes to be seen as manifestation of the immediate understanding set by the therapeutic situation itself, among the participants in the meaning of the relationship to be accomplished. Now the conflicting possibilities are divided between the partners of this situation. The doctor in being exposed to the suffering of the patient is first strengthened by the very suffering which is making the patient weak, strengthened in his healing understanding for the patient. But if he excludes himself from the discovery that understanding itself exposes him to an awareness of that suffering with which neither he nor the patient can hear, then to that degree he can inflict additional injury on other, in lieu of bearing that share which falls to him. Nietzsche's (and Husserl's) concern with the accomplishment of one's being is necessarily accompanied by the at first strange discovery that what does not destroy me can make me stronger. Or in terms relevant to this conference: what doesn't condemn me can make me more moral. In the therapeutic situation that is the knowledge which the events it makes possible must convey.

University of Kentucky,
Veterans Administration Medical Center,
Lexington

MONIQUE SCHNEIDER

LE VISIBLE ET LE TANGIBLE COMME PARADIGMES
DU SAVOIR

Pour assister au décollement de deux dimensions qui, dans l'oeuvre de
Merleau-Ponty, s'emboîtent et se recouvrent étroitement, je ne suis pas
tant partie d'une question posée par le texte lui-même, que d'une
question posée au texte. Question qui s'insère au départ dans une
problématique à la fois étrangère et parallèle.

En partant du couple de la représentation et de l'affect chez Freud,
on peut en effet être reconduit à une autre dualité structurant, non plus
le phénomène du devenir-conscient, mais celui de l'expérience sensible.
La problématique de la représentation, telle qu'elle est débridée par
Freud, s'arc-boute en effet sur des métaphores visuelles, corrélation que
soulignera d'ailleurs Merleau-Ponty en se proposant, dans ses "Notes de
travail", de "généraliser la critique du tableau visuel en critique de
la *Vorstellung*".[1] D'autre part, la thématique de l'affect induit des
métaphores se référant à la rencontre par le toucher. Analogie suggérée
par le langage lui-même qui, conférant un double sens au terme "être
touché par", propose une surimpression entre le registre affectif et
l'expérience tactile.

Apparier ainsi ces deux couples — représentation-affect et visible-
tangible — ne pouvait cependant manquer de provoquer un durcissement concernant l'aspect antithétique risquant de structurer de manière
trop rigide ces deux couples. Durcissement sans doute provoqué par la
fréquentation de la pensée freudienne, mettant volontiers l'accent sur les
clivages. Ce recours à l'opération de la coupure, de la séparation,
pourrait cependant induire en erreur, si on isolait ces diverses opérations pour conférer aux partages instaurés une dimension définitive. Les
clivages freudiens sont en fait mobiles et protéiformes, indéfiniment
reconduits sous des formes différentes, ne se superposant jamais
parfaitement aux clivages primitivement mis en place; ils doivent donc
être reçus comme instruments d'analyse plus que comme repères
délimitant un cadrage inamovible.

EN DEÇÀ DE LA COUPURE

Remettre en question le besoin de clivages fondateurs, telle est bien

297

A-T. Tymieniecka (ed.), Analecta Husserliana, Vol. XX, 297–311.
© 1986 *by D. Reidel Publishing Company.*

l'invitation inaugurale qui travaille la démarche de Merleau-Ponty. Remettre en question ces clivages, non pas pour les ignorer, mais pour retrouver une expérience située en deçà des oppositions classiques. D'emblée, l'expérience perceptive est approchée, non pas comme lieu où se combinent, où s'affrontent des modalités perceptives relativement hétérogénes, mais comme lieu de correspondances, lieu où les divers sens échangent leurs modes d'appréhension, conférant ainsi à l'expérience une dimension symbiotique:

Le dur et le mou, le grenu et le lisse, la lumière de la lune et du soleil dans notre souvenir se donnent avant tout, non comme des contenus sensoriels, mais comme un certain type de symbiose, une certaine manière qu'a le dehors de nous envahir, une certaine manière que nous avons de l'accueillir, et le souvenir ne fait ici que dégager l'armature de la perception d'où il est né. (P.P., 36)

L'expérience sensible adviendrait essentiellement comme rencontre à l'intérieur de laquelle il serait impossible de décider quel est l'élément agent et l'élément patient: "le dehors" ou nous-mêmes. Deux activités se conjugueraient, sans qu'il soit possible de regarder l'une ou l'autre comme porteuse d'initiative absolue. Inutile par conséquent de chercher à situer, comme dans une perspective classique — qu'elle soit idéaliste ou empiriste — le lieu du commencement; invasion et accueil opéreraient conjointement. Merleau-Ponty aura d'ailleurs recours, dans d'autres passages, à la métaphore de l'"accouplement". Mais, si parfaitement emboîté que soit cet accouplement, il laisse subsister une différence, une dualité: le monde, nous-mêmes; dualité sans doute héritée de la formulation traditionnelle du problème. C'est cette dualité supposée que Merleau-Ponty s'emploiera à dissoudre, pour retrouver une expérience quasi-extatique, où il deviendrait impossible de maintenir radicalement la séparation entre les partenaires: l'être du sujet et l'être du monde.

"Moi qui contemple le bleu du ciel, je ne suis pas en face de lui un sujet acosmique, je ne le possède pas en pensée, je ne déploie pas en avant de lui une idée du bleu qui m'en donnerait le secret, je m'abandonne à lui, je m'enfonce dans ce mystère, il "se pense en moi", je suis le ciel même qui se rassemble, se recueille et se met à exister pour soi, ma conscience est engorgée par ce bleu illimité." (P.P., 248)

Le registre choisi pour analyser cette rencontre située à la limite du fusionnel est celui qui, en principe, se prête le moins à cette expérience d'abolition des différences, puisqu'il s'agit du registre visuel. Mais de quelle vision s'agit-il? Merleau-Ponty n'écrit pas "moi qui perçois . . ."

ou "moi qui vois . . .", mais bien "moi qui contemple", terme qui n'est pas indifférent dans la mesure où il jouxte le domaine esthétique. Est-ce que la raison pour laquelle l'exercice visuel est ici présenté comme capable d'instaurer l'intimité habituellement dévolue au toucher? L'altérité présumée de l'objet — traditionnellement respectée dans le compte-rendu de l'expérience visuelle se réalisant dans la coupure — est ici engloutie pour faire place à un rapport d'identification: "je suis le ciel même qui se rassemble".

Comment lire ce texte? Toute velléité de commentaire se trouve sans doute réduite à l'impuissance si, aux séries de brouillages suggérés par Merleau-Ponty concernant les clivages traditionnels — affect-représentation, voir-toucher, monde-sujet — ne vient pas s'ajouter un autre brouillage concernant le partage entre modalités d'écriture: le partage, en l'occurence, entre écriture philosophique et écriture littéraire. Merleau-Ponty part en effet d'une question métaphysique — la place du sujet dans la rencontre perceptive — pour introduire un mode d'analyse ayant constamment recours au passage à la limite. En quel sens faut-il entendre "je suis le ciel même . . ."? Il serait tentant — tentation cédant d'ailleurs à la facilité — de lire ce texte en le situant dans un registre foncièrement littéraire ou poétique, comme s'il offrait une simple transcription métaphorique située dans les marges d'une saine approche analytique.

En face d'une telle hypothèse de lecture, deux réponses peuvent être élaborées, l'une latérale, l'autre plus directe. Il serait possible, d'une part, de prendre à la lettre les expressions de Merleau-Ponty et de les référer à une expérience perceptive originaire faisant le fond de toute perception. Les analyses de Piera Aulagnier concernant la rencontre originaire nous mettent en effet en présence d'une modalité de fonctionnement psychique se situant dans un en deçà radical par rapport à toute pensée de la dualité. Etudiant une instance située en amont du processus secondaire et du processus primaire, Piera Aulagnier dégage les linéaments d'une expérience perceptive originaire, le "pictogramme", expérience à l'intérieur de laquelle il serait impossible de séparer l'acte d'appréhension de l'objet appréhendé. "La représentation (. . .) de cette rencontre a la particularité d'ignorer la dualité qui la compose. Le représenté se donne à la psyché comme présentation d'elle-même: l'agent représentant voit dans la représentation l'oeuvre de son travail autonome, il y contemple l'engendrement de sa propre image."[2]

En dépit de la relative hétérogénéité des deux démarches, on pourrait opérer un détour et recevoir l'analyse du pictogramme, de la représentation propre à la rencontre originaire, comme livrant un des sens possibles de ce rapport d'identification instauré par Merleau-Ponty entre le sujet percevant et ce qui est traditionnellement posé comme objet. Dans la mesure où, selon Piera Aulagnier, "le représenté se donne à la psyché comme représentation d'elle-même", on pourrait supposer que c'est l'opération même d'appréhension qui se voit reflétée dans ce qu'un témoin extérieur désignerait comme son objet. Ainsi le mouvement de se diluer engendrerait-il, en une seule et même opération, le "bleu illimité" du ciel et le moi — ou plutôt l'infra-moi — n'existant que dans l'opération elle-même d'abandon. C'est ainsi sa propre extase que le sujet contemplant rencontrerait dans l'illimitation de ce dans quoi il se perd et se trouve à la fois.

Une telle analyse permettrait d'aller au delà des concepts de fusion et de symbiose, termes inadéquats dans la mesure où ils se réfèrent au mélange, à l'indistinction progressive, de deux entités pouvant être envisagées au départ comme distinctes. Or le rapport suggéré par Merleau-Ponty a une portée plus radicale, désignant, non pas une union, mais une identité d'être: "je suis . . .". Rapport d'identité qui va au delà d'un rapport d'accouplement, suggéré par d'autres passages, et qui pourrait renvoyer aux brèves analyses freudiennes concernant le "moi-plaisir". Lorsque le moi, selon l'analyse conduite dans "Pulsions et destins des pulsions", "prend en lui" l'objet qui se révèle être "source de plaisir", il traverse une expérience s'apparentant à un phénomène de transsubstantiation: il ne s'agit pas seulement d'éprouver du plaisir devant tel ou tel spectacle, mais d'abolir ce que la notion de spectacle comporte en fait de mise à distance; dans l'avènement du "moi-plaisir", le moi perd ses frontières virtuelles pour *devenir* ce qui est censé figurer comme spectacle.[3] Dans l'analyse freudienne comme dans ce passage de Merleau-Ponty, le propre du plaisir perceptif est d'abolir la situation de l'être "en face de . . .", pour découvrir une structure d'identité: "je suis".

Cependant, même si cette analyse nous reconduit au risque inclus dans ce que comporte de radical l'expérience perceptive, risque caractérisant ainsi une dimension effective de cette expérience, un tel mode de rencontre peut-il être présenté comme révélateur du rapport perceptif habituel? Ne s'agit-il pas d'une expérience-limite, jouxtant le domaine proprement perceptif, si tant est que la délimitation d'un tel domaine ne soit pas le fruit d'une construction réductrice et défensive: séparer des sphères d'appartenance idéales pour neutraliser à l'avance des phénomènes de contagion ou de transgression? La rencontre perceptive abolissant le sentiment de face à face pourrait en effet concerner, soit l'expérience esthétique du ravissement, soit certaines expériences psychiques se réalisant sur le modèle de la fascination, soit catastrophique, soit extatique? Une précision apportée par Merleau-Ponty permet en effet de déterminer le registre au sein duquel se situent des expériences de perception instauratrices d'un rapport d'identité:

Si je voulais traduire exactement l'expérience perceptive, je devrais dire qu'*on* perçoit en moi et non pas que je perçois. Toute sensation comporte un germe de rêve ou de dépersonnalisation comme nous l'éprouvons par cette sorte de stupeur où elle nous met quand nous vivons vraiment à son niveau. (P.P., 249)

"Cette sorte de stupeur . . .": expérience, qui, à la limite, ouvre sur l'atemporel, sur le *zeitlos*, expérience qui, en un sens et du fait même du gommage qu'elle opère sur toutes les inscriptions spatio-temporelles tentant de la circonscrire, ne pourrait jamais finir. Est-ce à dire que la perception la plus banale comporte en elle-même cette puissance de conflagration? La restriction apportée par Merleau-Ponty a une portée décisive: "quand nous vivons vraiment à son niveau". Par cette simple précision, ne quittons-nous pas le domaine de la psychologie pour aborder celui de l'éthique — une éthique d'ailleurs solidaire d'une ontologie et d'une esthétique? Une valorisation intervient nécessaire-ment, séparant la perception factice, déchue, d'une perception pleine, adéquate à son essence. La portée de l'ensemble de la *Phénoménologie de la perception* apparaît d'ailleurs sous un angle différent: l'essentiel ne serait pas tant de constater et d'étudier tels rouages d'un processus perceptif que de retrouver le chemin d'un savoir-percevoir. L'itinéraire suivi par Merleau-Ponty serait alors aimanté par la recherche d'une perception à découvrir ou à retrouver. La restriction introduite par la formule "quand nous vivons vraiment à son niveau" serait d'ailleurs corroborée par cette autre remarque, référant l'expérience perceptive à un destin de perte: "nous avons désappris de (. . .) sentir". "A la recherche de la perception perdue": c'est ainsi que peut être lue l'ensemble de la quête à laquelle se livre Merleau-Ponty tout au long de la *Phénoménologie de la perception*, texte qui prend alors valeur initiatique: nous reconduire à ce foyer perceptif à la fois vivant et renié.

Renié au nom même des principes qui régissent la nomenclature et le clivage entre concepts, entre registres et dimensions tenus pour différents.

Rapporté à cette nostalgie tournée vers la redécouverte d'une perception accomplie, le rapport entre voir et toucher prend un sens différent. Alors que, cédant au mirage d'une épiphanie perceptive, Merleau-Ponty insiste parfois sur la dimension de synergie inscrite dans notre corps — promesse d'un parfait emboîtement entre registres hétérogènes — l'allusion à la perte conduit à prendre acte des phénomènes de coupure partielle et de retard qui installent entre voir et toucher un rapport fait de tension:

> Dans l'expérience visuelle, qui pousse l'objectivation plus loin que l'expérience tactile, nous pouvons, au moins à première vue, nous flatter de constituer le monde, parce qu'elle nous présente un spectacle étalé devant nous à distance, nous *donne l'illusion* d'être présents immédiatement partout et de n'être situés nulle part. Mais l'expérience tactile adhère à la surface de notre corps, nous ne pouvons pas la déployer *devant nous*, elle ne devient pas tout à fait objet. (P.P., 365. Souligné par nous)

Au lieu d'être centrée sur le thème de la synergie ou de l'accouplement, l'analyse s'emploie ici à inscrire une brèche. Brèche non pas inéluctable, mais du moins menaçante, indiquant des directions à partir desquelles l'expérience perceptive peut conduire à l'avènement de paysages ou d'univers foncièrement différents. Le partage ne s'effectue d'ailleurs pas dans la neutralité, dans la mesure où des préférences, des solutions de moindre errance, sont indiquées. Ainsi l'expérience visuelle fait-elle l'objet d'une mise en garde, dans la mesure où elle est source d'une "illusion" possible. Illusion dont la portée culturelle fut décisive: toute la philosophie idéaliste de la connaissance peut être regardée comme dérivant d'une passion pour le voir, conduisant à ériger la vision comme le modèle de toute appréhension; passion qui serait à la source de l'ambition démiurgique située au coeur de l'entreprise idéaliste: "nous pouvons (. . .) nous flatter de constituer le monde". Cette prééminence du modèle visuel est d'ailleurs soulignée dans l'analyse par laquelle Alain Roger dénude l'articulation entre le privilège accordé au regard et l'orientation idéaliste:

> Si l'idéalisme est la tentation de l'Occident, c'est parce que l'Occident *regarde*, en dépit d'efforts infructueux, du moins jusqu'à une date récente, pour habiliter les autres claviers sensoriels. Nous sommes tous, quoi que nous pensions, ou plutôt parce que nous pensons (en termes de vision) des idéalistes invétérés. Lorsque Foucault distingue

un âge de la Ressemblance, un âge de la Représentation, etc., il se borne à décrire les avatars de l'idéalisme occidental. Il est donc vrai que l'aveugle est la réfutation de l'idéalisme, (. . .) parce que la cécité ne peut pas comprendre l'idéalisme, parce qu'il faut un regard pour le penser, pour y penser, comme à sa propre *théorie*. Dieu, d'ailleurs, est Pro-vidence, et qui ne voit l'impiété de l'affliger d'un goût, d'un odorat, pourvoyeur de sensations grossières.[4]

La corrélation entre le primat de la vision et la construction d'une philosophie de la représentation n'est cependant pas reçue par Merleau-Ponty sur le mode du constat; une confrontation se met en place, faisant rivaliser le modèle visuel et le modèle tactile. L'expérience tactile intervient en effet, non pas pour constituer une dimension parmi d'autres de l'expérience perceptive, mais plutôt pour indiquer la voie permettant de déjouer l'illusion visuelle, illusion créatrice de distance et de face à face défensif, illusion conduisant surtout au sentiment de toute-puissance. C'est à partir de l'expérience tactile, en promouvant cette dernière au statut de paradigme, que la philosophie pourra tenter d'échapper aux pièges inhérents à la tentation idéaliste ou, de façon plus générale, objectivante. Dans la mesure où le toucher nous enracine dans notre corps, il permet de réduire ce qui, dans la passion visuelle, peut apparaître comme *ubris*, ce qui installe en elle une menace perpétuelle de transgression ou de trahison: "illusion d'être présents immédiatement partout et de n'être situés nulle part". La désertion visuelle devrait ainsi se trouver endiguée pour que soit rendue possible une perception à la fois accomplie et originaire.

Pourquoi une telle suspicion planant sur la passion visuelle? Une fois de plus, le recours à une démarche latérale favorisera la mise en place d'effets de résonance. Dans l'insistance avec laquelle Freud revient à la métaphore visuelle, soit dans ses rêves, soit dans ses textes théoriques, on peut découvrir la mise en acte d'un projet destructeur: le rêve *non vixit* dote la vision d'une efficience dissolvante, comme si l'objet ne survivait pas au regard qui porte sur lui; absence de survie qui d'ailleurs est peut-être solidaire de son statut d'ob-jet: ci-gît la trace d'une manifestation vivante. Les réticences de Merleau-Ponty à l'égard d'un exercice visuel branché sur le goût de l'objectivation éclairent en outre par ricochet une remarque à la fois brève et décisive de Freud; dans les *Trois essais sur la théorie de la sexualité*, la mère est dite perdue au moment où l'enfant est capable de la "voir dans son ensemble". Il ne s'agit certes pas de la perte d'une personne, mais de la perte d'un régime ou d'un règne: règne de relative indivision, règne davantage centré sur le modèle du toucher que du voir. La méfiance de Merleau-Ponty à l'égard du pouvoir visuel, d'un pouvoir visuel allant jusqu'au bout d'une démesure objectivante, peut alors être regardée come une tentative pour préserver un premier règne.

Au thème de la synergie serait alors dévolue une fonction réparatrice: endiguer d'emblée l'éclatement qui menace au sein de toute expérience sensible totale, visuelle aussi bien que tactile. Une réponse pourrait être aportée, sur ce point, à la question soulevée par Jacques Colette, suspectant Merleau-Ponty d'avoir "introduit subrepticement une non-coïncidence"[5] au sein d'une présence pleine. La menace de déhiscence est en réalité pressentie d'emblée dans la démarche de Merleau-Ponty, mais pour se trouver, dans un premier temps du moins, aussitôt colmatée. C'est en effet à l'intérieur même de l'exercice visuel que Merleau-Ponty réintroduira ce qui fait le propre de l'expérience tactile; jouxtant les analyses sur la transgression visuelle − transgression il est vrai fondatrice − d'autres études présenteront la vision comme "palpation à distance", expérience permettant de rétablir un lien avec la dimension d'adhésion inhérente au toucher.

Lorsque Merleau-Ponty stigmatise "l'illusion" dont la vision peut être source ou, dans un renversement d'attitude, lorsqu'il incruste dans la vision un foyer de palpation, l'analyse se situe-t-elle sur le même registre? Elle ne se contente pas de répertorier des modalités hétérogènes égales en dignité, mais elle se fait attentive à ce qui, au sein de chaque expérience sensible, se situe en deçà ou au delà de la perte. Le modèle tangible figure alors l'équivalent du règne édénique à l'intérieur duquel Adam et Eve ignoraient aussi bien la honte que la possibilité de se *voir* nus. Nudité ablative et séparation seraient le lot d'une préférence visuelle, tandis que le modèle tangible préserverait sujet et objet de la catastrophe qui les pose comme séparés et voyants.

L'ÉMERGENCE VISIBLE

Si bien colmatée que soit, au début de l'itinéraire, la déhiscence menaçante − déhiscence entre registres différents, déhiscence entre soi et le lieu originaire − le mouvement de l'oeuvre vise à reconnaître comme nostalgie, comme rêve d'une unité perdue, ce qui pouvait paraître posé au départ comme emboîtement originaire. La déhiscence n'est donc pas seulement présente, dans l'élaboration de Merleau-Ponty, comme thème, mais tout aussi bien comme force en travail. Aucune conversion brutale, mais plutôt arrachement partiel, permettant à la dimension visuelle, d'abord noyée dans les autres dimensions ou partiellement tenue en bride, d'imposer ce qui fait sa spécificité. Autant la *Phénoménologie de la perception* paraît s'insérer dans une tentative d'annulation de la perte, autant l'oeuvre terminale, *Le visible et l'invisible*, parvient à entériner un certain travail de deuil: deuil d'une hypothétique symbiose posée comme donnée d'emblée. Le tabou n'est-il pas levé au niveau même du titre? S'il est vrai que "le visible" ne se réfère pas à une dimension purement psychologique de l'expérience, mais ouvre à l'ontologie, il n'est pas indifférent que seule cette modalité de l'expérience sensible soit capable de fonctionner comme point d'articulation. Avec la reconnaissance du privilège inconditionnel du visible se trouve consommé l'échec d'une tentative philosophique soucieuse de se situer uniquement au ras du vécu. Une telle tentative serait condamnée à faire l'économie du concept de monde, entité qui n'est rendue possible que par une transgression: s'éloigner hors d'un vécu postulé originairement pur. A la vision est octroyée la tâche consistant à consommer une telle transgression:

> La visioin est panorama; par les trous des yeux et du fond de mon invisible rédit, je *domine le monde* et le rejoins là où il est. Il y a une sorte de *folie de la vision* qui fait que, à la fois, je vais par elle au monde, et que, cependant, de toute évidence, les parties de ce monde ne coexistent pas sans moi; la table en soi n'a rien à voir avec le lit à un mètre d'elle − le monde est vision du monde et ne saurait être autre chose. (V.I., 105. Souligné par nous)

Texte étonnant si on le compare au passage de la *Phénoménologie de la perception* dénonçant dans la vision l'instrument possible d'une illusion. La dimension d'illusion n'est d'ailleurs pas contestée; tout au contraire, elle est portée jusqu'à ses dernières limites: la transgression ébauchée devient "une sorte de folie". Mais la déraison est ici accueillie dans son pouvoir fondateur: par elle un monde devient possible, la tentation idéaliste se trouvant du même coup, non pas cautionnée philosophiquement, mais réhabilitée comme illusion nécessaire, inséparable d'une instauration. Un tel retour en grâce de la vision et de son inévitable menace d'exil sonne du même coup le glas de la précédente nostalgie symbiotique. Ainsi l'idéal d'accouplement parfait ne peut-il qu'éclater. Nous ne sortons pas pour autant de toute promesse de correspondance, de recollement des parties disjointes, mais l'unité n'est plus posée comme absolument originaire; elle est déplacée, transférée du côté de ce qui n'est qu'annoncé:

Il faut nous habituer à penser que tout visible est taillé dans de tangible, tout être tactile *promis à* la visibilité, et qu'il y a empiètement, enjambement, non seulement entre le touché et le touchant, mais aussi entre le tangible et le visible. (...) Les deux parties sont parties totales et pourtant ne sont pas superposables. (V.I., 177. Souligné par nous)

Dans la mesure où tout tangible est "promis à la visibilité", la structure faisant correspondre tangible et visible devient structure d'écart. Ecart instaurateur d'une temporalité incluant une dimension de négativité: entre l'un et l'autre termes, entre les deux bords réunis par la seule promesse, s'installe la possibilité d'un délai indéfiniment reconduit. Dans le rêve des Parques, Freud donne à voir la fonction de l'attente; la mère figurée par la Parque demande à l'enfant d'attendre, de différer le moment de la satisfaction et l'autorité imposant le délai ne peut pas ne pas apparaître comme une figure de la mort. Remplacer une structure de coïncidence symbiotique par une promesse ne peut qu'instaurer, au sein de l'expérience perceptive, un mouvement de va-et-vient, de ricochet ou de différé.

Il ne s'agit d'ailleurs pas, dans l'itinéraire déplié par Merleau-Ponty, de passer d'une célébration nuptiale — la foi en un accouplement intégral et sans faille — à une célébration funèbre — la fête sombre où se disent la non-coïncidence irrémédiable, le manque abyssal — mais plutôt de poser les fondements de ce qui pourra s'instituer comme phénomène de chiasme, le différé ouvrant un circuit qui permettra un mouvement de recouvrement partiel: pas de clivage décisif entre les divers registres, ceux, entre autres, du tangible et du visible, mais une tentative toujours renouvelée de "réflexion sensible". L'expérience sera incapable toutefois de restituer, en un mouvement étalé dans le temps, la plénitude d'abord attendue de la rencontre sensible: "cette réflexion du corps sur lui-même avorte toujours au dernier moment." (V.I., 24) Toutefois, loin d'advenir comme échec pur et simple de la tentative initiale, cet avortement est porteur d'une dimension nouvelle; c'est grâce à lui qu'est rendue possible une émergence hors du *zeitlos* primitivement solidaire d'une perception "engorgée". C'est le rêve d'une toute-puissance originaire qui avorte, toute-puissance à l'intérieur de laquelle seraient fondus, interchangeables, le vouloir du sujet et l'offre prodiguée ou imposée par le monde. Mais, si blessante que soit la perte, elle ouvre l'espace possible d'une naissance — naissance située, datée — qui ne détiendrait pas les ressorts de son propre avènement.

LE CHIASME ORIGINAIRE

Circularité de la forme et du fond: l'oeuvre de Merleau-Ponty se propose elle-même comme étant en voie de déhiscence et d'enfantement, dans la mesure où elle s'efforce de reconnaître comme nostalgie ce qui était d'abord présenté comme donnée originaire. On débouche ainsi sur l'acceptation d'une double polarité. Il n'est pas question, d'une part, de s'installer dans le registre de la *Vorstellung*, d'accomplir jusqu'à son terme le mouvement d'objectivation et de coupure, mais il est tout aussi impossible de s'installer dans une expérience qui prendrait pour modèle le tangible, et ceci à un double niveau. Dès les analyses inaugurales de la *Phénoménologie de la perception*, le tangible apparaissait moins comme ordre séparé que comme paradigme d'un mode d'appréhension également présent dans la vision; la possibilité d'un toucher tactile était ainsi reconnue. La limitation du primat du tangible, telle qu'elle est conduite dans *Le visible et l'invisible*, va cependant plus loin: le visible n'est pas seulement réhabilité au nom de la germination tactile dont il porte en lui la possibilité, mais il est pris au sérieux dans ce qu'il est seul capable d'apporter: la dimension de coupure, de domination du monde. Dimension inséparable d'une déperdition au niveau du vécu. Mais il ne s'agit pas pour autant de valoriser cet aspect héroïque de renoncement à un espoir originaire d'intimité: ces deux dimensions ne constituent pas des aspects rigides, stabilisés, elles ne sont concevables que dans le mouvement au sein duquel elles passent l'une dans l'autre. D'où le thème de la "réflexion en bougé", réflexion mettant en place une coïncidence elle-même mobile. En dépit de la mise en place d'une bipolarité, nous resterions ainsi aux antipodes d'une pensée rigoureusement dualiste: les dimensions constituent autant de moments qui s'annulent au seuil même de leur accomplissement. Pensée résolument non sédentaire, soucieuse de capter, moins les conditions a priori de toute expérience possible, que des aspects ne pouvant être captés qu'à l'état naissant.

Le statut du connaître va d'ailleurs entériner ce processus de double polarité croisée. Ce qui était initialement dénoncé comme trahison deviendra exil inévitable. Les opérations de perte et de déni sont alors inscrites au coeur même de ce qui permet le travail de la conscience:

*Ce qu'*elle ne voit pas, c'est ce qui fait qu'elle voit, c'est son attache à l'Etre, (. . .) c'est la chair où naît *l'objet.* Il est inévitable que la conscience soit mystifiée, inversée, indirecte, par principe, elle voit les choses *par l'autre bout*, par principe elle méconnaît l'Etre et lui préfère l'objet, c'est à dire un être avec lequel elle a romppu, et qu'elle pose par delà cette négation, en niant cette négation (V.I., 302. Souligné dans le texte).

Cécité inaugurale de la conscience, ne pouvant à la fois assister au spectacle qui se déploie devant elle et prendre acte de l'ancrage à partir duquel s'inscrit cette ouverture. Dans cette analyse portant sur la cécité inaugurale de la conscience, cécité partielle, cette dernière est appréhendée sur le modèle de la vision; c'est d'ailleurs au lieu même de cette cécité fondatrice que viendra s'inscrire le foyer possible du rapport tactile: rapport non objectivant, mais actualisant l'ancrage à partir duquel se déploie l'émancipation visuelle. Impossible d'ailleurs d'atteindre, en une tentative de capture directe, cet en deçà de toute objectivation: c'est seulement par un travail d'analyse régressive que l'immédiat sera reconstruit en adoptant un cheminement qui accepte une complicité avec les processus de négation constitutifs de la mise en place de tout objet:

Si la coïncidence est perdue, ce n'est pas par hasard. (. . .)
Un immédiat perdu, à restituer difficilement, portera en lui-même, si on le restitue, le sédiment des démarches critiques par lesquelles on l'aura retrouvé, ce ne sera donc pas l'immédiat. (V.I., 162)

Les retrouvailles ne sauraient donc s'accomplir comme annulation rétrospective de la perte. Perte qui ne vient d'ailleurs pas inciser, engloutir l'expérience, mais qui est au fondement de son mouvement, dans la mesure où elle est inséparable de toute expérience de saisie:

L'idée du chiasme, c'est à dire: tout rapport à l'être est simultanément prendre et être pris, la prise est prise, elle est *inscrite*, et inscrite au même être qu'elle prend.
A partir de là, élaborer une idée de la philosophie: elle ne peut être prise totale et active, possession intellectuelle, puisque ce qu'il y a à saisir est une dépossession. (. . .) Elle est l'épreuve simultanée du prenant et du pris dans tous les ordres. (V.I., 319. Souligné dans le texte)

Dans la mesure où le couple du visible et du tangible se trouve apparié au couple de la possession et de la dépossession, le visible n'intervient pas pour articuler le dernier mot, mais il n'est pas pour autant annulé. La dépossession intervient comme destin de la prise, comme l'impossibilité dans laquelle elle se trouve de fonder sa propre émergence, de voir le mouvement qui lui permet de dominer son propre panorama.

Une fois reconnue la nécessité d'un tel chiasme, l'interprétation de l'aventure oedipienne se trouve radicalement renouvelée. Le tragique oedipien prend en effet naissance dans un choix résolu pour le modèle visuel, le service du logos, le culte apollinien; la dépossession ne peut ainsi être rencontrée que comme ce qui brise le mouvement originaire de la trajectoire, ce qui met en échec la volonté de pré-vision. D'où le heurt terminal avec une expérience de dépossession et l'entrée dans la cécité, dans l'univers de la palpation.

Or la geste dessinée par l'itinéraire de Merleau-Ponty consiste dans la réécriture d'un autre mythe. L'entrée dans le tactile ne vient pas d'une faillite de la vision. Inutile de se crever les yeux, mutilation marquant l'apogée du parcours tragique, puisque, en un sens, c'est toujours déjà chose faite. Partant de là, on pourrait estimer qu'Oedipe s'impose une cécité volontaire pour ne pas avoir à reconnaître une cécité instituée, originaire. Cécité inscrite dans les conditions de la prise de conscience vue comme phénomène d'éclairement: "Le avoir-conscience lui-même est à concevoir en transcendance, comme être dépassé par . . . et donc comme ignorance." (V.I., 250) L'ignorance ne serait donc pas échec. Rapportée au modèle tactile, elle atteste la consistance, la résistance de ce à quoi je me heurte.

On débouche ainsi sur un entrelacs de dimensions — à la fois disjointes, complémentaires, chevauchantes — qui nous reconduit à la manière dont Freud articule la dualité de la représentation et de l'affect. Dualité dans une certaine mesure parallèle à celle du voir et du toucher. C'est d'ailleurs à ce registre de l'affect que se réfère Merleau-Ponty en présentant le rapport au sens comme captif d'un phénomène d'investissement, l'affect étant approché par Freud à partir de la notion d'intensité d'investissement: "Le sens de la parole que je dis à quelqu'un, note Merleau-Ponty, lui 'tombe sur la tête', le *prend* avant qu'il ait compris." (V.I., 290. Souligné dans le texte)

La métaphore est significative et paradoxale, dans la mesure où cette expérience d'un sens qui nous "prend" — tout comme la grâce est censée tomber sur quelqu'un — correspond précisément à ce qui peut être éprouvé comme expérience de dépossession. La dépossession ne saurait donc être appréhendée comme pur phénomène de manque, puis-qu'elle est l'autre face d'un don; don qui est à la fois grâce et mutilation, puisque nous ne serons jamais contemporains de l'origine donatrice.

Une question reste cependant ouverte: dans quelle mesure l'aventure philosopohique n'est-elle pas intrinsèquement solidaire de la passion visuelle et de la croyance en la possibilité d'une "prise totale et active"? L'interrogation de Merleau-Ponty vient-elle sonner le glas de l'ambition philosophique, en dénudant les illusions sur les-quelles elle repose, ou permet-elle l'ébauche d'une autre approche philosophique? Approche qui, dans la mesure où elle est possible, ne peut s'effectuer que par le recours à un mode spécifique d'écriture. Non plus l'écriture qui étale, découpe et recompose, qui "résout"; mais une écriture qui sache accepter la complicité avec le mode tangible, avec son type de capture. Ecriture mitoyenne entre le champ philosophique au sens strict et le champ esthétique, dans la mesure où la parole doit, dans le même temps, capturer, toucher — par là même être touchée — et donner à voir. Or il est une figure stylistique qui correspond à cette double opération: la métaphore. Non pas figure immobile, mais véritablement "transport", où se conjoignent l'expérience du dévoilement et celle du dépaysement. L'écriture de Merleau-Ponty s'orientera d'ailleurs de manière de plus en plus manifese vers ce mode d'opération séductrice en quoi consiste la métaphore. Qu'il s'agisse du "transport" métaphorique ou de la séduction qui "dévoie" — ce qu'indique également le terme allemand de *Verführung* — nous sommes bien dans le registre d'une parole qui se fait non pas désignation directe, mais initiation. Or la métaphore agit précisément en nous impliquant tout comme dans le toucher, bien qu'elle ne soit rendue possible que par un décollement étranger au toucher; elle nous emporte, nous convertit, nous fait voir, sans nous indiquer dans le même temps les ressorts de sa propre opération. Mais reconnaître que seule la métaphore a le pouvoir de nous transporter dans l'originaire, c'est, abandonnant l'espoir de tout dévoilement de l'originaire dans sa virginité, reconnaître qu'il n'est d'originaire que rétrospectif, réinventé dans une écriture qui relève de l'art plus que de l'analyse portant sur un quelconque donné: "L'Etre est *ce qui exige de nous création* pour que nous en ayons l'expérience." (V.I., 251) Une phénoménologie de la perception n'est peut-être réalisable, en dernière analyse, qu'en effectuant une jonction avec une esthétique de la perception. Jonction qui ne saurait toutefois se présenter comme assimilation pure et simple, dans la mesure où le projet esthétique, pour l'essentiel, délimite son terrain du côté de ce qui peut s'offrir en fait de représentation. Le tangible, comme paradigme du non-représentatif, restera le témoin de ce qui, tout en fonctionnant comme ancrage de toute représentation, oeuvre ou connaissance, reste, dans une certaine mesure, inconvertible, impossible à monnayer en une représentation échangeable.

Centre National de la Recherche Scientifique

NOTES

[1] *Le visible et l'invisible* (Paris: Gallimard, 1964), p. 306. Les références à cette oeuvre (V.I.) ainsi qu'à la *Phénoménologie de la perception* (P.P.) seront désormais incluses dans le texte à la suite du passage cité.

[2] *La Violence de l'interprétation*, Paris, PUF.

[3] Dans *Freud et le plaisir* (Paris: Denoël, 1980), j'ai tenté de dégager le risque inhérent aux expériences de plaisir analysées par Freud: perdition ou régénération.

[4] Alain Roger: *Nus et paysages. Essai sur la fonction de l'art*, Paris, Aubier-Montaigne, 1978, p. 31.

[5] Exposé proposé au Colloque International de Phénoménologie, organisé par *l'Institut Mondial des Hautes Études Phénoménologiques*, Belmont, Mass. à Paris en 1981. L'étude en question est parue dans le volume: *Le Psychique et le corporel dans la pensée philosophique de M. Merleau-Ponty*, Travaux de recherches de l'Institut Mondial des Hautes Études Phénoménologiques, Fascicule I, Paris: Aubier-Montaigne, 1985.

PART V

THE LIFE-WORLD AND THE SPECIFICALLY MORAL SIGNIFICANCE OF THE COMMUNAL/SOCIAL WORLD

DALLAS LASKEY

THE CONSTITUTION OF THE HUMAN COMMUNITY: VALUE EXPERIENCE IN THE THOUGHT OF EDMUND HUSSERL; AN AXIOLOGICAL APPROACH TO ETHICS

I. THE CASE FOR AXIOLOGICAL ETHICS

The axiological approach to ethics has few followers in recent times because of fundamental disagreements concerning the characterization of the basic value terms needed to delineate the field, and the lack of a suitable method for resolving these value conflicts. Given these short-comings, it is hard to see how one could move on to specialized studies of ethical values and to work out a comprehensive theory of the ethical life. It is no wonder that some philosophers have not only turned away from value inquiry, but have condemned or discredited the entire project.[1] Frequently, such critics have recommended metaphysical or ontological alternatives to value inquiry, but so far substantial contributions to ethics have been singularly absent.

One of the virtues of the original project lay in the production of rich and varied new descriptions of the basic experiences that give rise to value claims. If for no other reason, the axiological approach broadened the scope of investigation beyond the favored studies of obligation, the virtues, and the good. The project foundered, I believe, because of certain methodological shortcomings; the constant and often hasty appeal to intuition was not universally accepted, for it tended to cut off debate and further investigation. Early phenomenological studies of value were also criticized as ahistorical, for while the relations between the valuing subject and his values were explored, the relations between the situational contexts and the subject were often ignored.

The traditional axiological studies of Brentano, Meinong, Scheler, and Hartmann were instituted with relatively early forms of phenomenological method. Even the more recent studies by Hans Reiner[2] and J. N. Findlay,[3] important and fruitful as they are, do not make use of the extensions and modifications introduced by Husserl in the later manuscripts. It is my contention that these later developments in Husserl's phenomenology are especially suited for axiological and ethical investigation. Few philosophers are aware of Husserl's contribu-

A-T. Tymieniecka (ed.), Analecta Husserliana, Vol. XX, 315–329.
© 1986 *by D. Reidel Publishing Company.*

tions to axiology and ethics since the bulk of the manuscripts dealing with these topics has not yet been published. This study does not make use of these materials specifically, but relies on the various scattered remarks on ethics and value made throughout his published writings.

I would like to claim that a return to the axiological approach can be justified not only in terms of the new descriptions of human experience it may bring, but also in terms of the distinctly different view of ethics and its tasks. In the everyday life it is regularly assumed that ethics has to do with the regulation of human conduct and that its chief question is simply — what ought we to do? But Husserl pointed out that our daily life is naive and that we live in a world that is already given to us. We thus take for granted the meanings embodied in our thinking, acting, and valuing and assume that reflection on our situation will produce an answer to the question. But we cannot answer the question as to what we ought to do until we have first determined what the moral situation really is. This latter task is often more difficult than it would appear, for it is often the case that we just do not know what is moral in the situation. Coming to see the situation as moral may be one of our central problems, especially if we are more inclined to take it as prudential, economic, technological, aesthetic, or some alternative perspective. The axiological approach, on the other hand, tries to clarify just those features of human experience which set it off as moral.

Along with the prospect of a different way of viewing the nature of ethics and its tasks, is a different method for certifying or legitimizing value claims. Husserl's doctrine of evidence was claimed to be the very center of phenomenological investigation,[4] and its role in the resolution of value conflicts is essential. Such conflicts have long resisted the efforts of investigators to achieve agreement, and if the axiological approach offers another way for resolving differences, then it should be most welcome.

In view of these considerations, I propose to reexamine the axiological approach to ethics and value in the light of an interpretation of phenomenology drawn from Husserl's later works. I propose to examine a small set of issues concerning the structure of axiological experience and show how they contribute to a different conception of ethics and its tasks. While metaphysical and ontological decisions may, in the last analysis, be unavoidable, I will try to avoid these in order to stay as close as possible to the structures of human value experience.

II. AXIOLOGY AS A FOUNDATIONAL STUDY FOR ETHICS

The study of values and valuational activity is claimed as a necessary foundation for ethics. Ethical value, as one class of value in a wide spectrum of values, can be more effectively investigated in the context of an axiological approach. Too often the study of ethics has been initiated with too narrow a focus, as though ethical decisions were confined to a narrow segment of cultural life. The axiological approach helps to broaden the scope of the investigation and seeks the originating conditions responsible for value; in other words, it seeks the foundations of value in human experience.

What is meant by the term "foundations" is not immediately clear. Prior to Kant, the search for foundations of ethics often took the form of a search for the origins of ethical distinctions. In the eighteenth century Hume looked for the foundations in human sentiment;[5] Kant, on the other hand, looked for them in practical reason.[6] After Kant the search for foundations shifted to a more epistemological emphasis and was focused on the justification problem and whether there were ultimate principles which could serve as stoppings points in the process of justification; e.g., the principles of utility, of self-realization, of freedom, etc. In some cases it was held that no single principle could serve as an ultimate ground and so a small set of independent principles were advanced in its place; e.g. W. D. Ross' prima facie obligations.[7] In more recent times, the search for foundations has been regarded with increasing suspicion, and antifoundationalist philosophers have tried to divert attention away from the problem of foundations to a consideration of alternative ways of organizing moral claims and with more emphasis on the systemic character of our beliefs.

In Husserlian phenomenology the question of foundations for ethics and value would appear to be ambiguous, depending on what phase of its development was being considered. In this paper I wish to interpret Husserl in the light of his later philosophy and to tackle the question of foundations from that perspective. Within the perspective of transcendental constitutive phenomenology the idea of foundations acquires a distinctive sense. It was not the search for a body of truths that would serve to ground other claims, nor was it a matter of finding a stopping place in a linear process of deduction. Rather, it was the search for the originating experiential conditions out of which emerged the sense and significance of value judgments. Such original experiential conditions were intentional features of experience which had to be phenomenologically clarified and legitimated through evidential performances appropriate to their horizons. Beyond a static phenomenology, a genetic inquiry was required to reach back into these originating conditions and to clarify the constitutional path from them to the current value claims. Foundational investigations would involve not just current axiological and ethical judgments in their present horizons, but their development from originating conditions through various historical epochs to the present. It had to show how the various levels of sense had been acquired or lost in the path of constitutional development. Given the changing social and cultural contexts from one age to another, it seems clear that our current judgments are complexes or sedimendations of meaning elements that have been built up over time. The task of unraveling all these elements is indeed formidable and leads one through wider and wider horizons. What is foundational at one level may thus be seen as derivative when viewed from a more basic level. The progressive disclosure of further foundations has the unexpected consequence of bringing into clearer focus the systemic nature of phenomenology as can be seen from its teleological character.

If the notion of foundations is thus found to be complex, it should be noted that the question of origins cannot really be separated from the notion of destination or goals (Husserl's "infinite tasks"). Because of the very nature of intentionality we are led through intentional links from foundations to the correlative notion of ultimate goals or tasks. Throughout his life Husserl saw his work in the context of a developing Western humanism.[8] The striving for a universal science which began with the Greeks and continued throughout history along with the belief in the universal norms of reason allowed Western man the possibility of transforming himself and his milieu and to pass on this commitment from generation to generation. Husserl claimed that this humanism provided the teleological context for his work, and it is important to see how our more local and particular norms are affeected by the larger defining norms of our civilization. This humanism was a far cry from that criticized by Heidegger in his *Letter on Humanism*;[9] this attack was directed to a fixed conception of man and his work — nothing could be further from Husserl's transcendental humanism.

In summary, it is clear that Husserl's search for foundations was unique in the history of philosophy. Set in the very widest possible context of Western humanism, the quest for foundations was a search for the origins of value judgments and norms in the primordial experiences of our culture. It was both a descriptive and a normative inquiry; descriptive in the sense of exposing the various intentional structures involved in the making and development of value judgments, and normative in the appraisal of such judgments through the evidence implicit in their horizons.

Would such a foundational investigation yield the set of basic terms which would be sufficient to delineate the field of axiology? While there is little doubt that this is the goal of Husserl's inquiries, it is clear that such an ambitious project was barely initiated. Further research on his unpublished manuscripts will reveal just how far Husserl traveled in this direction, but on the face of it it would seem implausible that he had really undertaken the extensive historico-critical studies needed for the task. It is more likely that the bare outlines were sketched, and that the completed task would require teams of phenomenologists exploring the social and cultural history and carefully sifting the results to severe evidential scrutiny.

III. THE NATURE OF VALUE EXPERIENCE

The clarification of the experiences which yield values proves to be far more difficult and frustrating than first imagined, and a cursory look over the results achieved by Brentano, Meinong, Husserl, Scheler, and Hartmann reveals a wealth of conflicting claims.[10] Such a disagreement poses a serious threat to the credibility of the phenomenological enterprise and a real challenge to any investigator. What is the root of the difficulty? If the phenomenological investigations were carefully carried out, then they should be repeatable with the same results. My hunch is that previous studies have been thwarted by methodological shortcomings which may have screened out important data. In the early part of this century, when enthusiasm for the axiological approach was at its height, relatively primitive forms of phenomenology were employed which were unable to handle the intersubjective contributions from social and cultural contexts.

It is an open question whether Husserl used an earlier methodological version in his studies of ethics and values. The study by Alois Roth[11] appears to ignore the role of transcendental constitutive phenomenology and is more concerned to emphasize analogies and links with the *Logical investigations*. One must wait until the manuscripts dealing with ethics and values have been published to decide this question. There is no doubt that Husserl lectured on ethics until the summer of 1924, so he would have had plenty of time to incorporate new developments. The limited references in Roth and Diemer[12] give no clear indication on this matter; a number of references to original manuscripts relating to Husserl's doctrine of material practice would suggest, however, that both static and genetic phenomenology were extensively involved.[13]

Let us now turn to some early statement of Hussrel concerning the character of value experience. In the first place values are there before us in the world of everyday experience; in his description of the thesis of the natural standpoint, Husserl notes that what appears before us is furnished with value characters "such as beautiful or ugly, agreeable or disagreeable, pleasant or unpleasant, and so forth."[14] Such values and practicalities were affirmed to belong to the constitution of the actually present objects as such. Value itself is characterized as the intentional correlate of an intentional act of feeling.[15] His remarks on this are quite clear:

> But in the act of valuation we are turned towards values, in acts of joy to the enjoyed, in acts of love to the beloved, in acting to the action, and *without* apprehending all of this. The intentional object, rather, that which is valued, enjoyed, beloved, hoped as such, the action as action, first becomes an apprehended object through a distinctively "objectifying" turn of thought. . . . in acts like those of appreciation we have an *intentional object in a double sense*; we must distinguish between the "subject matter" *pure and simple* and the *full intentional object*, and corresponding to this a *double intention*, in a twofold directedness.[16]

Clearly Husserl is speaking of valuation on the level of active synthesis and "in general, acts of sentiment and will are consolidated upon a higher level, and the intentional objectivity also differentiates itself accordingly."[17] Value is thus not a logical or an empirical quality, but an intentional object disclosed after the epoché has been activated. Values are intentional structures of experience and emerge in the intentional acts of feeling that we direct to our environment. Their properties can be described and have the same sort of autonomy as other constituted objects in logic, science, or art. As intentional objects they have horizons which need to be explicated, and are relatively stable features of experience which can be studied in their constitutional development.

In the early discussions, Husserl drew attention to the stratifications of consciousness and their particular modes of ordering, and how feeling, sentiment, will, decision, and valuation in every sense involve many and often varied stratifications in noetic and noematic phases.[18] What emerges in his discussion is the apparently dependent character of value predications on what is already presented to experience. Brentano's classification of intentional acts suggested much the same thing, for judgments and acts of love and hate always presupposed presentations. What is puzzling here is the priority placed on the doxic acts of belief on which acts of feeling and sentiment are predicated.[19] Value predications would thus seem to be supervenient on belief predications at least in the discussion in the *Ideas*; whether valuations play a role in primary synthesis is a question that was not brought up at this stage of his development. Scheler, on the other hand, appears to give priority to value predicates over belief predicates — at least in the "Ordo Amoris" essay.

Husserl places considerable stress on the fact that stratifications occur at the affective-volitional level; even when they are based on "presentations" they reveal a stratified formation.[20] Such stratifications are found in both the noetic and noematic phases so that "In the noema of the higher level the value as such is as it were a meaning-nucleus girt about with new thetic characters."[21] Husserl does not attempt to unravel these complicated structures and determine how "formative syntheses of value" are related to "those of fact."[22] But he does affirm that while living in the modifications of the valuing consciousness we do not need to take the doxic point of view, but can do so if we move from the suggestion theses to the thesis of modes of belief.

Granted that the doxic and valuational predicates are intertwined in a unitary whole, the question arises as to how such a unitary meaning is to be certified or legitimated. Husserl observes that the doctrine of evidence applies not merely to theoretical beliefs but also to the valuational claims as well:[23] "The 'theoretical' or '*doxological truth*,' or *self-evidence*, has its parallels in the '*axiological and practice truth or self-evidence*.' . . ."[24] In a most revealing footnote, Husserl asserts that while knowledge is a term applying to the correlate of self-evident judging, it also applies to "every kind of self-evident judging itself . . . ." Thus it seems clear that acts of sentiment and valuing can be appraised in terms of their evidence just as any other intentional act. This is confirmed in the *Cartesian Meditations* when he says that reason is "an all-embracing essentially necessary structural form belonging to all transcendental subjectivity."[25] Reason refers to possibilities of verification and verification refers ultimately to making evident and having as evident.[26] Thus I can safely reaffirm Roth's contention that the evidence problem is the uniquely central question of phenomenology and that in axiology and ethics the same severe evidential requirements hold good.

At this point I wish to continue with some observations of my own about the nature of affective and volitional acts and how they are to be legitimated. It was Brentano[27] who first raised the question of correctness and incorrectness regarding feelings, but he did not have all the intentional structures at his disposal to provide the necessary clarity. Husserl was critical of Brentano and accused him of failing to understand intentionality, particularly in its dynamic aspects.[28] One of the distinctive features of all intentional acts is their ability to be fulfilled in immediate experience or in sequences of immediate experiences. Intentional acts of feeling can thus be classified as empty or fulfilled, depending on whether the intentional object provides us with fulfillment of the noetic elements. Adequacy would be attained if every element of noetic structure were fulfilled or encountered in an *Erlebniss*, or where there is nothing in the noema that does not have its correlate in the noesis. Feelings can be directed to all sorts of things, real and unreal, posited and phantasized; they should be able to be appraised therefore in terms of their evidence. We do not need empirical objects to provide fulfillment, just as we do not need empirical objects to fulfill our logical or mathematical intentions. One is easily misled on this point, for in the case of perception in the natural standpoint, an object is required. However, within the reduced state, it is enough that the noematic structures do or do not fulfill their noetic correlates — and so it is with feelings.

It is a well known fact of human experience that people vary significantly in their capacity for feeling and emotion, and they are often judged in terms of the sensitivity or lack of it. It is also a well known fact that feelings can be cultivated and developed, and if Aristotle is right, they can be modified. This is one of the central pedagogical and ethical problems in life, for it is a difficult task to learn how to feel the right way about the right things. What is difficult to determine is whether or not the objects of our feelings confirm our intentional acts, for it is one thing in principle to admit that such verification is possible, but it is quite another thing in practice to be able to bring it about. Even more important is the ability to learn how to modify the noetic acts of feeling in such a way that they may have a greater chance of articulating their objects clearly and distinctly. Much work of concrete analysis remains to be done in this area, as Husserl has only sketched in bare outline how the project should be realized.

It is time now to reflect a moment on the question of the way in which values are apprehended. When insufficient attention is placed on the horizons of value or on the noetic contributions, and when the noematic phases are unduly separated from their noetic correlates and considered as autonomous and independent, then claims for the self-evidence of value judgments are liable to be made too hastily. This would seem to be the case for Scheler and Hartmann who affirmed a multitude of self-evident assertions about values and value hierarchies. It is no wonder that such affirmations have led to contradictions and led to criticism from all sides. A recent criticism of the axiological approach to ethics has been formulated by Anna-Teresa Tymieniecka in her work entitled "The Moral Sense."[29] Tymieniecka challenges the whole axiological enterprise for failing to provide a defensible account of moral experience. As I understand the argument, she objects to the conception of ethics in which moral experience is characterized as a kind of passive and contemplative viewing of values. She protests against the model of contemplative knowing of values, whether this be emotive or intellectual cognition. There is no doubt that Scheler and Hartmann occasionally write in such a way as to imply that moral deliberation is basically a question of the cognition of values in a given situation. According to this thesis, as long as one opts for the higher value, one acts or decides morally. Implicit in this conception is an assumption which dates back to Socrates, that if one knows what the good is, then one will automatically do it.

This contemplative model of moral experience is seriously wrong and I think that Tymieniecka is fully justified in her criticism. This intellectualist position is simply the result of insufficient attention to all the intentional factors involved; the whole dynamic of intentional life becomes distorted when the entire focus is directed to noematic structures alone. Values get hypotastized into something like Platonic essences and get totally separated from any concrete process of moral deliberation.

Tymieniecka's criticism of the axiological approach leads her to develop a new and strikingly original conception of the origin of moral values, namely, in what is termed "the moral sense." Objecting to the emphasis on the constitutional process which she claims is controlled by intellect, she explores the origins of moral values in a parallel constitutional system of considerable complexity. It is not my intent to appraise this novel contribution, which deserves a full-length treatment. My task is simply to determine whether moral experience and moral valuation can be accommodated within the axiological tradition and in a way that would be free from her criticism.

To repeat, I agree with the criticism of the model of moral experience as merely the cognition of values; such a model cannot account for the phenomenon of a deliberately made choice for a lesser good or for evil. But it seems to me that part of Tymieniecka's motivation for a new phenomenological method comes from an assumption that appears early in the paper[30] — the distinction between intellect and affectivity. Such a distinction appears to be raised to an ontological distinction in which the domains of intellect and affectivity are entirely separate. This seems to be phenomenologically unfounded. While it is certainly possible to describe the different functions of intellect and affectivity in phenomenological terms, yet it should be clear that these are simply different functions at work in the same self or ego. Their constant intertwining in our intentional life has already been referred to, and their constant interplay in our value judgments and ethical judgments is not to be denied but explicated. In the making of axiological judgments of any kind the whole person is involved and the complex dynamic of intentional structures must be carefully described; when such an analysis is undertaken, then moral experience will be seen to be much more complex than a mere cognition of values. This point will be elaborated in more detail in the following section.

The axiological tradition, if it is to command the attention and respect of philosophers, has much work on its hands — not the least being the phenomenological analysis of the basic value terms in a much more sophisticated manner than has been observed thus far. In my opinion, only a start has been made and Husserl's project for ethics and value is barely under way. There is a great need for the kind of historico-critical studies of particular values, such as that featured in the work of Hans Reiner in his Duty and Inclination.[31]

IV. THE NATURE OF ETHICS

Before embarking on the question of the nature of ethics, it might be useful to sketch the outline of Husserl's architectonic so as to provide a context for discussion. What are the requirements for a complete account of ethics and value? Husserl's answer was threefold: (1) one must first provide a phenomenological description of the intentioanl experiences responsible for the constitution of value, and then an account of the main kinds of value; (2) one must provide an account of the most fundamental generalizations or laws of values and a specification of the contents of such laws or principles; and, finally, (3) one must offer an account of the application of such values and principles in the lives and acts of men and society; this would led to an investigation into the kinds of person and kinds of community where such values could be realized.

As a consequence of these considerations the axiological approach to ethics would involve four major areas of investigation: a formal and a material axiology coming from the study of the noematic phases of intentional life, and a formal and material practice coming from the study of the noetic phases. Formal axiology is concerned with the laws and principles holding between values; material axiology has to do with the content of these laws and principles developed in formal axiology. Formal practice is concerned with the norms of correct willing and striving, the elements of objective duty, and the formal determinants of the highest ethical principle, the categorical imperative. Material practice represents the attempt to provide material contents for the norms of value and obligation and to account for the constitution of the ethical subject and the ethical community; it has also the task of finding contents for the categorical imperative. This completes the sketch of the architectonic.

This comprehensive division of labor represents for Husserl a complete picture of what is required for an account of ethics and value. It can be readily seen that the study of values is a necessary first step for the study of ethics, and that the task of comparing, ordering, and systematizing value claims leads to a consideration of the norms of correctness for conduct. What we normally think of as ethics would thus appear to fall under the domain of formal and material practice; the appraisal and ordering of ends and purposes is a distinctly ethical task, as is the determination of a self and the ethical community. A comprehensive theory requires all four studies however.

The architectonic sketched above seems quite appropriate for a static phenomenology, but a problem appears immediately — would not the results obtained from formal and material axiology merely reflect the values and morals in use in a given society? A further analysis seems

required of a genetic kind. A little reflection indicates that the values
and principles of a particular society may be simply a variant stage of
development, for a given society might appear as warlike and vindictive,
closed and oppressive, or as shortsighted and unimaginative. It is quite
unlikely that universal standards were embodied in the institutions and
practices of such a society. What is needed, then, is a series of genetic
studies on value analogous to the kind of critical studies instituted
by Husserl in the *Crisis of European Sciences and Transcendental
Phenomenology* and in *Erste Philosophie*.[32]

There is a further problem about the apodicity of the value claims
that might appear in a formal and material axiology. It is hard to see
how any value judgment formulated in any society up to now could be
fulfilled in an adequate manner. Some social forms do not seem to be
conducive to the development of those human sentiments and approvals
which have been admired in all ages; it is more likely that certain forms
of urban, commercial, and technological society stifle the expression of
our finer human sentiments of fellow feeling and concern for others. It
follows from this that it would be difficult to attain the results called for
in a static analysis with any degree of adequacy. A genetic analysis
leading back to the primordial conditions and their horizons and
evidence seems clearly called for, so that one could trace the constitutional development with its evidential appraisals from the past through
to the present. What sorts of results obtainable from this can scarcely be
imagined, yet the project seems eminently attractive and relevant.

A genetic analysis would be most useful in revealing the constitution
of both persons and communities. Husserl clearly saw the problem of
legitimacy as applying to the self and other people; we are fully responsible for such views and perhaps this may account for his almost
neurotic concern for claims to be rigorously established in evidence —
why he could not bear unexamined assumptions or posits or hypo-
thetical entities. In every phase of one's thinking, feeling, striving, and
willing there is the demand for rigour and precision; it is thus no
accident that he referred to phenomenology as an ethico-religious point
of view.[33]

The tasks for ethics involve the constitution of selves and the
constitution of society; this is an infinite task involving constant activity
on the part of every person in their entire life of thought, feeling, and
action. Far from a passive contemplation of values, this is a continuous
and ongoing struggle of heroic efforts. Husserl's ideal of the ethical

community, which is not far removed from Kant's third formulation of the Categorical Imperative, and which he took as a kind of *Liebesgemeinschaft*, required a society of persons with shared intentions, each one of which was capable of adequation, and each person operating with norms of correct willing, thinking, and striving. In the latter part of his life, more emphasis was placed on the concept of love (*Der Liebesbegriff*) as an essential factor in behavior. Love was an intentional act of feeling directed to other persons in such a way as to bring about a genuine ethical community. Progress toward such a community was not understood by Husserl to be in a straight line, but one which involved retrogression as well as advance. The task of living the moral life according to the norms of correctness in an immoral society was seemingly a vivid part of his own experience, and the interpretation and modification of Kant's Categorical Imperative along the lines suggested by Brentano, seem a reflection of that reality. Since we live in less than a perfect world, and often in drastically imperfect social situations, the ethical demands on the person are still clear — in all his acts and strivings, he is to make his conduct conform to the norms of correctness in the appropriate area. Duty requires the person to move toward the best possible development of himself and his community under the prevailing circumstances; in other words, to realize the most good that is possible in those conditions.

In retrospect, the tasks of ethics are of heroic proportions. It is not a matter of being able to read off the values in a particular situation, but a question of molding one's intentional life in such a way as to conform to the norms of correctness. What really marks off the ethical enterprise is the development of those feelings toward our fellowmen. In this paper, I did not mention the role of transcendental empathy which I think is essential not only for our understanding of other persons and their intentions, but for our participation in the realization of their aspirations where these meet the universal standards required from the problematic evidence. It is through empathic imagination or imaginative empathy that we are able to put ourselves in others' shoes — to see as they see, to feel as they feel, to strive as they strive — and this is an essential first step in coming to understand a situation in moral terms. Again, this is not passive contemplation of values but the active molding of a personality in accord with the highest possible standards. Lacking such an empathic knowledge of our neighbor makes genuine ethical behavior next to impossible. The phenomenological elucidation of the role of

empathy was for Husserl the key to the resolution of the intersubjectivity problem. Its role in morality is a project waiting to be realized.

Concordia Univesity

NOTES

[1] Martin Heidegger, "A Letter on Humanism," in *Phenomenology and Existentialism*, ed. Robert C. Solomon (Washington, D.C.: University Press of America, 1980).
[2] Hans Reiner, *Duty and Inclination: The Fundamentals of Morality Discussed and Refined with Special Regard to Kant & Schiller*, trans. Mark Santos (The Hague: Nijhoff, 1983).
[3] J. N. Findlay, *Values and Intentions* (New York: Macmillan, 1961).
[4] Alois Roth, *Husserls ethische Untersuchungen* (The Hague: Nijhoff, 1960).
[5] David Hume, *A Treatise on Human Nature*, ed. L. A. Selby-Bigge (Oxford: Clarendon Press, 1958).
[6] I. Kant, *Critique of Practical Reason*, trans. Thomas Abbott, 6th ed. (New York: Longman's Green, 1954).
[7] W. David Ross, *Foundations of Ethics* (Oxford: Clarendon Press, 1939).
[8] Dallas Laskey, "Husserl as a Humanistic Moralist," *Phenomenology Information Bulletin* 7 (October 1983).
[9] Heidegger, "A Letter on Humanism".
[10] Franz Brentano, *On the Origin of our Knowledge of Right and Wrong*, trans. Roderick Chisholm and Elizabeth Schneewind (New York: Humanities Press, 1969); Alexius Meinong, *On Emotional Presentation*, trans. Marie-Luise Schubert Kalsi (Evanston: Northwestern University Press, 1972); Max Scheler, *Formalism in Ethics and non-formal Ethics of Values*, trans. Manfred Frings and Roger L. Funk (Evanston: Northwestern University Press, 1973); J. N. Findlay, *Axiological Ethics* (London: Macmillan, 1970); Nicolai Hartmann, *Ethics*, 3 vols., trns. Stanton Colt (London: George Allen & Unwin, 1951).
[11] Roth, *Husserls ethische Untersuchungen*.
[12] Ibid.
[13] Ibid.., S49—54.
[14] Edmund Husserl, *Ideas: General Introduction to Pure Phenomenology*, trans. W. R. Boyce Gibson (London: George Allen & Unwin, 1958).
[15] Roth, *Husserls ethische Untersuchungen*, p. 124; Alwin Diemer, *Edmund Husserl: Versuch einer systematischen Darstellung seiner Phänomenologie* (Meisenheim am Glan: Anton Hain, 1956).
[16] Husserl, *Ideas*, p. 122.
[17] Ibid.. pp. 122—23.
[18] Ibid., p. 276.
[19] Ibid., p. 323.
[20] Ibid., p. 326.
[21] Ibid., p. 327.
[22] Ibid., p. 328.

23 Ibid., p. 389.

24 Ibid.

25 Edmund Husserl, *Cartesian Meditations*, trans. Dorion Cairns (The Hague: Nijhoff, 1960), p. 57.

26 Ibid.

27 Brentano, *Origin*, pp. 220−24, 244−47.

28 Edmund Husserl, *Formal and Transcendental Logic*, trans. Dorion Cairns (The Hague: Nijhoff, 1969), p. 245.

29 Anna-Teresa Tymieniecka, "The Moral Sense," in *Analecta Husserliana*, vol. 15, ed. Tymieniecka (Dordrecht: Reidel, 1983), pp. 3−78.

30 Ibid.

31 Reiner, *Duty and Inclination.*

32 Roth, *Husserls ethische Untersuchungen*, chap. 3.

33 Edmund Husserl, "Philosophy as a Rigorous Science," in *Edmund Husserl: Phenomenology and the Crisis of Philosophy*, by Quentin Lauer (New York: Harper Torchbooks, 1960), p. 71.

INTERSUBJECTIVITY AND THE VALUE OF
THE OTHER

The problem of intersubjectivity has been restricted traditionally to the epistemological issue of the knowability of the other.[1] Yet, the *value* of the other is also a central dimension of intersubjectivity, especially insofar as we perceive the other as autonomous of our own wishes and projects. A phenomenological analysis can reshape the field of intersubjectivity for both philosophy and the social sciences by showing that intersubjectivity can be examined not as an epistemological dilemma, but as a meaning-giving event in which the value of the other is constituted through interaction in the social world.

I. SELF-INTERPRETATION AND THOU-INTERPRETATION

Intersubjectivity presents us not only with the awareness of others, but the awareness that others are meaning-giving agents like ourselves. Others appear to us *as* others by virtue of the fact that their actions and beliefs are uniquely their own, yet overlap with and accompany our choices, motives, and projects. Each frame of reference, the other-directed and the self-directed, forms a primary level of interpretation according to which the value of the other is given to us through interaction in the social world. We will, therefore, examine the value of the other in terms of the categories of "thou-interpretation," the meaning assigned to the individual by others, and "self-interpretation," the self-assigned meaning of the individual.

Self-interpretation and thou-interpretation are relative frames of reference. The value of the other emerges in the interplay between both perspectives. Thou-interpretation is the external constitution of the value of the other. We, as observers — disinterested passers-by, acquaintances, social scientists, philosophers — determine for ourselves the meaning of others in the social world. Our observations continually shape and are shaped by the intersubjective context of our personal projects, social networks, historical epoch. Self-interpretation is the subjective meaning we as individuals assign to our acts from a position of being in the center of our own conscious world. Our situation is seen

A-T. Tymieniecka (ed.), *Analecta Husserliana, Vol XX*, 331–338.
© 1986 *by D. Reidel Publishing Company.*

directly in reference to our goals, needs, and projects. From our position within the social world we "constitute" our own value, we design a course of action and carve out life-plans.

The work of two phenomenologists, Alfred Schutz and Anna-Teresa Tymieniecka, is of particular relevance for analysis of the problem of intersubjectivity in terms of the value of the other.[2] Schutz and Tymieniecka both reject the epistemological approach of transcendental phenomenology and emphasize, instead, the dynamic constitution of meaning through interaction in the social world. Schutz, however, stresses the role of thou-interpretation, especially in his examination of the stock of knowledge and systems of relevance operant in the common-sense attitude. Tymieniecka develops extensively the function of self-interpretation, with a focus on transaction, the human condition, and the moral sense.

Schutz locates his discussion of intersubjectivity within the common-sense attitude. The common-sense attitude is comprised of those beliefs about the social world which we take for granted.[3] Unless a problem arises that calls into doubt a specific belief, our trust in the world goes unquestioned. In the common-sense attitude the existence of the other is simply assumed. Each day we awaken to a world of others reporting the news, driving to work, opening umbrellas, waving to friends. That these hatted and cloaked beings are human, much like ourselves, is accepted as a matter of fact. A second glance at Schutz' account of the pregivenness of the other suggests that the other is always given as a valued being. As we go about our daily lives we are careful not to bump into, disregard, or intrude on others. We do not move others out of our way as we might in the case of objects. We expect that others will be treated humanely even if we do not know them directly, or have a vested interest in their situation.

Although Schutz claims that our knowledge of the other is indirect, and based on the immediate awareness we have of ourselves, Schutz' approach to the other is primarily from the perspective of thou-interpretation. The structures basic to Schutz' examination of intersubjectivity, i.e., stock of knowledge, schemes of typification, systems of relevance, are external and seemingly normative measures of characteristics we might reliably expect the other to exhibit, especially as a member of a designated social group. Perception of the other is interpreted through a standardized stock of knowledge that is acquired, for the most part, by enculturation and that supplies the interpretational patterns for making sense of the other.[4] The stock of knowledge organizes our perception of the other according to schemes of typification that classify others in terms of social roles, kinship practices, etc.[5] We select, from a variety of possible typifications, the scheme of typification that is most relevant to the problem at hand.[6] The stock of knowledge is always in flux. Yet, although schemes of typification may change in content and systems of relevance may shift, our perception of the other is always relative to general sociological patterns.

Tymieniecka, in contrast, regards intersubjectivity as the dynamic field in which we develop the subjective, or personal, meaning of our existence. Each individual creates an existential script of life meaningfulness that blends into the intersubjective interlocking of individual projects. We engage not in a process of "self-enclosed individualizing,"[7] but in a complex self-differentiation by which we call the social world into being. Through self-interpretation we break out of any strict opposition between ourselves and the lifeworld. Others enable us to expand our life meaningfulness and to constitute a common field, or social world, the existential arena in which our humanity unfolds.

Tymieniecka's approach to intersubjectivity emphasizes the evolutive process of self-interpretation. The explanatory concepts that are central to Tymieniecka's analysis of intersubjectivity, i.e., transaction, consent, and relevance, show how we shape our existence as living individuals immersed in the changing and unpredictable flux of social currents.[8] Transaction regards the social world as a place of evolving existential subjects and endeavors, through the development of pluri-individual interpretive systems, to generate new sources of meaning. Through transaction, individual interests coalesce into a series of socio-existential complexes which, in their totality, constitute the texture of the social world. The foundational network of transaction is marked by a mutual consent, or agreement, that recognizes the interests of others. Such consent is at once deliberate, rational, and subjective, sentimental. Intersubjective consent has as its goal the attainment of transactional relevance: the collectively shared aim to be achieved by the overlapping of inter-individual survival interests. The value of the other emerges amidst existential action by which we promote self-interest through mutual engagement in the social world.

The thou-interpretation of Schutz and the self-interpretation of Tymieniecka describe the value levels of the other in the social world. In thou-interpretation, we identify the social roles of the other through recognition of indexical patterns by which we are positioned in social interaction. In self-interpretation we discover the existential script of the other as a meaning giving individual in pursuit of inter-individual projects. Both thou-interpretation and self-interpretation insist upon and show the constitution of intersubjectivity as an *operative category* that pervades all action in the social world.

II. THE VALUE OF THE OTHER

At the common-sense level, human rights offer the most basic evidence that the other has value in the social world. Irrespective of the various interpretations of human rights, we commonly assume that there is a value intrinsic to human beings, a value that ought not be abrogated. In the common sense attitude we straight-forwardly assume that the other, like ourselves, has a certain dignity and the right to respect.

Human rights are constituted in the common-sense attitude through the process of thou-interpretation and self-interpretation. From the perspective of thou-interpretation, our external regard of the other, we see our self-worth intimately connected to the value of others and in safeguarding the value of others, we affirm our own value. From the standpoint of self-interpretation, the recognition that all persons are self-interpreters and have self-relevance points immediately to the rights and value of all individuals. The constitution of human rights and the value of the other pervades all our intersubjective dealings and gives grounds for specific ethical practices.

A philosophical inquiry into the seemingly automatic recognition, or assignment to the other of human rights shows that straight-forward description, by itself, is insufficient. A constitutive analysis of human rights and, especially, one that can recast the problem of intersubjectivity, necessitates the explicit development of a philosophical methodology that can identify relevant phenomena and thereby elicit how the value of the other is embedded in daily life. The constitution of the formal system of human rights in laws and institutions, which are often explicitly political in nature, must be articulated. The informal network of human rights, which focuses on the emergence of human rights through each individual's inner dynamic and spinning of projects, must also be analyzed.

A phenomenological framework for common-sense constitution of the formal system of human rights is developed by Schutz in "Equality and the Meaning Structure of the Social World."[9] Schutz' analysis of equality treats complexities in external meaning structures neither as metaphysical issues grounded in meta-ethics or epistemology, nor as mere matters of social praxis in which specific formulations of equality must be judged just or unjust, but as meaning structures constituted in the common-sense world of social interaction. On a Schutzian model, to speak of equality is already to assume that there is sufficient similarity in the value of others, and in the rights to which others are entitled, to warrant assessments of equality and inequality. Although Schutz himself does not make this point explicit, the social construction of equality makes sense only on the premise that individuals are equal in dignity.

Schutz' methodology emphasizes conflicts in the construction of equality and human rights and continues his approach primarily from the standpoint of thou-interpretation. The precise understanding of equality, for Schutz, is shaped within concrete social groups, each of which has distinct circuits of meaning and acceptable ways of enacting that meaning. The intersubjective world of the "in group" has its own common situation, common problems, typical means and ends. Similarly, members of the "out group" have their own central myths, common valuations, and typical resolutions. Members of the "out group" do not share the same world view as members of the "in group" and may constitute entirely different meanings for equality.[10] Domains of relevance are defined differently by each social group such that, Schutz maintains, the concept of equality is always relative. The United Nations Declaration of human rights, for instance, guarantees some forms of equality, but not the equality of material goods.

Issues of equality are measured by Schutz in terms of external standards of thou-interpretation, i.e., administrative and legislative measures, tax laws, etc.[11] Schutz' common-sense attitude determines both what is and what ought to be the value of others according to the existing systems of relevance that order the practices of a specific culture. The role of self-interpretation in regard to issues of equality is, in effect, presented by Schutz as reactive rather than creative, as limited to ensuring that one is not deprived of any of the rights or goods that one's social group accepts as standard.

The methodological approach developed by Tymieniecka in "The Moral Sense: A Discourse on the Phenomenological Foundation of the Social World and of Ethics"[12] brings to light the harmonious interweaving of individual actions in the human conditions. Like Schutz,

Tymieniecka presupposes the value of the other and the human rights of the other. Tymieniecka, like Schutz, also seeks to free moral experience as valuation from its direct submission to objective factors, focusing on the meaning structure of human rights rather than the formulation of specific criteria for judging concrete social and sociopolitical frameworks. In marked contrast to Schutz, however, Tymieniecka's phenomenological investigation of the informal network of human rights grounds the value of the other not in the common-sense but in the moral sense.

The moral sense and human rights are viewed by Tymieniecka from the perspective of self-interpretation, for which the lived, or existential significance of human rights is of utmost relevance. Human rights, Tymieniecka claims, delineate an experiential course of action, and "express the *existential significance of the Human Condition.*"[13] Human rights thereby function as self-evident axioms that underlie specific endeavors in the social world and that manifest "virtualities,"[14] or the potentiality, of the human condition.

Tymieniecka affirms that the condition of constitutive genesis of shared meaning lies in the moral sense, and not in the intellective operations of the transcendental ego.[15] The moral sense is the virtual giver of meaning in the field of intersubjectivity and pervades all aspects of our personal lives. As the virtual element of the human condition, the moral sense is present in and gives rise to the social world. The progressive unfolding of the moral sense expresses itself, as well, in societal movement toward intersubjective harmony.

III. ACTUAL AND VIRTUAL LEVELS OF INTERSUBJECTIVE CONSTITUTION

The philosophical status of the value of the other is contingent on whether that value is held to be an actual, or a virtual element of the human situation. Whereas the common-sense constitutes the value of the other in terms of what actually is, the moral sense constitutes the value of the other in terms of the virtuality of possibilities that might be. Schutz' constitutive analysis claims that the value of the other arises through the cognitive operation of the common-sense. The value of the other is typified by the common-sense attitude according to the actual systems of relevance in practice in a specific cultural group. Tymieniecka's constitutive analysis maintains, however, that the value of the other emerges through the unfolding of the moral sense, the constitutive element whose potentiality is expressed in the existential movement of each individual's life. The value of the other rests on a transcendental, non-normative ground that is manifest in the unpredictable progress of an individual's life script.

Whether the value of the other should be constituted at the actual or the virtual level is at once a matter of polemical debate and of philosophical discretion. Systems of relevance inevitably come into play in any response to questions such as whether constitution of the other through Schutz' common-sense leads necessarily to a static or closed philosophical system, or whether constitution of the other according to Tymieniecka's moral sense may entail unwarranted metaphysical presuppositions. A principle of philosophical discretion may be proposed in light of the endless character of such debate: to accept and develop whichever approach, or combination of approaches deals most adequately with the full spectrum of issues concerning intersubjectivity. While "adequacy," used in this sense, may suggest a philosophical criterion that is as much moral as epistemological, immediate disclamer of the possibility of a moral base to epistemological concerns may not be warranted. Indeed, throughout our recasting of the problem of intersubjectivity we have shown that value and epistemology may be more closely related than we often anticipate.

University of Southern Maine

NOTES

[1] Cf. Edmund Husserl, *Cartesian Meditations*, tran. D. Cairns (The Hague: Nijhoff, 1960), especially the "Fifth Meditation"; René Descartes, "Meditations," in *The Philosophical Works of Descartes*, trans. E. E. Haldane and G. R. T. Ross (Cambridge: University Press, 1972), pp. 131—200.

[2] Cf. Alfred Schutz, *The Phenomenology of the Social World*, trans. George Walsh and Frederick Lehnert (Evanston: Northwestern University Press, 1967), and "Common-Sense and Scientific Interpretation of Human Action," *Collected Papers*, vol. 1 (The Hague: Nijhoff, 1973), pp. 3—47; Anna-Teresa Tymieniecka, "The Phenomenology of Man and of the Human Condition in the Human Sciences," *Analecta Husserliana*, vol. 14, ed. A-T. Tymieniecka (Dordrecht: D. Reidel, 1983), pp. 21—50; and "The Moral Sense: A Discourse on the Phenomenological Foundation of the Social World and of Ethics," *Analecta Husserliana*,vol. 15, ed. A-T. Tymieniecka (Dordrecht: D. Reidel, 1983), pp. 3—78.

[3] Cf. Schutz, "Some Structures of the Life-World," in *Collected Papers* vol. 3 (The Hague: Nijhoff, 1970), pp. 116—132.

[4] Ibid.

[5] Cf. Schutz, "On Multiple Realities," *Collected Papers*, vol. 1, pp. 283–287.

[6] Ibid.

[7] Tymieniecka, "The Moral Sense," p. 54.

[8] Ibid., pp. 49–54.

[9] Schutz, "Equality and the Meaning Structure of the Social World," *Collected Papers*, vol. 2 (The Hague: Nijhoff, 1971), pp. 226–269.

[10] Ibid., pp. 243–245.

[11] Ibid., p. 255.

[12] Tymieniecka, *op. cit.*

[13] Ibid., p. 71.

[14] Ibid., p. 54.

[15] Ibid., p. 11.

JOSEPH J. KOCKELMANS

PHENOMENOLOGICAL CONCEPTIONS OF THE LIFE-WORLD[1]

I

Herbert Spiegelberg wrote in 1960 that "The most influential and suggestive idea that has come out of the study and edition of Husserl's unpublished manuscripts thus far is that of the *Lebenswelt* or world of lived experience."[2] Ten years later David Carr quoted the same statement with approval;[3] and in the same year Aron Gurwitsch joined Husserl in calling a phenomenology of the life-world a scientific task of the first order of importance.[4] I fully agree with these authors. Many phenomenologists today are convinced that the life-world notion is one of the most influential ideas of Husserl's later philosophy. This influence has been enormous and evidence for this can be found in a number of different areas of theoretical concern. We find the influence of Husserl's ideas about the life-world first of all in the realm of phenomenological philosophy proper. Suffice it here just to mention a few names of phenomenologists in whose works the life-world idea has found a positive reception: H. Conrad-Martius, J. Patocka, R. Ingarden, E. Fink, L. Landgrebe, J.-P. Sartre, M. Merleau-Ponty, G. Berger, S. Breton, A. Gurwitsch, H. Spiegelberg, and A. Schutz.

A second area in which Husserl's conception of the life-world appeared to be extremely fruitful and productive is the realm of philosophical anthropology. Here too a great number of authors can be mentioned, such as G. Brand, C. van Peursen, W. Keller, A-T. Tymieniecka, N. Mohanty, J. Sinha, S. Strasser, R. Kwant, and W. Luypen.

A third domain where this influence has been considerable is obviously empirical psychology and psychiatry, where the names of E. Minkowski, E. Straus, J.-P. Sartre, F. Buytendijk, J. van den Berg, J. Linschoten, C. Graumann, and A. Gurwitsch immediately suggest themselves.

Another important area in which Husserl's ideas about the life-world appeared to be of prime importance is that of the social sciences. Here

339

A-T. Tymieniecka (ed.), Analecta Husserliana, Vol. XX, 339–355.

the works of Schutz, Berger, Luckmann, Natanson, and many others can be mentioned.

Finally, Husserl's ideas about the life-world have also been applied in reflections on the philosophy of science (A. Gurwitsch, T. Kisiel, and myself), in investigations concerned with the historical disciplines (Hohl and Carr), and in literary criticism (G. Poulet, R. Ingarden, and the early Miller).[5]

Yet the notion of the life-world was affected by serious difficulties from the very beginning. Some of these problems are intimately connected with Husserl's conception of transcendental phenomenology and its relation to the so-called "mundane" phenomenology of the life-world, difficulties which led Scheler, Hartmann, Heidegger, Sartre, Merleau-Ponty, Ingarden, Fink, De Waelhens, and many others to develop other conceptions of phenomenological philosophy.[6] Other difficulties are connected with the precise function of the life-world itself within Husserl's transcendental phenomenology as a whole. Does the life-world doctrine of Husserl's later works open up new paths of investigation, or is it only a new and clearer outline of the programmatic intentions which his phenomenology had from the very beginning?[7]

Then there are the difficulties which result from the fact that the concept of the world in general and that of the world of perception in particular, which had already played an important part in Husserl's earlier conception of phenomenology, somehow are to be related to the life-world of the later works, as well as from the fact that in the twenties Husserl had introduced still another concept of world, namely that of "the world of immediate experience." What is particularly confusing in this case is that it is not totally clear how the world of immediate experience is to be related to the world of perception on the one hand, and to the life-world of the later works on the other.[8] Another set of difficulties is connected with the fact that during the last years of his life Husserl seems to have been using the concept of the life-world in more than one meaning, so that there also is a serious ambiguity in Husserl's notion of the life-world itself.[9] Then there are also difficulties connected with the fact that Husserl distinguishes between the life-world and certain *Sonderwelten;* it is not totally clear here whether the priority is to be given to these sub-worlds, or to the life-world as a whole.[10]

Finally, let us assume that Husserl indeed was correct in suggesting that a reduction of the world of the sciences to the life-world is necessary and, thus, that the bracketing of the scientific worlds can be justified by the validity of the life-world as an original dimension of self-givenness. Let us assume also that it is possible to develop an eidetic science of the life-world which can give a foundation to all other sciences, including logic, on the ground that in its basic validity it precedes every other science. Then it seems to follow that this "ontology of the life-world" is really the new fundament of all truth so that reflections on the function of the transcendental subjectivity become superfluous. This is so particularly in view of the fact that the world-constituting ego itself is an integral part of the life-world itself.[11]

In making these remarks it was not my intention to suggest that these and other difficulties cannot be resolved from the perspective of Husserl's transcendental phenomenology. I merely wished to substantiate the thesis that the doctrine of the life-world from the very beginning was affected by problems and ambiguities, although Husserl may very well have been aware of at least some of these problems. Yet, even though these difficulties have been discussed in detail in the secondary literature, some of the problems still remain, and ambiguity often still prevails. Before attempting to discuss some of the issues critically I shall first attempt once more to reconstruct Husserl's own development in regard to the notion of world, because I am convinced that the very complexity of this development is in part the "cause" of much of the ambiguity and confusion.

II

Although the idea of the life-world had occupied Husserl in one form or another at least since 1920,[12] Husserl had decided not to communicate his insights concerning the life-world as long as he was still unable to adequately solve the basic problems involved. When toward the end of his life he became convinced that he had succeeded in answering the most important questions which the life-world poses to the philosopher, he made the life-world the central theme of the first part of *The Crisis*. This part of the work appeared in 1936 as a first installment in the journal *Philosophia*. Husserl's ideas, however, did not become known widely until Landgrebe (in *Experience and Judgment* and in articles) and Merleau-Ponty (in *Phenomenology of Perception*) had introduced them in their publications on the basis of their knowledge of unpublished manuscripts.[13]

These publications at first created the impression that during the last years of his life Husserl had decided to replace his original studies of the transcendental subjectivity with new investigations concerning the life-world. Nothing, however, was farther from the truth. When the remainder of the incomplete work was edited by Walter Biemel in 1954, it appeared that the investigations about the life-world in *The Crisis* form only one of the four different ways in which the constituting activity of the transcendental subjectivity can be brought to light. In other words, the world of our immediate experiences is not the last level which phenomenological analyses can uncover; this world itself is also constituted, and the clarification of its constitution must precisely discover the anonymous functioning achievements of the transcendental subjectivity.[14] This is already quite evident from the fact that the title of the section in which Husserl deals with the life-world in *The Crisis,* explicitly speaks of the way into transcendental phenomenology from an inquiry into the life-world.[15]

Furthermore, one must not forget that Husserl's conception of the life-world as found in *The Crisis* was really the "logical" outcome of his earlier investigations concerning the world of immediate experience, the first traces of which can be found already in manuscripts which date from 1905. Originally these investigations were conducted in two different directions: first, in connection with the problem of the phenomenological reduction;[16] secondly, in the course of inquiries meant to clarify the very essence of perception.[17] Later, trying to lay the foundations of the general theory of science, which was already anticipated in the first volume of *Ideas,*[18] and to delineate the subject matter of psychology in particular, Husserl focused again on the world just as it is given to us in immediate experience.[19] Only in the latter context do we find the motives which, in the period beginning about 1920, led Husserl to the problem of the life-world as we now find it in *The Crisis.* Seen from this perspective Husserl's conception of the life-world as it is presented in *The Crisis,* appears as the harmonious synthesis of his view on the phenomenological reduction found in *First Philosophy* and *Cartesian Meditations* on the one hand, and his so-called "mundane phenomenology of the world of immediate experience," which was outlined briefly for the first time in *Phenomenological Psychology,* on the other.

Although it is my intention in the following reflections to focus on the most important insights in regard to the life-world as Husserl describes

them in *The Crisis,* I wish first to make a few brief comments on the
most significant results of Husserl's investigations concerning the everyday world, the world of perception, and the world of immediate
experience.

1. *The Everyday World of Common Experience and the*
   Phenomenological Reduction

Husserl usually describes the phenomenological reduction as a disconnection of the so-called general thesis of our natural attitude. According
to Husserl one must make a radical distinction between the natural and
the philosophical attitudes and, thus, correspondingly between the non-
philosophical and the philosophical sciences. One of the most
characteristic features of the non-philosophical sciences is the fact that
they do not basically question the fundamental thesis underlying our
natural attitude. They deliberately put aside all skepticism and every
kind of critique of knowledge in order to describe and explain the "facts
themselves" as accurately and objectively as possible as they immediately manifest themselves to us. Philosophy, on the other hand, cannot
presuppose anything whatsoever and, therefore, is obliged to pursue
critical inquiries into the very possibility of human knowledge and
science.[20] Thus on the one side, we have the sciences of the dogmatic
attitude, which are oriented to the things themselves and are not
concerned with epistemological problems. On the other side, we have
the sciences of the specifically philosophical attitude. They explicitly try
to consider all the skeptical problems concerning the possibility and the
limits of our knowledge and of science. The former remain in the realm
of the natural attitude; the latter bracket the basic thesis underlying this
attitude by means of a special phenomenological reduction.

According to Husserl human beings generally and ordinarily live in
the natural attitude. In this attitude they find continuously present to
them and standing over against them the one spatio-temporal fact-world
to which they themselves belong, also. As members of this "community
of people" they find this fact-world to be out there; and they also take it
always just as it gives itself to them as something that exists out there.
Any doubt, negation, or uncertainty of particular data of the natural
world does not affect the general thesis of the natural standpoint,
namely that there is a real world out there that basically and in principle
at least is in the manner in which it gives itself to them in their

immediate experiences. If they wish to know more about the world and if they wish to know it more truly and comprehensively than their everyday experiences enable them to do, and if they are to solve all the problems which manifest themselves on this basis, they have to turn to the sciences. But these, too, continue to presuppose the general thesis of the natural attitude. If they wish to become philosophers, they must bracket this general thesis of the natural attitude and turn their attention to the realm of phenomena immediately given to "pure and trancendental consciousness" which constitutes the "phenomenological residuum" which the reduction brings to the fore.

Thus in this way, while developing the doctrine of the phenomenological reduction, Husserl was led to a first conception of the world: the world is the all-encompassing doxic basis that as an all-encompassing horizon includes every particular subject matter of all our positing activities.[21]

2. The World of Perception

In his analyses of perceptual acts Husserl was again led to the world as a phenomenological theme. For analyses of the various syntheses in which the perception of material things comes into being show that one cannot restrict himself to the perception of a particular thing as an isolated phenomenon, if he intends to discover the concrete and full meaning of the thing perceived. The perceived thing always manifests itself in a certain (outer) horizon, in a background of things which are consciously and more or less explicitly "meant" along with the particular thing in question. Every material thing appears to manifest itself in perception as having its outer horizon which is not only spatially, but also temporally extended indefinitely. When I perceive a tree I always co-perceive a number of other things which are equally there perceptually in the field of perception. But as long as I am turned to the tree I do not explicitly apprehend them. They appeared, but they are not concretely singled out. Every act of perception has such a zone of background "intuitions" which together constitute my consciousness of all that which as a matter of fact lies in the co-perceived background. Taken in its full concretion as correlate of our concrete perceptual life, this "spatial" horizon is our surrounding world. In this surrounding world in which we live, open possibilities of further experience manifest themselves and thus refer us to ever wider horizons. That is why we experience our surrounding world as part of "the" world, that part, namely, which is now directly accessible to us.

But this world does not only possess a spatial order; it has also a temporal order and a horizon in the succession of time. The world now present to me has its own temporal horizon which extends unendingly in two directions, its known and unknown past and future.[22]

3. The World of Immediate Experience

These ideas occupied Husserl again in the twenties when he began to look for a deeper insight into the relationship between phenomenology and psychology on the one hand, and the problem of the transcendental reduction, on the other. The first concern would lead to his conception about a "general ontology of the world of our immediate experience," whereas the second concern would lead to the conception of the life-world as found in *The Crisis*.

In 1925 and subsequent years in a series of lectures, entitled *Phenomenological Psychology*, Husserl tried to delineate the subject matter, methods, and function of a new eidetic science, called "phenomenological psychology," and especially to determine its relation to empirical psychology on the one hand and to transcendental phenomenology on the other. Now it appeared that it is impossible to define the "pure psychical" which is the subject matter of phenomenological psychology, if one does not take his starting point first in a general ontology of the world of our immediate experience.[23]

In the brief outline of such a general ontology of the world of immediate experience, Husserl deals first with the problem, already raised by Kant, according to which the world can never be an object of any possible experience. Husserl admits that obviously only individual entities are immediately experienced; yet he maintains again that in these experiences which indeed immediately relate to individual things, the world is co-experienced as their necessary horizon. Here Husserl thus extends his view first developed for perceptual experiences to all other immediate experiences.

The "general ontology of the world of immediate experience" must try to bring to light the essential structures without which such a world could not possibly be what in fact it appears to be. In so doing the investigator must make use of the method of free variation. The descriptive and eidetic analyses of the most general structures of the world which are to be performed with the help of the method of free variation, must take their starting point in the consideration of exemplary, real, individual things which must be taken in their essential relations to the horizon which in fact is the world.[24]

When we, while studying these phenomena, make use of the method of free variation to focus attention on the essential structure of experienced beings as such, it becomes clear soon that what we usually call "the" world, is not the world as it is immediately experienced at all; this world is rather always already covered over with layers of meaning which have their origin in our conceptions and opinions which, themselves, are the results of our theoretical and practical activities. On closer investigation it becomes clear that our contemporary view on the world is especially influenced by two factors: the insights made available by the social sciences and the conviction that our scientific conception of the world is the true conception. In Husserl's opinion, it is not difficult to show that the determinations given to the world on the basis of the accomplishments of the sciences, cannot be considered to be essential structures which necessarily belong to the world of immediate experience as such. On the contrary, one must turn from the world as it always is already there for us, with its layers of meaning attributed to it by our tradition and particularly by the sciences, and return to the world as it manifests itself to us prior to the sciences, i.e., the immediately lived world with its original givenness which is the underlying basis for every scientific determination. Correlatively we must go back from our scientific view on the world to our pre-scientific experiences of the world.[25] The danger flowing from the fact that our world is only one among many worlds can be overcome by applying the method of free variation, which helps us to focus our attention not on what is specific for any given world, but on the invariant structures that must be found in any world which is livable for human beings.

Now to the general structure which necessarily belongs to the very essence of our and every conceivable world there belongs the distinction between different regions of being, of which the regions of living and non-living beings are the most striking examples. Furthermore, the region of consciously living *subjects* appears to occupy a privileged and central place in the world of our immediate experiences. Although every concrete substance in the world possesses necessarily and essentially spatial, temporal, and causal aspects, the consciously living subjects manifest themselves as, in addition to these characteristics, having certain traits all of which belong to the order of the "psychical."[26] Phenomenological psychology is the regional ontology which concerns itself with the essential and invariant structures of the psychical as such.

4. The Life-World in Husserl's "Crisis"

During the last ten years of his life Husserl realized that phenomenological investigations of the world in which we live can form an excellent introduction to the fundamental problems of transcendental phenomenology. For reflections on the life-world appear to prepare the phenomenologist in an eminent way for the study of the universal and transcendental constitutions of transcendental subjectivity. This idea had already occurred to Husserl in the investigations which were to lead to the lecture course "Phenomenological Psychology," but it had not yet been developed there in a systematic fashion. Returning to the idea in the early thirties Husserl wrote in *The Crisis* that under the influence of modern science since the time of Galileo, the life-world, i.e., the "immediately intuited world," has been replaced by the objective world of the sciences which, in the opinion of most Western scientists, is to be taken as the "true" world, so that the totality of all the objects which are, or at least can be, studied by these sciences constitute "reality" in the strict sense of the term. In Husserl's view, it is not too difficult to show that such a view is completely unacceptable. For first of all, no object of any science is available to direct and immediate experience (perception), whereas the things and events of the life-world offer themselves, actually or at least potentially, in direct experiences.

Furthermore, the universe of the sciences proves to be a network of ideal constructs, a theoretico-logical superstructure; its apprehension and conception, thus, are of the same nature as those of the world of the mathematical sciences. For the construction of the world of the empirical sciences, too, implies procedures on the part of man in which idealization plays an essential part. Idealization, however, presupposes that there is something that can be given directly and which can then be idealized. Therefore, because of its intrinsic meaning as a superstructure, the world of the empirical sciences must have a firm basis upon which it rests and upon which it is built. This foundation cannot be anything other than the life-world and the immediate evidence of our lived experiences, i.e., experiences in which the things and events themselves immediately present themselves as such. Thus all scientific truth, whether logical, mathematical, or empirical, has its final justification in evidences which concern events and happenings in the life-world.[27]

In *The Crisis* Husserl thus first explains in what sense the scientific conceptions of the world necessarily presuppose the life-world. He then sets out to develop an outline of a general ontology of the life-world, while paying special attention to the typical reduction involved in the development of such an ontology as well as to the methods to be used. Husserl stresses here that the life-world is not a purely theoretical world, but rather the concrete world in which we live; thus the life-world encompasses the intentional correlates of all human activities which we perform as individuals and as members of a community. Husserl also tries to establish the precise relationship between the life-world and the world of the sciences. Finally, he attempts to show that the life-world itself is the product of the constituting activities of the transcendental subjectivity.

The life-world itself is thus by no means immediately accessible as such, since everyone has already undergone the influence of his culture, and because Western man particularly is deeply influenced by the scientific interpretation of the world. A special reduction, a temporary suspension of our culture and of the sciences, is therefore indispensable if one is to uncover the life-world itself and its essential structures. Once we have performed this reduction, we are then in a position to study the life-world itself in the ontology of the life world which has the form of a "mundane," i.e., a non-transcendental, phenomenology.

Toward the end of his life Husserl thus believed that the "mundane" phenomenology of the life-world is an important propaedeutic for transcendental phenomenology. For, on the other hand, it is evident that the analyses of "mundane" phenomenology itself have no final meaning if they are not complemented by analyses in the transcendental sphere toward which the transcendental reduction gives us the proper access. But, on the other hand, only after the inquiries of "mundane" phenomenology have been completed does the transcendental reduction have a sound basis and a proper guide. Thus we must first return from the world of culture and science to the original life-world by means of a first reduction; then the transcendental reduction must lead us further back from the structures of the life-world to the hidden achievements of the "functioning intentionalities" of the transcendental subjectivity. The systematic discovery of these achievements and their correlates allows us then to describe the original constitution of the characteristic features of the life-world as such and all the objective structures later based upon them.[28]

In my book *Edmund Husserl's Phenomenological Psychology* (1967)[29] I compared the life-world with its correlative mundane phenomenology of *The Crisis* with the world of our immediate experiences and the ontology of the world of immediate experience of *Phenomenological Psychology* and stated there that there is a perfect correspondence between these two. I concluded these observations with the claim that what Husserl in *The Crisis* calls the "original life-world" is doubtless what he earlier in *Phenomenological Psychology* had referred to as the world of our immediate experience, but that this observation should not blind us to the fact that the formulation of *The Crisis* is by far the more comprehensive and desirable description of the world. In 1970 Carr[30] correctly pointed out that these claims as they stand are unacceptable in that in *Phenomenological Psychology* the world of immediate experience is understood not as a social, historical, and cultural world, but as the world of strictly perceived things. What I should have claimed was that, like Gurwitsch, I had come to the conclusion that what the expression "the world of immediate experience" in *Phenomenological Psychology* stands for is still an integral *part* of what Husserl later would call the *eidos* life-world. One can then say that the formulation of *The Crisis* is the more comprehensive one; it is also the more desirable one precisely because it takes the socio-historical aspect of the life-world seriously.

III

I started this paper with the remark that many phenomenologists today are convinced that the life-world idea is one of the most influential ones that has come out of the study of Husserl's unpublished manuscripts. Yet it should be noted also that although this claim is undoubtedly true, nevertheless virtually nobody has followed Husserl's concern with the life-world without introducing basic and essential modifications in his original conception. Of those who stayed most closely to Husserl's own position, Gurwitsch in my opinion must be mentioned specifically. In *The Field of Consciousness* Gurwitsch maintained Husserl's transcendental phenomenology in most of its essential elements. On the basis of this philosophical framework he then developed a consistent theory of perception which again remains faithful to Husserl's own position. He finally employed the insights so gained to develop his own conception of the organization of consciousness, as found in modern psychology in general and in Gestalt psychology in particular, into a consistent phenomenological theory of the field of consciousness and its basic structure of organization: theme, thematic field, and margin. Yet already in this book Gurwitsch pointed to aspects of Husserl's transcendental phenomenology which were unacceptable to him (the *hulē-morphē* doctrine of *Ideas,* the egological conception of consciousness, and certain aspects of Husserl's conception of intentionality and attention).[31]

Virtually all other authors mentioned rejected Husserl's transcendental idealism. Some authors seem to have been of the opinion that a strictly "mundane" phenomenology is philosophically adequate. Most authors, however, have realized that a strictly mundane phenomenology is philosophically without any foundation or justification. Some have therefore attempted to combine phenomenology with some form of metaphysics (Ingarden, Scheler, Hartmann, Sartre). Others have tried to give a transcendental foundation to phenomenology by means of some form of hermeneutic philosophy (Heidegger, Gadamer, Ricoeur). I plan to return to this attempt shortly.

Some authors, partly inspired by both Husserl and Heidegger, have thought that one can avoid the dilemma of Husserl's transcendental idealism and Heidegger's hermeneutic phenomenology, by giving the reflections on the life-world a foundation in some form of philosophical anthropology.[32] Yet it seems to me that Schrag recently has shown convincingly that such an approach is not very productive in that a philosophical anthropology itself is without ultimate justification as long as one refrains from what Schrag calls "radical reflection," i.e., reflections of the kind in which both Husserl and Heidegger precisely were engaged.[33]

But even if we do not focus primarily on underlying philosophical difficulties, but instead turn to Husserl's own analyses of the life-world, we find very few scholars who have accepted Husserl's analyses without major modifications. Marx has defended the view that there are serious difficulties connected with Husserl's attempt to relate the various sub-worlds, which constitute the regions with which the various regional ontologies and the corresponding empirical sciences concern themselves, to the concept of the life-world itself.[34]

Furthermore, Carr has shown convincingly, I think, that in Husserl's later philosophy the term "life-world" is used to refer to phenomena which really belong to different strata, and that Husserl failed to properly relate the world of culture, the world of our immediate experience, and the worlds of the sciences.[35] Finally, Gadamer has pointed to the fact that Husserl fails to relate theory and practice in an acceptable manner, and continues to subscribe to the illusion that from science rational decisions can be derived which then can constitute a universal practice.[36]

Finally, if we limit our reflections to the contributions which Husserl made to the human sciences, it must be said here too that although a mere description of the universal and invariable structures of the life-world, of every possible life-world, of the *eidos* life-world, may perhaps be helpful for the formulation of a broad, a priori framework of meaning from which psychological and social phenomena can be studied empirically, such a framework of meaning nevertheless is obviously totally inadequate for the explanation and understanding of these phenomena. Thus even in the realm of the human sciences Husserl's theory of the life-world can only play a relatively limited part. In addition to descriptive analyses of the life-world, interpretative and critical analyses of concrete life-worlds must be developed if the scientists concerned with human phenomena are to formulate meaningful and relevant problems and give equally meaningful and relevant explanations for them.[37]

In view of the fact that Heidegger's interpretation of phenomenology has had a deep influence on the manner in which many other thinkers such as Sartre, Merleau-Ponty, Fink, Breton, Gusdorf, van Peursen, Keller, Strasser, and many others, have tried to re-interpret Husserl's transcendental phenomenology, I wish to conclude these reflections with a brief comparison of Husserl's conception of the life-world with Heidegger's description of the world which is constitutive of the very Being of man as *Dasein*.[38]

Both Husserl and Heidegger appear to agree that the basic concern of philosophy consists in the attempt to answer the question concerning the meaning and Being of being. Both also agree that in trying to achieve this goal the philosopher should not be concerned primarily with the ultimate cause of all finite things, but rather with answering the question of *how* the Being of beings and the Being of the world are to be constituted. Finally, both agree that in answering the question concerning the meaning of Being a privileged position is to be given to the subjectivity, i.e., that being which asks the question concerning the Being of beings. "We agree that being taken in the sense of what you call 'world' cannot be clarified in its transcendental constitution by means of a return to a being of the same kind."[39] The disagreement between Husserl and Heidegger is to be found mainly in their different views of the meaning and function of the transcendental reduction. Heidegger denies the possibility of this reduction, whereas in Husserl's opinion there can be no philosophy without it. Referring to Heidegger's re-interpretation of phenomenology Husserl wrote: ". . . one misunderstands my phenomenology backwards from a level which it was its very purpose to overcome; in other words, . . . one has failed to understand the fundamental novelty of the phenomenological reduction and hence the progress from mundane subjectivity (i.e. man) to transcendental subjectivity; consequently . . . one has remained stuck in an anthropology . . . which according to my doctrine has not yet reached the genuine philosophical level, and whose interpretation as a philosophy means a lapse into 'transcendental anthropologism,' that is 'psychologism.' "[40]

Many other points of disagreement are intimately connected with this basic difference: disagreement about the idea that philosophy is to be a strict and rigorous science, about the question of whether or not philosophy can reach apodictic evidences, about the interpretation of the meaning of immanence and transcendence, etc. In all these issues Heidegger fundamentally agrees and equally fundamentally disagrees with Husserl. The source of the basic disagreement between the two authors consists here in the fact that the root of all philosophical truth (which for both authors is to be found in the transcendental subjectivity) consists for Husserl in a subjectivity which originally is world-less, whereas for Heidegger the "ultimate" starting point of philosophy is to be found in man as Being-in-the-world. Husserl determines his conception of the subjectivity by the demands of the transcendental reduction; Heidegger argues that a genuine understanding of the human subjectivity makes such a reduction impossible. ". . . what is the mode of Being of that being in which the 'world' becomes constituted? That is the central problem of *Being and Time,* that is, of a fundamental ontology of *Dasein.* One must try to show that the mode of Being of human *Dasein* is completely different from that of all other beings and that this mode of Being, as that which it in fact is, precisely contains the possibility of the transcendental constitution. Transcendental constitution is a central possibility for the ek-sistence of the *factical* self. Thus the *concrete* person is as such (that is as a being) never a 'mundane real fact,' since man is never merely present-at-hand, but rather ek-sists. And the 'marvel' consists in this that the understanding of Dasein's ek-sistence makes the transcendental constitution of everything which is possible, feasible."[41] Thus for Heidegger the world is obviously constituted, but in constituting the world man always finds the world already there, because it is a constitutive component of his own mode of Being of which he always already has a pre-ontological understanding. This is the reason why every attempt to let things manifest themselves from themselves and in themselves on this level must take the concrete form of a hermeneutic phenomenology and critique.[42]

The Pennsylvania State University

NOTES

[1] An earlier draft of this paper was presented at the Program of The World Phenomenology Institute held at the annual meeting of the Eastern Division of the American Philosophical Association, Boston, in December of 1980.

[2] Herbert Spiegelberg, *The Phenomenological Movement. A Historical Introduction*, 2 vols. (The Hague: Nijhoff, 1960), vol. I, p. 159.

[3] David Carr, "Husserl's Problematic Concept of the Life-World," in Frederick Elliston and Peter McCormick (eds.), *Husserl. Expositions and Appraisals* (Notre Dame: University of Notre Dame Press, 1977), 202—212, p. 202.

[4] Aron Gurwitsch, "Problems of the Life-World," in Maurice Natanson (ed.), *Phenomenology and Social Reality* (The Hague: Nijhoff, 1970), 35—61, p. 35.

[5] For bibliographical details of the publications of the authors quoted here, see: Peter McCormick and Frederick Elliston (eds.), *Husserl. Shorter Works* (Notre Dame: University of Notre Dame Press, 1981), pp. 381—430.

[6] Cf. Joseph J. Kockelmans, *Phenomenology. The Philosophy of Edmund Husserl and Its Interpretation* (Garden City: Doubleday, 1967), pp. 183—193, 221—236, and passim. See also Spiegelberg, *The Phenomenological Movement*, passim.

[7] Hans-Georg Gadamer, *Philosophical Hermeneutics*, trans. David E. Linge (Berkeley: University of California Press, 1976), p. 182.

[8] Cf. Joseph J. Kockelmans, *A First Introduction to Husserl's Phenomenology* (Pittsburgh: Duquesne University Press, 1967), pp. 250—280, 281-314; *Edmund Husserl's Phenomenological Psychology* (Pittsburgh: Duquesne University Press, 1967), pp. 161—177, 264-301. See also the articles by Gurwitsch and Carr quoted above.

[9] David Carr, *art. cit.,* pp. 203-211.

[10] Werner Marx, "The Life-World and the Particular Sub-Worlds," in Maurice Natanson (ed.), *Phenomenology and Social Reality,* pp. 62—72. Cf. also *Vernunft und Welt. Zwischen Tradition und anderem Anfang* (The Hague: Nijhoff, 1970), pp. 45—77.

[11] Hans-Georg Gadamer, *op. cit.*, pp. 190—195.

[12] Hubert Hohl, *Lebenswelt und Geschichte. Grundzüge der Spätphilosophie E. Husserls* (Freiburg: Karl Alber, 1962), pp. 24ff.

[13] Herbert Spiegelberg, *op. cit.*, vol. I, p. 159.

[14] Walter Biemel, "Les phases décisives dans le developpement de la philosophie de Husserl," in *Husserl*. Actes du troisième Colloque international de phénoménologie. Royaumont: 23 au 30 avril 1957. Ed. M.-A. Bera (Paris: Minuit, 1959)., pp. 32—62; cf. also *Ibid.*, p. 65 and p. 85.

[15] Edmund Husserl, *The Crisis of European Sciences and Transcendental Phenomenology. An Introduction to Phenomenological Philosophy*, trans. David Carr (Evanston: Northwestern University Press, 1970), p. 103.

[16] Edmund Husserl, *The Idea of Phenomenology*, trans. William P. Alston and George Nakhnikian (The Hague: Nijhoff, 1964), pp. 13—21.

[17] Edmund Husserl, *Ideas. General Introduction to Pure Phenomenology*, trans. W. R. Boyce Gibson (New York: Collier Books, 1962), pp. 91—100, 105—107, 113—139, 218—219, and passim.

[18] *Ibid.*, pp. 45—71.

[19] Edmund Husserl, *Phänomenologische Psychologie*, ed. Walter Biemel (The Hague: Nijhoff, 1962), pp. 64—105 (passim).

[20] Edmund Husserl, *The Idea*, pp. 13—21.

[21] Ludwig Landgrebe, "The World as Phenomenological Problem," in *Philosophy and Phenomenological Research* 1 (1939) 39—53. Cf. Joseph J. Kockelmans, *A First Introduction*, pp. 261—262.

[22] Edmund Husserl, *Ideas*, pp. 105—107. Cf. Joseph J. Kockelmans, *A First Introduction*, pp. 262—263.

[23] Edmund Husserl, *Phänomenologische Psychologie*, pp. 55—64.

[24] *Ibid.*, pp. 93—97.

[25] *Ibid.*, pp. 52—57.

[26] *Ibid.*, pp. 99—100. Cf. Joseph J. Kockelmans, *A First Introduction*, pp. 263—267.

[27] Edmund Husserl, *The Crisis*, pp. 48—53, 103—114, 121—135, 142—143, and passim.

[28] *Ibid.*, pp. 123—143. Cf. Spiegelberg, *op. cit.*, vol. I, pp. 160—162; Aron Gurwitsch, "The Last Work of Edmund Husserl," in *Philosophy and Phenomenological Research*, 16 (1955—1956), 370—399.

[29] Joseph J. Kockelmans, *Edmund Husserl's Phenomenological Psychology*, p. 288.

[30] David Carr, *art. cit.*, p. 212, note 25.

[31] Aron Gurwitsch, "Problems of the Life-World"; cf. *The Field of Consciousness* (Pittsburgh: Duquesne University Press, 1964).

[32] This is the case not only for some followers of Scheler, but also for many "existential phenomenologists" who reject the metaphysical position of Sartre. A similar view is sometimes defended by phenomenologists who reject Husserl's transcendental idealism. As for the latter see Alfred Schutz, *Collected Papers*, vol. I: *The Problem of Social Reality*, ed. M. Natanson (The Hague: Nijhoff, 1967), pp. 97—203 (passim).

[33] Calvin O. Schrag, *Radical Reflection and the Origin of the Human Sciences* (West Lafayette, Ind.: Purdue University Press, 1980), pp. 29—51, and passim.

[34] Werner Marx, "The Life-World and the Particular Sub-World," *loc. cit.*, pp. 62—72.

[35] David Carr, *art. cit.*, pp. 202−212.

[36] Hans-Georg Gadamer, *Philosophical Hermeneutics*, p. 196.

[37] Cf. Joseph J. Kockelmans, "Some Reflections on the Meaning and Function of Interpretative Sociology," in *Tijdschrift voor Filosofie*, 43 (1980), pp. 294−324.

[38] Martin Heidegger, *Being and Time*, trans. John Macquarrie and Edward Robinson (London: SCM Press, 1962), pp. 91−148.

[39] Edmund Husserl, *Phänomenologische Psychologie*, p. 601 (letter of Heidegger to Husserl).

[40] Edmund Husserl, *Ideen zu einer reinen Phänomenologie und phänomenologischen Philosophie*, vol. III: *Die Phänomenologie und die Fundamente der Wissenschaften*, ed. Marly Biemel (The Hague: Nijhoff, 1971), p. 140.

[41] Edmund Husserl, *Phänomenologische Psychologie*, pp. 601−602.

[42] Cf. my essay "World-Constitution. Reflections on Husserl's Transcendental Idealism," in *Analecta Husserliana*, ed. Tymieniecka, 1 (1971), 11−35, pp. 31−33.

ALEKSANDER GELLA

CONTROVERSIES ABOUT HUMANISM IN SOCIOLOGY

The problem of humanism appeared late in the development of sociology. The founders of sociology — Saint Simon and August Comte, as well as their antagonist Herbert Spencer — despite all the other differences that divided them, were so deeply impressed by the nineteenth-century achievements of the natural sciences that they had no hesitation in applying the models and methods of the natural sciences to the new field of investigation: sociology.

First, a small digression concerning the terms "sociology" and "humanism." The first was coined by August Comte. By itself the very terminology reflects a departure from traditional humanistic training. No nineteenth-century intellectual with a humanistic educational background would dare create a new term by borrowing one part from Latin and a second from Greek. Perhaps in the future a community of people living on space stations and speaking only in Esperanto will call their studies of societal problems "spolechnoy-science" (from Russian and English), or "scientpolechnoy" (from English and Russian). Such terms would present their author's ignorance of historical languages in the same extent that August Comte (an alumni of École Polytechnique), was insensitive to the usage of Latin and Greek in one term.

The term "humanism" was coined by F. I. Niethammer in 1808,[1] and since then it has been commonly used to designate the theoretical and practical postulate of education, based on the studies of Greco-Roman culture and languages. However, the concept itself is sometimes traced back to pre-Socratic times. The religious center in Delphi "joined the elements of universal morality, sympathy, and compassion, with the aristocratic pride and self-consciousness of its own value."[2] The ancient Greeks realized the universal laws of human nature and, therefore, not individualism but humanism was their intellectual principle. This principle was further refined under the pen of Varro and Cicero.

During the formative period of sociology there was no room for humanistic reflection on either the methodological or the subjective level. Sociological interests were reduced to utilitarian problems studied with a positivistic concern for exactness and precision regarding answer-

A-T. Tymieniecka (ed.), Analecta Husserliana, Vol. XX, 357—367.

able questions only. This development in sociology marked a significant break with the Greek roots of the Western epistemological tradition. Humanism had no place in sociology until Wilhelm Dilthey launched his vehement criticism of the new discipline which had declared itself a scientific study of human affairs while omitted the most essential elements of social life, that is, those which did not exist in the natural sciences.[3]

Dilthey understood the damaging results of the divorce which this new discipline, fighting for the status of a science, took from its closest sisters: history and philosophy. The effort of sociologists to win for sociology a citizenship within the realm of science crippled the study of human societies. The problem of humanism in sociology appeared under the impact of this criticism that arose from the so-called "German humanistic revolution in philosophy," started by Dilthey, Windelband, and Rickert.

The controversies around humanism exist, however, not only in sociology, but also in many other fields of modern sciences. There is not one but two basic concepts of humanism which must be clarified before any consideration of the fate of humanism in sociology.

Modern controversies about humanism, which have behind them twenty-five centuries of tradition, began with two opposite concepts of man: those of Protagoras and Socrates. The first proposed an anthropocentric and naturalistic humanism, while the second, influenced by ethical intellectualism, proposed the suppression of the human-animal and the elevation of man's life over nature. In the first case humanism is a philosophical attitude most simply defined by a slogan "man is the measure of all things." In the second case it is an intellectual discipline which demands the union of wisdom and morality, knowledge and virtue.

Protagorean man is a thinking being who, knowing that everything is relative, in a natural way prefers pleasure and relaxation over effort and pain. Socratic man, who assumes that there is a hierarchy of values, sees the virtue of united knowledge, justice, and courage as the highest human value. Thus this man exerts himself in order to achieve the higher values rather than mere pleasures.

In the ideas of these two thinkers are the roots of modern humanism. Though there are many varieties of contemporary humanism, they can be divided into three basic groups. One is anthropocentric and naturalistic mixed with the liberal political tradition of the Western world;

the second is also anthropocentric and naturalistic but is adapted to Marxist philosophy and under the name "socialist humanism" is used as a mere decoration in Communist politics. The third humanism stands in opposition to the first and second to the degree Socrates was to Protagoras. It is transcendent and nonnaturalistic. Its favorite conception of a human-being is not as a "thinking animal" but as a "human-person" who has to control the animal part of itself.

The third type of Western humanism found expression in the form of Christian humanism.[4] But one can find the attitudes which deserve to be included to the third type of humanism in other non-Christian parts of the world as well. For example, the effort of Confucius "to educate the self-controlled man who through a permanent self-improvement can achieve the sense of justice and respect for human dignity in himself and others,"[5] also deserves the name of humanism. Contemporary sociology, however, borrowed its social ethics from Protagoras and not from Socrates and the Christian tradition. How strange! Sociology deals with human societies that differ from animal societies by virtue of their cultural progress and development. However, progress through change is not achieveable without the transcendent attitude of at least some members of human groups. Thus, the sociologists of the naturalistic orientation have been unable to present a sound explanation of the cultural development of mankind.

The anthropocentric and naturalistic humanism is often treated as the liberation of men from religious burdens, and is supported by those organizations which are fighting against monotheistic religions. "An increasing number of Americans — in their capacity as parents, lawyers, teachers, judges" — wrote Thomas Molnar — "are worried today about the sudden appearance in schools, media and legislatures of terms like humanism, secular humanism, humanistic religion, autonomous values, self-creation, situation ethics, secular city, sensitivity training and so on."[6]

This anthropocentric, naturalistic fashionable fad plays a significant role in the development of sociology today. It penetrates successfully even Christian churches. The present abuse of the term humanism is dangerous because it affects the clarity of sociological analyses of current cultural developments. Under pressure of this fashion we are inclined to forget that the basic elements of humanism exist in the assumptions of all great monotheistic religions, and, therefore, these religions are defending the ideas of human rights, human dignity, and human brotherhood.[7] Among them Christianity seems to be the most human and humanized religion, if it is taken in its real and genuine sense.[8]

However, those who defend this view have to remember that transcendent, nonnaturalistic humanism does not need to be religious. Any effort to elevate the human being to a higher level of personality belongs to what we call humanism. In popular usage, however, humanism is often taken as an equivalent of humanitarianism. The efforts of humanitarians to make people happier, freer, or healthier can be seen as humanistic acts of persons who sacrifice their time and energy for these goals, but not when these goals are declared to be only for the actors' material and hedonistic benefit. In other words, if the humanitarian acts impose some moral discipline on the actor and by this help him to rise above his earthly cravings — they present humanistic phenomena.

In sociology, controversies about humanism developed on two levels. On one level they concern philosophical and anthropological problems of human nature, the origin of society, the evolution of the human race, the nature and character of moral phenomena, etc. On the other level they concern methodology.

While at its beginning the philosophical base of sociology was formed by positivism on the one hand, and British empiricism on the other hand, later on the main stream of sociological research was determined by pragmatism, neo-positivism, and behaviorism. These philosophical trends (if behaviorism can be generously treated as a "philosophy") determined both the subject matter of sociology and its methodology.

The subject matter and scope of interest of sociology had been highly limited by the naturalistic credo of empiricism and positivism. Thus, all phenomena which could not be reduced to, or in an evident way derived from, nature had to be rejected from sociological studies. As a result, all that which is uniquely human could not be taken into the workshop of the "scientific" sociologist.

Nevertheless, the temptation of rigorous application of natural science methods to sociology dominated the humanistic, or at least independent strictly sociological, approach. Most of contemporary sociologists blindly accepted B. F. Skinner's argument: "The methods of science have been enormously successful wherever they have been tried. Let us then apply them to human affairs."[9] However, he was unable to show us how, even with the methods of natural science, one can explain the acts of human cultural creativity.[10]

There is not one acceptable definition of sociology. Otto Neurath, a member of the Vienna Circle (who knew more about sociology than other members of the Circle), believed that physics was the only true science and that it produced statements that were controlled by "observational sentences." He therefore demanded that in sociology all "What appears in statements as 'mental,' 'personality,' or 'social,' must be expressible as something spatiotemporal or else vanish from science."[11]

However, those who took this direction limited sociological studies and rejected from the kingdom of science all historical and cultural disciplines that could not reduce their statements to Neurath's demand. Thus, he and his associates in the Vienna Circle, who were trying to present the unity of science as a "logical necessity," had to build up a modern myth of science as the only rational way of cognition. It was a radical answer to Dilthey and all those who defended humanistic cognition as equally valid with physical cognition. But, soon after, Carnap himself came to the conclusion that "at the present time, laws of psychology, and social science cannot be derived from those of biology and physics."[12]

At the same time, when sociologists influenced by philosophers of neo-positivism tried to form a scientific, empirical science of social behavior, some influential sociologists pointed out the impossibility of sociological development without taking into account those strictly human elements that have no room in the natural sciences. They stated that values, meanings, motivations, understandings, and the influences of historical events on the present, which were unnecessary in the natural sciences, were the most essential coefficients of sociology.

Controversies about methods had an impact on the scope of interest and the subject matter of sociology. The more independent they are from rigorous demands of the methodology of natural science, the larger, if not also deeper, is the scope of sociological interest. In other words, the attempts to develop sociology on the pattern of natural science hampered the expansion of its scope. Naturally, some methodological directives borrowed from biology and physics are helpful, but only insofar as they respect the fact that the aim of sociology is not only to explain a given social phenomena, as it is in the natural sciences, but also to understand them.

Max Weber was the first author who paid attention to sociology "as a science concerning itself with the interpretative understanding of social action and thereby with a causal explanation of its course and consequences."[13] In addition, the concept of causality — which according to Bertrand Russell should be erased from all philosophical dictionaries[14] — in sociology plays quite a different role in our studies of nature. Robert McIver, one of the few who occupied a humanistic position in American sociology, assumed that

> There is an essential difference from the standpoint of social causation between a paper flying before the wind and a man flying from a pursuing crowd. The paper knows no fear and the wind no hate, but without fear and hate the man would not fly nor the crowd pursue.[15]

Pragmatic-minded empiricists rejected this argument, and polemics have continued. The most rancorous enemies of humanistic sociology and other studies of cultural phenomena as a science with their own concepts and methods were philosophers of science from the neo-positivistic school. It is enough to recall Hempel, Popper, Neurath, and the youngest, Thomas Kuhn. Hempel rejected the possibility of explanation in historical research. Popper exercised a great influence on the acceptance of the methodology of physics in sociology by labeling those who opposed the methods of physics in sociology as "proto-naturalists." Kuhn contributed to the inferiority complex of social scientists in general, and sociology in particular, since he treated sociology as a "proto-science." Sociology did not fit his concept of science based on the idea of the "unity of science."

That naturalistic concepts and related postulates for sociology appeared in the works of the nineteenth-century founders of sociology — Comte, Spencer, Durkheim — is understood as a result of beliefs in the omnipotence of science and the dominant philosophies of that time. It is more difficult to explain how, after Dilthey, Windelband, and Rickert in philosophy, and Max Weber, Simmel, Sorokin, and Znaniecki in sociology, authors like Skinner and Dodd have still had a larger range of followers. It is worth mentioning that Skinner is even treated by some authors as a humanist (!), and one of his articles appeared in a collection titled *The Humanist Alternative*.[16] But in the same collection Floyd Matson sharply condemns behaviorism: "The behaviorist view of man as a helpless pawn in the fell clutch of circumstance is paralleled in its anti-Humanism by the currently popular doctrine which traces human conduct to blind instinctive urges, notably that of aggression, arising from primordial and predatory ancestry."[17] The newsletters of America's

Humanist Sociology Association show that this group does not have a definite image of humanism. Therefore, all who are against the political, social, economic, or academic status quo are treated as humanistic. It would be nice if all of them were at least actual humanist sociologists. However, the president of the Humanist Sociology Association, MacLung Lee, even included William Graham Sumner as a humanist. How strange! Sumner was a scholar whom an American historian called "prime minister in the kingdom of plutocracy."[18] Sumner was the leading figure of the American version of Social Darwinism, responsible for the legitimization of economic exploitation in the period of so-called ragged capitalism.

Thus, in sociology we notice two fields of controversies about humanism: one, ideological in character, the second, methodological. Naturalism and anthropocentrism formed the philosophical platform on which liberals, leftists, Marxists and racists could cooperate together. Neo-positivism, behaviorism, and logical empiricism formed the platform for methodological discourse. While entrance onto the first platform required only a formal declaration of one's viewpoint, the elevation to the second platform was possible exclusively to those qualified to discuss modern methodology's issues.

Humanistically minded sociologists at the beginning of the twentieth century were too close in time to the famous and highly prestigious philosophers of neo-positivism to have a proper perspective on the entire controversy.

The neo-positivisitic critics of sociology finally met an interesting response from those sociologists who themselves were trained in analytical philosophy. It is interesting that some of the most important answers to positivistically minded critics of sociology as well as to the sociologists blindly faithful to the methodological demands of logical positivism, empiricism, and/or behaviorism, were offered by sociologists in "Marxist" Poland.

Soon after Pitirim Sorokin launched his attack on American sociology,[19] Stanislaw Ossowski published his criticism of biological patterns and the methodology of natural sciences borrowed by sociologists.[20]

The critical views of some outstanding younger Polish sociologists cannot be explained by the influence of Marxism, which, despite Marx's hatred of Comte, is deeply infected by positivism.[21] It has been rather a result of the opposition toward Marxism that caused postwar students of sociology to immerse themselves in those trends which promised scientific objectivity. However, having a compulsory training in Marxist criticism on the one hand, and the memory of Znaniecki's important contribution to the humanistic sociology on the other hand, they were able to offer highly interesting contributions to the present discussion on the nature, methods, and relation of sociology to the natural sciences. They slowly discovered the weak points in the concept of the "unity of science as a logical necessity." Edmund Mokrzycki (one of the youngest of Ossowski's students) made an excellent contribution to the study of the "impact of the philosophy of science on sociology over the last fifty years, i.e., from the rise of Vienna Circle, which has resulted in a deep-reaching and basically undesirable methodological orientation in sociology."[22] Mokrzycki is right in noticing that methodology

became the link between "true science" and sociology, an interpretation of the scientific method that was both professional and accessible to non-professionals. Methodology thereby not only provided an additional and deeper justification of the pro-naturalist tendencies, but also practical possibilities for putting them into effect.[23]

The domination of sociology by empirical sociological methodology is one of the reasons that humanistic trends are still dominated by the nonhumanistic. This domination is rooted in the fact that humanistic sociology requires from its followers not only knowledge of methodology in general and humanistic methods in particular, but also a larger educational background containing at least some philosophical and historical knowledge.

This may suggest why the giants of humanistic sociology — Weber, Znaniecki, Sorokin, Mannheim — had extensive knowledge beyond sociology itself. Weber studied law, history, and economics; Znaniecki was a philosopher with a solid background in general humanistic education before he turned to sociology; Sorokin wrote first *Leo Tolstoi as a Philosopher*, and had a considerabe historical knowledge; Mannheim had been trained in various philosophical schools with teachers that included George Lukacs, Heinrich Rickert, and Edmund Husserl.[24]

On the other hand, we have those who contributed to nonhumanistic sociology by prescribing "scientific methods" for sociology. As a contemporary critic wrote, they had neither a serious knowledge of philosophy nor a larger humanistic background:

George A. Lundberg did not know the publication of the Vienna Circle, Paul Lazarsfeld blamed the philosophers of science for ignoring the real methodological problems of social sciences, and Hans Zetterberg, when writing about the axiomatization and the empirical verification of sociological theories, almost exclusively quoted sociologists and psychologists and his knowledge of the texts on the subject by philosophers of science was extremely modest.[25]

The fate of humanistic sociology is determined, however, not only by the development and trends in a given historical period, but also by the appealing promises and technical character. One can assume that it will never have the same appeal as the nonhumanistic. By the term "non-humanistic" should be understood all those sociological trends and schools that have neglected or rejected the most humane elements of social existence: values, meanings, motivation, and understanding. This nonhumanistic sociology can help us to solve or explain thousands of small-scale social problems, situations, and processes of collective behavior. Despite all its weaknesses, it could slowly contribute to the welfare of society. The relative easiness of research, according to its distinctive methods and technics, gives a quantitative victory over the more ambitious tasks of humanistic sociology.

The later, which draws its inspiration from the entire heritage of Western philosophy, is vulnerable to many kinds of criticisms because of lack of mathematical exactness on the one hand, and because of an elitarian ethos on the other hand, since it is like transcendent, anti-naturalist humanism available rather to the few than to the masses.

The twentieth century has been an age of specialization in all fields of intellectual inquiry. "Teachers" became "instructors." Most philosophers became logicians and/or methodologists; most sociologists became social researchers. The room for humanistic approach has been shrinking.

There is, however, one hope: the significance of humanistic sociology should grow up parallel to the needs of international cooperation. The cooperation and mutual understanding among nations is the *sine qua non* of the biological survival of mankind. Those who believe that human intellect can play a role in the political evolution of the global community of nations will agree that the social sciences in general have a significant place in this process of cooperation. The humanistic approach then apears to be most significant because of its principles of "interpretive understanding," "humanistic coefficient," and search for "meanings."

In this historical circumstance humanistic sociology also needs a phenomenological perspective, since this sociology, as well as phenomenological philosophy, fights against the minimalism of positivism, against naturalism and narrow empiricism. It searches for that which is "eidetic" in social process.

NOTES

[1] F. I. Niethammer, *Der Streit des Philanthropismus und Humanismus in der Theorie des Erziehungs-Unterrichts unserer Zeit* (Jena, 1808).

[2] S. Jedynak, "O humanizmie przed Protagorasem," in *Humanizm Socjalistyczny* (Warsaw, 1969), pp. 13—14.

[3] Dilthey discussed this problem in his essay, "Normen und Naturgesetze" (1882).

[4] T. S. Eliot wrote that: "the humanistic point of view is auxiliary to and dependent upon the religious point of view" (*Selected Essays of T. S. Eliot* [New York: Harcourt Brace, 1932], p. 427).

[5] Irving Babbitt, *Democracy and Leadership* (Boston: Houghton Mifflin, 1924), p. 3. Babbitt maintained also that: "Religion indeed may more readily dispense with humanism than humanism with religion . . ." ("Humanism: An Essay at Definition," in *Humanism in America*, ed. Norman Foester [New York: Farrar & Rinehart, 1930), pp. 43—44).

[6] Thomas Molnar, *Christian Humanism: A Critique of the Secular City and its Ideology* (Chicago: Franciscan Herald Press, 1978), p. vii.

[7] Wincenty Granat, *Upodstaw humanizm chrzescijanskiego* (Pozman, 1976), p. 53.

[8] Molnar, *Christian Humanism*, pp. 4—5.

[9] B. F. Skinner, *Science and Human Behavior* (Chicago: Free Press, 1953), p. 5.

[10] See Arthur Koestler, *Act of Creation* (New York: Macmillan, 1964).

[11] Otto Neurath, "Empiricism and Sociology," in *Empirical Sociology: The Scientific Content of History and Political Economy*, ed. M. Neurath and R. S. Cohen (Dordrecht: D. Reidel, 1973), p. 325.

[12] R. Carnap, "Logical Foundations of the Unity of Science," in *International Encyclopedia of Unified Science*, p. 61.

[13] Max Weber, *Economy and Society*, ed. G. Rath and C. Willich (New York, 1968), p. 4.

[14] Bertrand Russell, *History of Western Philosophy*.

[15] Robert McIver.

[16] Paul Kurtz, *The Humanist Alternative: Some Definitions of Humanism* (Buffalo: Prometheus Books, 1973).

[17] Floyd W. Matson, "Toward a New Humanism" in ibid., p. 95.

[18] Upton Sinclair, *The Goose Step* (Pasadena, 1924), p. 123; quoted by Richard Hofstadter, in *Social Darwinism in American Thought* (New York: Macmillan, 1964), p. 63.

[19] Pitirim Sorokin, *Fads and Foibles of Sociology and Related Sciences* (Chicago: H. Regnery, 1956).

[20] Stanislaw Ossowski, *O osobliwosciach nauk spolecznych* (Warsaw: PWN, 1962).

[21] According to Seymour M. Lipset, "Karl Marx was a rationalist and positivist who believed that political actions must be based on scientific knowledge" ("Are Rationality and Reason Dead?" *Humanist*, March—April 1975, p. 10).

[22] Edmund Mokrzycki, *Philosophy of Science and Sociology: From the Methodological Doctrine to Research Practice* (London: Routledge & Kegan Paul, 1983), p. 2.
[23] Ibid., p. 71.
[24] Aleksander Gella et al., *Humanism in Sociology: Its Historical Roots and Contemporary Problems* (Washington: University Press of America, 1978); see the chapters on Weber, Znaniecki, Sorokin, and Mannheim.
[25] Mokrzycki, *Philosophy of Science and Sociology*, p. 74.

EDWARD VACEK

THE FUNCTION OF NORMS IN SOCIAL EXISTENCE

Philosophical and theological debates over the past few decades have often centered on the proper place of norms in the moral life. Utilitarians have argued among themselves over act and rule-based theories, and they have argued with deontologists over exceptions to norms. Christian theologians, stirred by "situation ethics," have struggled over the proper relation of norm to context. The goal of this paper is to bring some clarity to this debate by examining where norms are inappropriate, where useful, and where necessary. The general question we put to ourselves is, "Why do we have norms at all?"

For purpose of clarity, I want at the outset to give what I think is a paradigmatic example. Marriage is for many people the central focus of much of their moral existence. Still, there are and there can be no universal norms prescribing who shall get married or whom one shall marry. There are, however, universal norms that apply once one has entered the state of marriage, e.g., "Thou shalt not commit adultery." There are, furthermore, norms in marriage which are useful guides, such as the norm that both partners should be equally involved in the rearing of children.

In the first half of this paper I hope to pursue this triple division of not-appropriate, useful, and necessary. In the second half, applying some themes from Max Scheler, I want to examine the function of norms in the light of three forms of human knowing.

I. THE MORAL LIFE AND MORAL NORMS

A common confusion in ethics has been to understand the moral life and moral norms as coextensive.[1] The impulse in civil society that gives birth to the protest, "There ought to be a law," finds its twin brother in the moral arena: "There ought to be a moral norm for every moral action." Once born, the child then restricts the identity of the parent: whatever is not covered by a moral norm is treated as if it were morally indifferent.

A-T. Tymieniecka (ed.), Analecta Husserliana, Vol. XX, 369–391.
© 1986 *by D. Reidel Publishing Company.*

A. Insufficiency of Universal Norms

The first thesis of this paper is that the moral realm is not coextensive with any set of existent moral norms. Human beings constantly face new challenges, whether it be to freeze human embryos or to develop neutron bombs. Such activities do not suddenly become moral or immoral only after norms have been devised. We often admit we are not sure whether a certain activity is morally good or evil, without thereby denying that a moral evaluation is appropriate. In fact, we frequently try to formulate an appropriate moral norm because we are sure that some moral evaluation is requisite.

If the moral realm is not straightforwardly coextensive with any existent set of norms, is it perhaps coextensive with a set of norms which in principle could be formulated? In other words, are there any morally relevant ways of being and acting which in principle fall outside every possible set of norms? The answer to this question is complex.

Some, such as Schüller,[2] Rahner,[3] or Meilander,[4] have argued that not all moral activities are able to be subsumed under universal norms. They reject the claim that any action, in order to count as moral, must entirely submit to the test of universalizability. That test holds that I ought not will an act unless I am willing that every other agent similarly situated also will the same act.[5] Those who object to this constraint point out that moral activities have two kinds of irreducibly particular features: the individuality of the agent and the Kairos element of a fitting time and place. In at least some acts one or both of these make a relevant difference. The racial situation in the United States in the 1960s was undoubtedly immoral, as it had been in varying ways throughout the previous 300 years. But only to a Martin Luther King, Jr., and a few others did it fall as a moral possibility and moral demand that structural reform be initiated. There had to be the right person in the right position at the right time. Others may have been acting within the bounds of the minimally decent Samaritan when they avoided racist acts. But only a few could have acted as these men did. In the concrete world where moral action takes place, theirs was a moral calling not possessed by all. Theirs was a time laden with possibilities not previously present.

Keen[6] has further argued, rather forcefully, that to balance the Appolonian tendency which seeks universal order we need also to allow for another authentic human tendency, namely, the Dionysian. This tendency is guided more by the creative impulse and fosters genuine novelty in history. The Dionysian tendency appears most clearly in those participative experiences wherein persons unite with a force or movement that is greater than themselves and therein transcend themselves. The ultimate form of such experience is, of course, mystical experience. We shall return to this topic at the end of this paper.

There are three ways in which one might try to continue to make moral life synonymous with what falls under universal norms. Because of the nature of judgment and language, an individual's action in a kairotic moment can be given an ostensibly universal form:[7] "If one's name is Socrates and one has learned that one is the wisest of persons, and if all around in Athens there are Sophists offering sophistry, then one ought to expose those sophists." Such apparent universals can be multiplied so that they exhaustively encompass the unique, by introducing into the norm an essence which has one or at most a limited number of possible exemplifications.[8]

Scheler has argued[9] that the agents and all the value-relevant factors involved in moral activity each have an intuitable essence or intelligible form. Some of these essences in principle apply to a range of objects, acts, human beings, etc., while others apply to one and only one.[10] In the case of human persons, both the essence of uniqueness and the essence of a common human nature combine.[11] In this sense, then, we might grant that, if a moral norm is allowed to contain both common and individual elements, it would be possible to formulate a norm for each and every action. But this is a far cry from what we mean by a universal norm.

A second tack can be taken to show that universal moral norms and moral existence are coextensive. One can choose sufficiently formal norms. "Do good and avoid evil" is one example. Even the unique can apparently be encompassed in this fashion, for example, "Everyone must follow his or her unique vocation."

Scheler has pointed out that a system of norms is usually established along the model of Porphyry's tree. One norm will be formal to those below it and material to those above it, though even the most formal will have some material or content in it. Something, however, will be concealed in such a system. The unique factors will stand not within but outside such a system. In opposition to Kant and the medievalists, Scheler and Wojtyla[12] after him, have held that the person is not an indifferent X of rational nature, a mere instance of a universal nature. When one moves intellectually from this person to human beings in general, the *positive* content of being "this" person is omitted. Similarly, Scheler and others have held that the kairotic moment, once lost, may never return and thus it is not strictly replicable or universalizable. One need not be perplexed by the presence of thoroughly individual elements in a moral decision. Only those who equate universality with objectivity and individuality with subjectivism need be disturbed.[13]

A third tack is to note that the standard case of a moral norm is one with sufficient generality so as to encompass a variety of particular cases. The norm says, "Thou shall not kill," and all cases of the unjustifiable taking of the life of a human being are envisioned. The universal norm stands. It then falls to practical judgment or wisdom whether a given case is a case of murder.

It might immediately be noted, however, that such encompassing norms usually *presuppose* that the individuality of the agent and the kairotic moment are *not relevant*, e.g., that the agent is not a civil servant deputed to execute at the designated time a criminal in an era where alternatives to capital punishment are not available. According to Scheler,[14] philosophers such as Kant have been so concerned to overcome the idiosyncrasies of personal desire, that they have envisioned persons simply as universal agents, each not really a concrete individual person. Hence Kant's categorical imperatives are also only for the universal aspects of persons, not for that which is most unique in them; and thus, such categorical imperatives regulate actions *without real agents.* If one wants to build a long bridge, careful engineering is required. Building bridges is not essential to the moral life of human beings in general. Building bridges, however, is a moral activity for one who *is* as part of his or her identity a bridge-builder, and such a person ought to take advantage of adequate engineering knowledge. To fail to do so makes one not only a bad bridge-builder, but also a failure in that person's way of being a moral being. Of course, to nonculpably fail as a bridge-builder is not to be a failure as a moral person. Kant rightly saw that success in one's projects is not a criterion of moral being. The same, however, can be said of any other activity that might be named in a norm. We shall return to this observation in a moment.

From what we have observed, we can see that the more formal a norm is, the more incapable it is of giving guidance concerning the particular, while the more particular it is, the more it approaches the judgment of conscience that in *these* here-and-now circumstances *I* may, must, or must not do this one deed. Further, we see that the attempt to see the moral realm as coextensive with moral norms either leads to the loss of the universalizability of the norms or it omits from the moral realm crucial elements that are constitutive of the moral act. That Saul run for public office seems a paradigmatic moral event, but it is one that cannot be simply deduced from universal norms.

Independent of these theoretical difficulties, it remains true that in many moral situations no moral norm has been formulated and that a complete set of moral norms could be formulated only with difficulty and applied with even greater difficulty.[15] What moral norm, we might ask, tells a bereaved wife when to give up grieving and get on with life? What moral norm declares decisively after how many children a vasectomy is permissible, proscribed, or even prescribed?

B. Rudiments of a Moral Theory

Underlying these questions and remarks, of course, is a theory of the moral life. In a theological view, as I shall develop in the final section of this paper, the moral life is ultimately one of cooperating with God in the enhancement of being. One thinks of such Christian metaphors as helping to build the Kingdom of God or of responding to what God is doing in the world. In less explicitly theological terms, the moral life is the life dedicated to maintaining and enhancing the realization of value. The moral life primarily refers to what befits the human, but goes beyond to include acting in concert with nature in the preservation and increase of the good. One might well call such a theory teleological, if that term itself were not so ambiguous. For the purposes of this paper, perhaps it is serviceable enough.[16]

From the teleological viewpoint, it is clear that Kant's starting point is ill-chosen. As Scheler[17] has trenchantly argued, the primary moral experience is not one of a sense of duty and the basic moral act is not one lived in conformity to self-legislated norms. We do many acts simply out of a love for the good.[18] Most of the time we tell the truth (such an ominous phrase!) just because we want to say something to another person. We are generous because we like to help others or because they are in need of us. Scheler perceptively pointed out that the sense of duty arises not out of virtue, but rather is founded on an experience of our moral deficiency. There is a disorder in us that wants to avoid the truth, and so we are tempted to lie. The sense of duty, often accompanied by the consciousness of an obligating norm, rises as a resistance to this temptation. By contrast, the great-souled person, the *belle âme*, is one who spontaneously and with ease does the good. Unlike the person who labors under the rational compulsion of duty, the maturely moral person sees the good to be done and simply does it.[19] Scheler[20] lamented that love and joy had been obscured as the primary fonts of moral behavior. The Bavarian Catholic blamed the Protestant professor of Königsberg for the cloud of sternness that hovers over popular conceptions of the moral life.

This identification of the moral realm with duty has had the further consequence of reducing morally alternative actions into morally indifferent actions. This idea of moral alternatives is frequently overlooked, so some explanation is called for. If two or fifty-two alternatives each promote the good in varying degrees, we may not have a duty to perform any one of those alternatives, but that lack of a sense of duty does not strip these alternative deeds of their moral significance. To be sure, the ancient maxim, "De minimis non curat lex," can be left intact. Many actions are morally trivial, but triviality does not in itself require the conclusion that these actions are morally indifferent. Put simply, every action that enacts a person's freedom is a moral act, either morally good or morally bad. That which is trivial should not be given exaggerated importance, of course; still, reflection on a love relation makes it vividly clear how even the trivial — a pink rose or an unkind remark — can add to or subtract from a relationship.

The basic moral experience, as Maguire[21] has tried to show, is that of the befitting. An agent experiences a gap between what is and what could be in given circumstances. What befits is what would close this gap.[22] At times a person may experience such a gap as an offer of an opportunity to realize new value, what some call an ethics of invitation or calling.[23] When there is resistance to this *invitation*, the gap is experienced as a lack of perfection in the world or in the agent. At other times, as Dyck[24] has argued, agents may experience such a gap as a *demand* confronting them with the obligation to achieve at least minimum standards of goodness. When there is resistance to this demand, the sense of requiredness will be experienced as duty.

The disparity between what is and what could be must be perceived not only in itself; it must also be perceived as a live possibility for the agent's action.[25] The case is not just that "things" could be better (always true); nor that this particular object, relation, act, etc., could be improved; but further, that the maintenance or enhancement of this set of values is "ours" to achieve. If we neglect to act, we neglect a relation of ourselves to an enhancement of this value-complex, and thereby we are less than we could be. By saying no to this set of values, we may, in the absence of other, compensating value affirmations, become less than we were, perhaps even reversing symbolically the various virtues and tendencies (*Gesinnungen*) toward the good that belong in every human heart.

Thus far we have seen that the moral realm is larger than the set of moral norms. In this section we have tried, however briefly, to locate or describe ordinary moral experience. Our task here is to indicate the usefulness of moral norms in moral life. Before doing so, we need to propose at least a preliminary statement of what we mean by "moral norms."

The morally good life, as we have seen, consists in the personally or socially enacted maintenance and enhancement of value. Every object, actual or potential, has positive or negative value. There is a value, or usually, a set of values, that pertains to each person, act, relation, structure, and so forth. Each of these items, whether real or ideal, is a bearer of values, and the values they bear can be intuited as distinguishable though interrelated phenomena. Some intuitable values are realizable in this world, while others are not, e.g., universal harmony among creatures. The values resident on any item are generally multiple. In particular, the values of all human actions are indefinitely complex since minimally they include a reference to the person who acts, the object of the act, the circumstances, and the possible consequences.

Acts of a particular kind generally have regularly recurring features. They regularly realize a maintenance, a diminution, or an increase of value. The Decalogue is an obvious example of a set of such acts. In addition to their central function as *loci* of obedience to Yahweh, these Semitic laws rather primitively pointed out that certain human activities regularly enhanced or devalued human and divine relationships. Over the millenia, these rough laws of moral behavior have been greatly expanded, clarified, and refined into many other prescriptions and proscriptions. As new values came to consciousness or as new patterns of values came to be intuited, and as these values and value-clusters were seen to be regularly threatened or enhanced by a particular act or a particular way of being, that act or that way of being became proscribed or prescribed.

When a proscription or prescription is conceptually formulated, we have an exlicit norm. The proper formulation of such norms and the relation of one norm to another is, of course, a major task of ethicians. Such norms need never be formulated.[26] Often they are only implicit in the parables and stories by which people guide their lives. They may be as quaintly stated as the maxim that "good girls don't kiss on the first date." They may be as unrefined as "You can do whatever you want as long as you don't hurt anyone." Whatever their sophistication, explicit moral norms are conceptual formulations of ways of being or ways of acting which are required generally for the maintenance and enhancement of value. Values are not founded on norms; rather, norms rest upon the values that they attempt to preserve, banish, or augment.[27] Norms are criticizable not only in terms of their consistency with one another, but most especially in terms of their adequacy in articulating the moral task.

Moral norms are of two major types, which the medieval tradition called material norms and formal norms,[28] but which more contemporaneously could be termed action norms and virtue norms. Action norms rather straightforwardly indicate actions that are to be performed or avoided. These norms are usually stated apodictically, though, as Cahill has pointed out,[29] it can be argued that their form hides some implied conditions. "Do not tell a lie" might be seen as shorthand for "Do not tell a lie, unless some greater harm would result," where such harm might be either to the practice of truth-telling or to some greater value. As the history of ethical reflection on truth-telling reveals, the person who always told the unvarnished truth whenever asked to do so would in fact threaten human communication. In their usual form, action norms indicate patterns of human behavior that regularly lead to a surplus of disvalue or the absence of value. When freely engaged in, such behavior is generally immoral. The norms for such behavior have, in contemporary terminology, a *prima facie* validity. These norms envision circumstances where normally there is no live possibility of doing some other high-valued act, e.g., save a life, that might demand the otherwise proscribed behavior.

Virtue norms take two closely related forms. (We shall not include here the tautological formal norms such as "Do not murder," or "Do not be excessively preoccupied with money.") Virtue norms either dictate the kind of attitude which should inform our actions, or they state the kind of person we should become. Since the action of a person reciprocally flows from and forms who that person is, the dividing line between these two kinds of virtue norms is not always clear. Thus the norm, "Be kind," can refer either to the attitude which should inform our dealings with people and animals, or it can be a statement that we should so act as to make kindness a characteristic of our selfhood. In the first form, the norm "Be courageous" might tell us how to go about waging battle, while in the second form it might tell us to develop, perhaps through fear-overcoming activities, a character that might only occasionally, if ever, be called into fear-surpassing action.

To complete this section, we must finally say something about the mutability and variability of moral norms. Any question of the usefulness of norms would largely be moot if it were denied that moral norms continue to originate and disappear in response to changes in the history of an individual, group, or the human community. Even that supposed defender of eternal and universal norms, Thomas Aquinas, found it necessary to discuss the dispensability of the God-given precepts of the Decalogue. Thomas recognized that some circumstances make following the letter of the law inappropriate, thereby setting the stage for the distinction between the norm and the values that the norm is intended to protect.[30] Where new complexes of values occur, say, through the addition of new or unusual circumstances, new moral norms may be formulated and new moral practices are legitimated. As an example, we might cite the various, often inadequate proposals that have sprung up in recent decades to foster a healthy self-love in response to the discoveries of contemporary psychology.

That there is a diversity of norms between cultures is, I take it, firmly established by anthropologists. That there has been an evolution of norms within a given culture can also be taken as established by historians. One need only think of such standard examples as slavery, divorce, adultery, or birth control. Sometimes new norms arise in response to genuinely new possibilities, e.g., a proscription of nuclear war. Sometimes new norms arise as insight is achieved (e.g., a proscription of sexual discrimination), or as insight is lost (e.g., the common good gives way to an almost absolute primacy of private property). Sometimes the same formulation of a norm will assume various meanings, e.g., the norm to honor one's parents changes as one matures into adulthood. In all these changes, moral norms function to point out certain value-complexes and to resist the loss of other value-complexes. We shall return to this theme in the second half of this paper.

Moral norms are *useful* for progress in the moral life. It is possible for the great-souled person to perceive a value or complex of values as to-be-realized and then immediately respond without any explicit thought of the relevant norms. In fact, such explicit thought may hinder moral action, similarly to the way that reflection on just how we walk or move our arm actually impedes the walking or throwing.[31] Nonetheless, as we shall see in the second half of this paper, moral norms are often helpful because they help us get clear on what values are present; they help us to resist temptation; and they indicate likely ways of cooperating with God. They are not indispensable. In fact, they are secondary and derivative from a sensitive value consciousness. But still they can be useful.

C. Moral Institution and Success

Christians, among others, have religious reasons for being suspicious about norms. They have, however, seen the usefulness of norms in pointing out areas of moral failure and in providing instruction. Norms are, however, more than useful; they are indispensable for social life. We proceed now to this third thesis of the paper.

Certain moral norms function to *constitute* a moral *institution*.[32] While social practices or institutions may not be essential to a perfectly moral *individual*, they are essential to a society whose members can only interact in the context of a set of stable expectations.[33] The act of driving down one or the other side of a street is morally indifferent *until* a *practice* of one or the other is established. Thereafter, normally it is morally required to drive down, say, the right side. An institution sets up a system of expectations which demand conformity.[34] Polygyny may or may not be what marriage should be, but once the practice is accepted, the husband ought to sleep with each of his wives in a fair way. Moral criticism here should be directed not to acts within the practice, but to the practice itself. One does not criticize slaves for not voting, but the system that prevents them from voting. Conversely, one ought not, at least normally, ask whether in this or that instance, an act of adultery might produce more good than sexual fidelity. In general, norms establish an institution, but then the institution, once entered upon, demands conformity to the rules.

The point we are making here is not that one ought never violate the system of norms that create and regulate a social institution.[35] The problems involved in the evaluation of institutions are similar to those faced by proponents of a rule-based morality. A rule morality occasionally proscribes an individual action that of itself would increase the good, because in the long run such actions would decrease the good.[36] Conversely, occasionally the violation of a rule better serves the values the rule is intended to protect, but in so doing it jeopardizes the rule. Sadly for the rationalist and perhaps the deontologist in each of us, norms and the institutions they constitute often enough conflict with one another, and usually no clear way of adjudicating between such conflicts is given.[37] The norm of obtaining informed consent may yield to the norm of beneficence in the hospital emergency room but not in the patient's room on the day before surgery.

The full range of moral norms, as we have noted, includes more than rules which function deontologically to specify the minimum standards for living social life.[38] But even at this level, the richness and complexity of human life contains within it the basis of what can only be called the tragic. That is, a conflict of basic goods and evils occurs. Kant's famous example of truth-telling to the homicidal inquirer reveals the conflict. The disagreement that has arisen over the strict applicability of Rawl's lexical ordering of basic goods indicates the same point.

Institutions share several features with games.[39] Games do not exist without the rules that constitute them.[40] Moreover, one must be conscious of those rules and intend to act in conformity with them or, at least, be willing to be criticized or penalized for not following them. The child who mindlessly moves a chess pawn four spaces diagonally is not playing chess, but neither is the child who unwittingly makes what in fact is a brilliant check-mate move. The rules of a game are what Shwayder calls restricting norms. That is, they take an activity such as play that we might otherwise engage in, and they specify one system of rules as proper to playing this game. Without the rules, this game does not exist, even though play continues.

Similarly, the norms that constitute social practices are necessary for the very existence of those institutions. People could mutually benefit one another without a contract. But once some interchanges are entered upon in the form of contracts, then one is bound by the norms which create and regulate contracts. Much of human life is built upon such informally and formally agreed upon rules of social life.

Analogous with the minimal standards formulated in deontological rules, there are in games rules which *constitute* the game, rules, for example, that forbid bringing a hockey stick onto the football field.

There are also in a game strategic rules which have to do not with determining which game we play, but how we may succeed at the game. Thus in football one should pass on third down and long yardage, unless one thinks that a draw play or even a quick kick might catch the opponent unawares.

Teleological moral norms generally are more flexible and in this they share a feature with these subordinate strategies of a game. In teleological systems, moral norms indicate general patterns for attaining happiness, personal fulfillment, or fullness of being. Moral norms and practices thus sometimes define aspects of human interaction and at other times serve as calculated strategies for succeeding at teleologically given goals.

To be sure, success is not the sole criterion for the morality of one's actions. Tennyson was right when he said that it is better to have loved and lost than never to have loved at all. Nonetheless, the aim still is love, successful love. The project to improve myself or my neighborhood is a moral endeavor, even if I die before the task is completed. When the self has committed itself to the value-to-be-realized, it has already begun the formation of its own moral goodness, though greater goodness lies in the full engagement which comes through persistence in realizing the goal.

In the first half of this paper, we have examined the moral life and found that there are distinguishable areas of that life in which moral norms may be essential, useful, or not ultimately decisive. In the second half of this paper we shall examine three different mind-sets that give rise to the formulation of norms.

II. THE FORMULATION OF NORMS

In the second part of this paper we ask why norms, especially moral norms, are ever formulated at all. Here we will not answer the extremely important question concerning how, psychogenetically, norms arise. The developmental psychologies of Kohlberg and Gilligan are illuminating probes into that area; so too the older work of those psychologists and sociologists who study the gradual internalization of a community's moral system. These questions and approaches are extremely valuable. But, even if completed, they would still be subject to the philosophical criticism of the genetic fallacy. That is, the fact that someone has learned geography, say, in order to please a teacher gives no grounds for saying that desiring approval is the meaning of what is learned.

We will focus on three origins for norms, three relations between consciousness and value that may eventuate in the birth of a formulated moral norm. Scheler classified knowledge into three types: *Wesenswissen* or *Bildungswissen, Herrschaftswissen,* and *Erlösungswissen.*[42] The first is knowledge of essences arrived at through eidetic intuition and passed on through education. The second is knowledge which "serves our ability to exercise power over nature, society, and history." The third is our "contact on the level of knowledge with a reality intuited as supremely powerful and holy and at the same time the highest good and the universal ground of being." We shall use this framework to understand the reasons for formulating norms.

A. Knowledge of Essence

(1) *The Desire to Know.* The first source of norm-formulation, though I suspect the least common, is the pure desire to know. As reflective people, we ask questions of reality not simply in order to get on better with the business of life, but rather because we want to know what we are doing and why. We might say we are uncomfortable with just being good and doing good; we also want to know in a conceptual way what it is that we should do. Usually it is enough just to get an inkling of the appropriateness or inappropriateness of our moral behavior.[43] Sometimes, however, we want clarity of mind on that issue or we are interested in the consistency of our moral being and doing; and so we explicitly formulate our norms. The formulation of norms brings clarity to our minds and makes it possible to expose inconsistency, thus also satisfying the rational desire for system-building. Let us look further at this issue of consistency and then at that clarity which is self-knowledge.

Every culture, I presume, is filled with maxims. These maxims are formulated to help in recurring decisions.[44] The maxims are often contradictory: "The squeaky wheel gets the grease"; "The quacking duck gets shot"; or, "No one is going to blow your horn for you"; and, "Beware of those who go about parading their good deeds." Human beings, I suppose, have a rough way of figuring out when to apply one maxim and when another. But the rational mind, the mind that has undergone some sort of intellectual conversion, sees the conflicts between these maxims and desires to harmonize them.

Consistency is not the only or even the most important virtue of the human mind. Being well-integrated also is not the sole norm of morality. But consistency and integration are demands of the rational mind. And so the pure desire to know leads one to go beyond unthematized actions and social practices and beyond popular maxims toward a set of moral norms that is both consistent and harmonious with the run of one's moral life.

The desire to be consistent is not the only motive behind this pure desire to know which leads to the formulation of norms. As von Hildebrand[45] has argued, the sense of moral agency is heightened in reflective behavior. It is satisfying to explicitly know what one is doing and why one acts. The moral life essentially requires as its prerequisite a rational agent. The ancients distinguished between acts of a human being and human acts. Such a distinction can be (and was) too rationalistically conceived. Nonetheless, there is a heightened sense of agency which would find its epitome in Sartre's vision of pure freedom or in Robert Solomon's descriptions of the emotions. Both positions are illusory and rationalistic, but they are attractive at the same time. They feed upon the heightened sense of agency we have when we act reflectively (though von Hildebrand goes too far when he insists that the fully moral act must possess this reflective clarity). Most human acts flow spontaneously from the "graced" or the relatively freed self without needing to be processed through a thematizing consciousness. Still there are moments when we come to greater consciousness of who we are through reflection on our deeds.

Scheler[46] insisted that we can be under illusions about ourselves as much as about any external objects. The self does not necessarily know itself. The creation of moral norms greatly facilitates this process of self-knowledge. Moral norms provide for us an object outside ourselves in view of which we can grasp ourselves as moral agents.[47] This object outside ourselves, the moral norm, can be alienating.[48] It can lead to self-righteousness or what Scheler and the Christian tradition unfairly call pharisaism. Nonetheless, through conceptualization of the norms, as through any conceptualization of the self and its activities, the self is able, as I think Hegel tried to show, to more clearly grasp itself in its very nonobjectifiable being. It sees itself in the categorized act as that which transcends its immanence in the given act. Thus it can grasp itself, however obscurely, as a moral agent through the moral norm. This desire for self-possession is one crucial reason for creating the mirrors which are moral norms. The mirrors lead to self-righteousness when we act so as to appear good to ourselves or others. The mirror leads to self-knowledge when it reflects back to us who we are as free agents.

It should be clear that this process of formulating norms can go haywire. One can accidentally formulate good norms based on faulty or incomplete insights into the moral good or the moral evil. Conversely, one can formulate poorly the good insights one has. Each mistake clouds the mirror of self-knowledge. Moreover, the possession of a set of good moral norms can itself lead to illusions, leading one to think that one is morally good since one's mirror is so precious. Persons may forget their personal center while they take their lofty moral norms as pictures of themselves. Liberals who pride themselves on their nonracist positions, but who never hire a black or a Latino, might properly ask themselves if they see themselves through their norms or if they see merely the norms themselves.

Of course, things need not go so badly. Norms can actually be the beginning of new behavior. Just as painters fashion their inspirations through working on the canvas, so one who strives to fashion moral norms may create new possibilities for moral action, action that once formulated may then hover over the person as a beckoning or demanding possibility. The moral genius is one who has the passionate interest to discover such moral possibilities, and the moral theorist, however untrained, is one who formulates such possibilities into moral norms. Both enable people better to know and to realize their authentic selves.

(2) *Educational Function.* Closely linked with this desire to know is the desire to teach. There is the sense, just noted, in which we can teach ourselves, but we will speak here only of teaching others. Moral norms name typically required actions and ways of being. They point out the value in a way of being or kind of action. Much as a guide in an art museum enables us to see features in a Rembrandt that we would otherwise not easily see, so moral norms are fashioned as concise ways of communicating the value and disvalue of various actions or ways of being. We are not speaking here of their psychological function of indicating to others what will win approval or disapproval, but rather of indicating to others what is in itself approvable or disapprovable so that they too might see the value.[49] Value is the ground for desire, not desire the basis for value. Much confusion in popular and professional ethics could be eliminated if this cardinal insight were more fully grasped.[50]

For Scheler,[51] the primary role of an authority must be to lead others beyond their narrow vision into new possibilities. They obey in the hope that in the performance of a commanded deed new values will be discovered.

The desire to teach can, of course, have many motives. Here we need only mention two. The first is the perceived gap in other persons between the world of values they presently appreciate and the larger world of values that they could intuit. We may want to protect others from their blindness or we may want to open them to a richer world. Second, the desire to teach is motivated by a desire to create a shared world. Human solidarity is promoted by a shared world, and norms can facilitate shared perceptions of the world.

Scheler[52] insisted that mere training never brings a person to moral insight. The education of others, which offers insight and appeals to freedom, is made possible by the objectivated otherness of the moral norms. Only when freedom is elicited can genuine education take place, and that freedom is respected when norms are presented as objects independent of any coercive force, e.g., the promise of acceptance or reward on the part of the one who presents the norms. These norms, though they are the product of human minds, nonetheless develop an independence of the minds that generate them. The desire, noted in the previous section, to make our various maxims and norms consistent with one another depends on the quasi-independence of these norms, once they have been formulated.

This quasi-independence of moral norms gives them an appearance of existing apart from the subjectivity that originates them. Thus they are more readily perceived not merely as demands of an authority or even as desires of a beloved, but rather as demands or invitations of being itself, which in fact they are or ought to be. By concealing the originating subjectivity of the one who first formulates them, explicit moral norms thus are apt vehicles for explicating the objective demands of the realm of value.[53] They become useful tools for educating others into the objectivity of moral existence.

B.   Knowledge for Control

(1) *Resisting Temptation.* The most common reason for formulating norms, it seems to me, lies not in the pure desire to know or even to teach. Rather, according to Scheler, norms are formulated to resist the inclinations that spring from the moral incompleteness in all of us. We

have just spoken of that form of incompleteness which is ignorance. In the face of ignorance, a community teaches a member to expand his or her awareness and sensitivities. There is another form of incompleteness, moral incompleteness.[54] There are in each of us tendencies that inhibit us from becoming or doing the good we perceive. For example, I know I should tell the truth, but I am afraid of losing face. Or, it would be good for me to be generous, but I really do not feel like it. Moral norms function in such cases as a counterweight to these wayward tendencies. It is the awareness of our evil tendencies that motivate formulating moral norms.

Emmet asked,[55] will there be any need for moral norms in heaven? As an exercise of philosophical imagination, her question is a good one. If, for the sake of argument, we are in a state of being where we spontaneously do the good and intuitively know the mind of others, we will have no need for norms. We might want to spend mornings in heaven exercising our pure desire to know by thinking up the norms that guide our good behavior just as we might calculate the laws that govern the trajectory of a golf ball, but in the heavenly afternoons we just hit the golf ball and we just treat our neighbors honestly and with respect. Here on earth, matters are different.

Most often, then, moral norms originate out of our felt sense of weakness. The person who intuits, perhaps suddenly, perhaps gradually, an "ideal ought" often finds that he or she does not immediately spring into action on behalf of that value. Our inner life is complex, and so we need gradually to restructure our attitudes so as to incorporate the new intuition. Moral norms serve as reminders to counter any tendency to slide back into previous attitudes and behaviors.

Moral norms usually are publicly formulated. The more common of the public norms counteract basic sinful or immoral tendencies of human beings. We can safely presume that "Thou shalt not kill" or "Thou shalt not commit adultery" were devised because people did kill, commit adultery, and so forth, or at least were sorely tempted to do so.[56] The virtue norms such as be courageous, be honest, be reasonable were devised because human beings easily give way to fear, dishonesty, and foolishness, often with greater ease than to their virtuous opposites.

Here we can note that we also formulate norms because of their usefulness in helping us to evaluate and correct.[57] When we say that we (or, more commonly, others) ought to have known better, we commonly refer to norms as the basis for our accusation. By appeal to a commonly accepted set of moral norms, we are able to indicate an objective standard that an offender knew or should have known. Such standards possess, as we have seen, an objectivity, an otherness, that stands over against the misbehavior. Such standards, which usually have been tested by community experience, provide clearer evaluative standards than the more formal judgment that "good people would not do that sort of thing." Thus a crucial reason for evolving norms is to provide an apparently impersonal forum for making judgments on one's own or another's character or behavior.

There are significant dangers when these evil tendencies are not acknowledged as the primary origin for formulating prohibitory norms. At least four reasons can be given why one should recognize one's evil inclinations before submitting to the norms and why, accordingly, a community should not multiply norms. First, we should recognize that prohibitory norms actually can instruct the truly innocent in evil. Just as telling a child not to look in mommy's closet near Christmas time piques the child's curiosity, so moral norms indicate new possibilities and thereby may evoke corresponding inclinations. Second, as Scheler[58] has pointed out, the truly good person will resent the prohibitory norm and be moved toward violation of it. Since the norm commands what such a truly good person would otherwise readily do, the command character interferes with the spontaneity of the act. The truly good person is tempted to become the sinner deserving such a command, since the essence of the command in a proscription is to oppose an evil tendency.

Third, there is the problem of externalism, wherein a person will model proper behavior without developing the heart that should issue in that behavior. Unless one is aware of one's own inclination, the norm quickly becomes at best a device for training, not education. Finally, as a corollary, the moral blindness of self-righteousness arises when moral norms are embraced by those who have not experienced their own inner tendencies. Self-righteous persons are those who conform their behavior to moral norms and then judge that they are good, free from evil tendencies, by the mere fact of that conformity. Self-acknowledged sinners frequently compare their action with moral norms and thereby judge that they have acted immorally. Both are liable to mistake making the moral life synonymous with fulfillment of moral norms, but self-righteous people are more prone to this mistake because they infer their goodness from behavior without examining the dividedness of their hearts.

(2) *Social Existence.* The need to order societal relations is, as we have seen, another source for formulating moral norms. More than monads, humans are social beings; and social existence is enacted in social encounters. These social encounters necessitate a system of expectations, institutions, or practices.[59] Each social system has its own set of expectations. Most of these expectations are learned through identification, observation, and imitation. Out of the vast variety of ways that human beings could interact, each society — a nation, a church, a club, a family — selects some as appropriate to its identity, and then it reinforces those patterns by social approbation. These expectations may be unformulated or formulated, but either way they combine to form what has been called the "social construction of reality."

The social origin of norms, then, lies in the need that people in a society have for predictable interactions. A society cannot function without some set of shared expectations. Shared expectations indicate what actions and ways of being are consonant with being a member of that society.[60] To violate those norms is to weaken the shared subjectivity that such a community possesses, and it invites criticism or exclusion in response. Societal laws take on a moral significance because in general they promote the common good.[61] Some of these norms may be in themselves merely arbitrary decisions where some decision is needed. Others will directly flow from the nature of being a person in communion with other persons. Together they form the fabric of social life.

The norms of a given society form a loose whole. Thus they modify one another. The existence of private property created the need for norms concerning stealing, privacy, and autonomy. The development of ecological concerns brought forth norms that modify this original set of norms. As histories transpire, one culture will evolve norms that are quite different from those of another culture. The result is inevitably a morality that is relative to the given culture, a morality that can be corrected only by looking to the values underlying all systems of moral norms.

To say that these norms may be corrected in the light of underlying values should not be taken to be an idealist desire that all moral systems should be the same. Genuine personal and societal diversity will always lead to a different selection from among the whole range of possible values.[62] As these individuals and societies evolve, of course, new norms will be formed as ways of preserving the greater insights into the value-possibilities or as ways of correcting for the hazards which occur when value-insight is lost. In short, moral norms are formulated in order to provide a system of expectations in a society wherein social harmony is fostered and social disharmony is minimized. The norms serve our need to control and advance the various communities of which we are a member.

C. Knowledge for Salvation

The last reason for formulating moral norms is their role in augmenting religious knowledge, the knowledge by which we participate in the holy and good ground of being. In moral *essential* knowledge, we desire to know the ideal essences of the world in itself. In moral *control* knowledge, we desire to order ourselves, our society and history in such a way that there is both harmony and progress in value. In moral *religious* knowledge, we desire to know God so as to be united with God's activity and God's being, and thus be saved.

In most theisms — Deism is an obvious exception — God is in some way presently active in the world. God's activity may be that of creating, sustaining, correcting, enhancing, and so forth. In a religion which views this activity as purposeful or at least as nonarbitrary, the believer is one who responds to or cooperates with this divine activity and intention.[63] Union with God's activity is union with God. Scheler[64] borrowed an Augustinean phrase, "amare Deum et mundum in Deo," to assert that we co-perform with God's own activity.

The third reason for formulating norms, is to grasp where and how God is acting and what God's intention is for the world. If it were discovered that God is a malevolent God, the religious response — under this hypothesis, contrasting to essential or control responses — would be to bring about a decrease of good in the world. For Christians, the loving God is building the universe, and hence response to God means to cooperate in building the kingdom of God.

Knowledge of God's activity typologically takes three different types, though usually there is an intermingling of these types, as, of course, there should be. In the first, emphasis is placed on the naked will of God. Believers seek to know the will of God, and the fact that God has decreed a certain behavior as God's will is reason enough for that behavior. Moral norms, often revealed in sacred scriptures, are perceived as the decrees of God and our actions are responses in obedience to that will.[65] In the second type, emphasis is placed on the mind of God. Believers ground their action more on what is in the mind of God than on the fact that God has decreed an action or law. Here essential knowledge is utilized, but it is utilized as a step toward knowing essences as reflections of the ideal essences in the mind of God. Moral norms describe patterns of being and acting which realize the ideal essences and essential connections which are thought to be in God and according to which it is thought that God creates the world.

In the third type, emphasis is placed on the heart of God. Believers seek to know the heart of God. That is, they seek to know the direction of the creative and unifying activity of God. Moral norms provide, on the one hand, warning signs against those actions and ways of being that would reverse that direction and, on the other, as guide posts pointing in the direction of God's love toward the possible discovery of new ways of being and acting.

These three types have led both to systems of fixed and to systems of revisable moral norms. When God's will is seen as immutable, when God's ideas are seen as fixed, and when God's love is simply unitive, the systems are fixed. There are, however, systems of revisable norms built on these same three types. Protestants have emphasized God's will, but some such as Barth have insisted that God is free to violate moral norms.[66] Other Christians have contended that God's wisdom can lead God to override even revealed essential norms. Those who hold divine love to be both unitive and creative argue that ethics must be open to new possibilities. For each of these positions, the values underlying norms can come to be seen with greater depth or wider extension, and thus the norms which formulate these values may have to be revised.

Throughout this paper we have proposed two simple themes. The moral life is larger than the moral norms which formulate some of the values that found the moral life. Nonetheless, the search for moral norms enables us to grasp more clearly these values, to perceive our obligations, to organize a moral community, to direct our lives, and ultimately to unite us with God's activity.

NOTES

[1] Richard B. Brandt, *A Theory of the Good and the Right* (Oxford: Clarendon Press, 1979), pp. 194–99; Dietrich von Hildebrand, *Ethics* (Chicago: Franciscan Herald Press, 1953), pp. 182–84; R. M. Hare, *Moral Thinking* (Oxford: Clarendon Press, 1981), p. 63.

390 EDWARD VACEK

2 Bruno Schüller, *Gesetz und Freiheit* (Düsseldorf: Patmos Verlag, 1966), pp. 14, 66.
3 Karl Rahner, *The Dynamic Element in the Church* (New York: Herder & Herder, 1964), pp. 13—29.
4 Gilbert Meilander, "Is What Is Right for Me Right for all Persons Similarly Situated?", *Journal of Religious Ethics* 8, (Spring, 1980): 125—34.
5 Hare, p. 42.
6 Sam Keen, *Apology for Wonder* (New York: Harper & Row, 1969), pp. 152—64.
7 Dorothy Emmet, *Rules, Roles, and Relations* (New York: St. Martin's Press, 1966), pp. 63—64.
8 Stanley Hauerwas, *Character and the Christian Life* (San Antonio: Trinity University Press, 1979), pp. 30—31.
9 Max Scheler, *Formalism in Ethics and Non-Formal Ethics of Values* (Evanston: Northwestern University Press, 1973), pp. 100—104.
10 Ibid., p. 76
11 Karol Wojtyla, *The Acting Person* (Dordrecht: D. Reidel, 1979), pp. 77—79.
12 Ibid., pp. 73—84.
13 Scheler, Formalism in Ethics p. 191.
14 Ibid., p. 373.
15 Richard A. McCormick, S.J., "Notes on Moral Theology," *Theological Studies* 43 (Spring, 1982): 89.
16 William K. Frankena, *Ethics* (Englewood Cliffs: Prentice-Hall, 1973), p. 14.
17 Scheler, *Formalism in Ethics*, pp.. 184—95, 210.
18 McCormick, p. 90.
19 Scheler, *Formalism in Ethics*, pp. 227—28; Schüller, p.166.
20 Scheler, *Formalism in Ethics*, pp. xxiii, 288.
21 Daniel C. Maguire, *The Moral Choice* (Garden City N.Y.: Doubleday, 1978), p. 71.
22 Von Hildebrand, pp. 244—56.
23 Scheler, *Formalism in Ethics*, p. 222.
24 Arthur J. Dyck, "Moral Requiredness: Bridging the Gap between 'Ought' and 'Is.'" *Journal of Religious Ethics* 9 (Spring 1981): 131—34.
25 Scheler, *Formalism in Ethics*, pp. 206—11.
26 D. S. Shwayder, *The Stratification of Behaviour* (London: Routledge & Kegan Paul, 1965), p. 97.
27 Scheler, *Formalism in Ethics*, pp. 203—12.
28 Timothy E. O'Connell, *Principles for a Catholic Morality* (New York: Seabury Press, 1976), pp. 157—61.
29 Lisa Sowle Cahill, "Teleology, Utilitarianism, and Christian Ethics," *Theological Studies* 42 (December 1981): 612—17.
30 Bernard Häring, "Toward an Epistemology of Ethics," in *Norm and Context in Christian Ethics*, ed. Gene Outka and Paul Ramsey (New York: Charles Scribner's Sons, 1968), p. 211.
31 Max Scheler, *Von Umsturz der Werte* (Bern: Franke Verlag, 1972), p. 303.
32 John Rawls, *A Theory of Justice* (Cambridge: Harvard University Press, 1971), p. 55.
33 Emmet, pp. 7—9.
34 R. S. Downie, *Roles and Values* (London: Methuen, 1971), pp. 35—39; Shwayder, pp. 252—59.

35 William A. Luijpen, *Existential Phenomenology* (Pittsburgh: Duquesne University Press, 1969), p. 329.
36 Paul Ramsey, *Deeds and Rules in Christian Ethics* (New York: Charles Scribner's Sons, 1967), pp. 134—44.
37 Robert M. Veatch, *A Theory of Medical Ethics* (New York: Basic Books, 1981), p. 297.
38 Ibid., pp. 298—305.
39 Georg Henrik von Wright, *Norm and Action* (New York: Humanities Press, 1963), pp. 13—16.
40 Emmet, pp. 57—59; Shwayder, pp. 265—68.
41 Shwayder, p. 265.
42 Alfons Deeken, S. J., *Process and Permanence in Ethics* (New York: Paulist Press, 1974), pp. 225—37.
43 Shwayder, p. 96.
44 Hare, pp. 40—43.
45 Von Hildebrand, p. 323.
46 Scheler, *Von Umsturz der Werte*, pp. 213—28.
47 Schüller, p. 17.
48 William Korff, *Norm und Sittlichkeit* (Mainz: Matthias-Grünewald, 1973), pp. 126—27.
49 Schüller, p. 41.
50 Brandt, p.32.
51 Scheler, *Formalism in Ethics*, pp. 327—28.
52 Ibid., pp. 547—75.
53 Hare, p. 86.
54 Scheler, *Formalism in Ethics*, p. 211.
55 Emmet, p. 11.
56 Luijpen , pp. 336—40.
57 Shwayder, pp. 260—63.
58 Scheler, *Formalism in Ethics*, pp. 213—14.
59 Shwayder, pp. 252—59.
60 Ibid., p. 254.
61 Ibid., pp. 279—80.
62 Emmet, pp. 102—3.
63 James Gustafson, "Moral Discernment in the Christian Life," in *Norm and Context in Christian Ethics*, ed. Outka and Ramsey, p. 27.
64 Scheler, *Formalism in Ethics*, pp. 212, 498—501.
65 Schüller, p. 32.
66 Hauerwas, p. 2.

MARK C. THELIN

CHINESE VALUES: A SOCIOLOGIST'S VIEW

I. INTRODUCTION

Within the moral perspective of the main cultures under discussion here, the Chinese approach stands out especially. In the following paper, I wish to give a short analysis of six core values in Chinese society. I will be doing this as a sociologist rather than as a philosopher, the substance of what I say being selected from the world of everyday experience. If I have any qualification for doing this, it is because I was born in China and have lived more than half my life either on the Chinese Mainland or on Taiwan. While there is an extensive literature on values, and Chinese values, in particular, I have chosen to concentrate on direct personal experience of life in a Chinese setting. Furthermore, since most of us are academics, I hope it will seem reasonable if I draw heavily upon the academic environment with which I am most familiar: the community which is Tunghai University on Taiwan. While there are certain distinctive features about this particular academic community, I am convinced that, in the main, it is typical of other such communities on the island and, perhaps, of Chinese communities elsewhere in South or Southeast Asia, as well. It goes without saying that my selection of core values may not necessarily be identical with a selection made by a typical *Chinese* scholar, though I presume that the degree of overlapping would be considerable. In any event, my purpose here is to stimulate thought and discussion. Questions may be raised as to how representative Taiwan may be of "all of China," and I would be quick to admit that significant differences in values may well exist, given the divergent development streams which have occurred on both sides of the Strait of Taiwan during the past three decades. At the same time, if these "core values" are really "core," then the underlying commonality should not have changed too much in thirty years.

One final caveat before we begin: In comparing two societies or cultures, frequently what you find in one you will also find in the other, but there is a difference in balance, and it is this difference in balance which makes all the difference.[1] This point was brought home to me on

A-T. Tymieniecka (ed.), Analecta Husserliana, Vol. XX, 393–405.
© 1986 *by D. Reidel Publishing Company.*

the first occasion when I made a presentation of this sort in India three years ago. In the discussion which followed, a Hindu scholar exclaimed, "You weren't describing *Chinese* values . . . you were describing *Indian* values!" He would have been the first to admit, however, that the two value systems are not the same in their patterns of outward expression and emphasis.

II. THE BASIC CONCEPTS

The first point in our consideration of this subject concerns the question, What do we mean by value? Numerous definitions exist,[2] of which I shall select one which is distinctively sociological: A *value* is an idea or concept revealed in (inferred from) a person's overt behavior. The term encompasses interests, likes, duties, moral obligations, desires, wants, and needs.[3] People who hold dearly to the values of individualism, democracy, and equality, for example, behave differently from those who do not. This is admittedly an ostensive definition,[4] yet I think it will suffice for the purposes of our discussion.

Given the broad range of behavior which can be encompassed by such a definition, how can I be justified in selecting only six representative values to be thus treated? This brings in the concept of core value, a value which is widely recognized by the members of a specific society as being central to the system of values typically found in that society. In this case, "centrality" may be "defined" as the percentage of people holding the value, i.e., if, let us say, ninety percent of the people hold to the value, then it can be considered a "core value." Beyond this, if most people in the society under observation reveal a deep emotional attachment to the particular value, this can also be considered grounds for classifying it as a "core value." My selection of six of these is, admittedly, somewhat arbitrary, especially in the light of the constraints of time and space. Suffice it to say, however, I feel confident that most of my Chinese colleagues would support my selection, though differences in balance (emphasis) undoubtedly exist.

III. SELECTED VALUES

1. *Familism*

This means that, above all else, the family and relationships among its members are preeminent. In traditional Chinese society it was not uncommon for one's friends to be drawn largely from the circle of one's relatives. This was particularly true in a village setting. Familism entailed a mutuality of rights and duties that delimited the scope and content of relations which a person could form with other people. Parents were obligated to rear their young children as best they could, while grown children were obligated to care for their elderly parents. This was a "natural" system of social security up to the present time. Today on Taiwan this is still true, though due to modernization the constraints are less heavy. In many ways, filial piety has been the "cement" binding the family members together. At present, the forms of piety are changing, though many firmly believe that the substance has not changed. Unhappily, for the older generation there seems to be a confusion of form and susbtance so that it is not altogether uncommon to hear an elderly person deploring the loss of filial piety in the younger generation.

While increasingly, friendships, professional, and commercial ties, etc., are formed outside the family circle, as a general rule there is no question as to which takes precedence in instances of relational conflict. Children have the right to parental care, but within limits determined by the parents. The phrase, "within limits determined by the parents", means precisely that. A child's right to nurture as an individual in traditional Chinese society was assessed in the context of the rights of the other children in the family. Birth order was particularly significant. Hence, the youngest (or the youngest two or three children) in situations of dire poverty might be denied food, affection, etc., so that the elder children would have enough. This is an instance of a reinterpretation of respect for the elderly (see below) purely in terms of birth order within a single generation.

Even today, among young children the eldest is expected to have certain rights and privileges denied to those siblings coming after him. The "other side" of these rights and privileges is that the eldest child has more responsibility over the younger ones. Hence, a friend once remarked, "Taiwan is the land where the small carry the tiny!"

As the child grows up, the set of obligations and rights vis-à-vis the other children becomes even more binding, particularly at the time the father retires or dies, and the eldest son must either take his place or act for him in decision-making matters affecting the family. Some leeway is allowed as to according the child advantageous or prejudicial treatment within the family. I recall one family in which the second of three sons was seen as clearly the intellectual inferior of his brothers. At the death of the father, he was the one chosen to stay at home in order to take care of the ailing grandmother and mother. To do this he had to drop out of school. His brothers went on to college and eventually overseas for graduate study, where both acquired Ph.D.'s. With the passing of the mother and grandmother, he rejoined his brothers overseas. It now became their responsibility to find work for him and to support him fully (including support for his wife and children) until a job came his way.

It goes without saying, of course, that if the eldest child is mentally retarded or otherwise handicapped the privileges and responsibilities are reassigned to the next oldest child, though preference is usually given to boys.

As we have suggested, familism implies complete fulfillment of the obligations attendant upon one's position within the family. In fact, there is a range of acceptable behavior, depending upon the age and condition of both parties — the parent and the child. The parent is clearly dominant when the child is very young. Under normal conditions this dominance diminishes with the child's increasing age and maturity, though the limits within which the child may be free to develop his own interests and skills are still determined by the parent. In many ways the child is regarded as "the passive receptacle" for information, skills, attitudes, opinions, etc., instilled in him by the parent. He may be indulged or severely punished at the whim of the parent. An extreme (and, simultaneously, pathetic) case is one in which the parent, someone known to me personally, would beat his children at the end of every marking period in school if they did not stand first in their respective classes. As one of my neighbors, himself educated at the graduate level in the United States, once put it, "I know from the books I have read that I should not beat my children . . . but I am Chinese, so I must beat them . . ." — a view that differs little from the biblical injunction about sparing the rod and spoiling the child!

2. *Respect for the Elderly*

Generational respect ramifies out beyond the family circle to include, at least overtly, respect for all elderly people. I clearly recall repeated instances of this expressed value on crowded buses, for example, in the Taiwan of twenty years ago. A young mother might board a bus holding onto a small child with each hand, a smaller one on her back, and herself obviously pregnant with a fourth. The chances were good that she would be allowed to stand until she got off at her destination. If a stooped, wizened little old lady or man got on, however, at least half a dozen young men would spring up to offer their seats.

Respect for the elderly is intensified when applied to someone in the position of teacher. This individual has broad powers over students, particularly in the elementary grades. Not uncommonly a possible difference of opinion between parents and teacher over the latter's treatment of their child is resolved in favor of the teacher. In other words, "Teacher knows best!" In contrast to an American setting, teachers in China may beat their pupils as punishment for poor performance on examinations, and the teacher who is especially severe may be granted high prestige accordingly. Even in nondisciplinary matters, the teacher's influence over a student or former student may still be very broad. While sitting in on a committee meeting one day, a meeting convened to arrange an international program for Chinese and American academics in Taipei, the name of a potential speaker was presented for consideration. He was a former student of mine at Tunghai University, had returned from graduate work in this country, and had been named as one of the outstanding ten young men in the Republic of China a year or so earlier. When the convener of the meeting learned that the man in question was one of my former students, he exclaimed, "You don't have to request, you can command!!!"

Before moving on to consideration of the value of personalism, let me comment briefly on something closely related to respect for the elderly: the idea of earnestness. In a Western context, this term implies sincerity, as when a person earnestly seeks the well-being of someone else. In a Chinese context, the outward appearance of sincerity or "being earnest" is emphasized, though if the "outward appearance" coincides with the "inner substance," all the better. A singularly apt illustration for this point is the case of a retired development expert who became an advisor to one of the ministries of the central government of the Republic of China, after almost twenty years of service with an international development agency. He was royally wined and dined, invited to make major addresses, including keynote speeches, at a host of seminars and conferences, toured numerous projects entirely at government expense — all of this as befitted someone in his exalted status of advisor to the central government. His recommendations for what the government should do, however, were almost totally ignored. He took this kind of treatment for about a year and then resigned. No one could have accused the officials at the relevant ministry of being deficient in "earnestness" in their treatment of this professional expert, yet, in fact, that was precisely the way they were.

3. *Personalism*

The Chinese term for this value is *rén=chíng*, human feelings, and it implies the effective utilization of existing personal relationships between people. As a friend, an American married to a Chinese, once put it, "Personal relations are everything out here!" More often than not in my direct experience of the Chinese milieu, who you know is much more important than what you know — though, to give this value "a Chinese twist," let me add that a fundamental component of what you know is who you know. It is often implicitly assumed that an individual will use his personal relationships to further his own or his group's ends. To do otherwise, is to behave stupidly. By extension, this means that knowing someone's name and vice versa is extremely significant in developing a relationship. A British friend once commented, "This society is so personal that it's basically impersonal!" Thus, if you don't know someone personally, i.e., if you don't know his name and he or she doesn't know yours and you have not met face-to-face, then that individual is a nonperson to you. The late writer, Lin Yu-tang, has explained that Confucian ethics say nothing whatever about the relationship between strangers.[5]

I have often thought about this personalism when being elbowed out of the way at the ticket counter in a train or bus station, though I must confess that, with the increasing whiteness of my beard and the diminishing hair on my head, such behavior has been directed at me less and less as the years have gone by.

Many Westerners, I feel, do not take seriously enough the constraints placed upon someone in a personal relationship. In other words, relationships in a personalistic society cut both ways. While it can be argued that all good things come to those possessing the relevant personal connections, the other side of the proverbial coin is the great emphasis placed on reciprocity. If I do something for you now and if you accept the favor, then you are obligated to me at some future date. It follows that the more significant the present favor, the weightier the future obligation.

Personalism, of course, is centered in the family, but it ramifies outward into all areas of society. In addition to "shirt-tail relations," one's schoolmates, friends, and even professors are not immune to its influences. The fact that you have taught a student is usually interpreted as your having established a personal relationship with him. Frequently, such a relationship is openly recognized as one that may be utilized in assisting the student to the attainment of some personal end or ambition.

To illustrate: shortly after returning to Taiwan following completion of my graduate studies, I was approached one day by a former student who had been enrolled in one of my English courses six years earlier. During that time we had not kept in contact and, for all practical purposes, our relationship had become nonexistent. (I vaguely recalled that he was in the broad middle-band of students in a sophomore English course.) There he was, unannounced and on my doorstep. I invited him in and, after exchanging a few pleasantries, he turned to the reason for his visit. He was applying to a graduate school in the United States and needed a reference letter assessing his English ability, preferably a letter written by a former professor who was a native speaker of English.

I demurred. I explained that I was not the one to write the language evaluation simply because our contact during the intervening six years had been nonexistent. Until that moment I had no idea as to how his English facility had developed, impressions gained from a few minutes' conversation were, indeed, very superficial, and, hence, for purely professional reasons I was not qualified. The former student initially became crestfallen, his disappointment shortly changing to what I would call "muted anger." We continued on other subjects of conversation for a few minutes more and then he left. I have neither seen nor heard from him since. It was painfully obvious that I had violated his expectations of our relationship — a relationship which he had defined as a personal one. I had experienced "the near edge" of "the sword that cuts two ways"!

Thus, one can revel in one's personal relationships, but he may also find them very constraining. By becoming personally involved with someone, one also becomes vulnerable, at least potentially, to exploitation.

4. *Propriety*

In *My Fair Lady* Henry Higgins observes, "The French don't care *what* you say, so long as you say it *properly!*" The same very much applies to the Chinese. They possess one of the world's oldest and most sophisticated languages, such that, when one writes well, considerations other than style are involved. The writer should be a fine calligrapher, as well as a verbal stylist. In few countries is more prestige accorded fine penmanship than in China — where good calligraphers may become public figures and reap rich rewards for their labors, both monetarily and in terms of high status.

By the same token, when one speaks, accent, appropriate forms of address, phrasing, and cadence are "properly emphasized." I am convinced that Chinese has one of the world's simplest grammars — and one of its most elaborate systems of politesse. Take the commonplace matter of forms of address. If a person wishes to be truly courteous in conversation, he will address you face-to-face either by title alone or in the third person, using your full name, as in, "Will Dr. Thelin be going down to the city this afternoon?" If the person addressed is elderly or apparently of a status superior to the speaker's but whose name is not known, he or she may be addressed only by title, as in "Will the department chairman be going down to the city this afternoon?" or by the word you in its polite form, as with *Nin* used in place of *ni*, the generic word for the singular second-person pronoun. Given names tend to be assiduously avoided except among equals who know each other well, in certain forms of very polite correspondence, or when a superior addresses an inferior, e.g., a household head to a servant, a professor to a student, etc.

The incident I am about to relate occurred some years ago. It remains to me one of the clearest examples of propriety with which I am familiar. (I might add that it also clearly illustrates the concept of 'face.')

An American visiting professor of psychology presented me one day with a list of Chinese translations of psychological terms in English. As the list had been published many years previously, the professor wanted to know if someone could check the translations to see if they were still up to date. We had no Chinese professor of psychology in the department at the time, and, in casting about for someone to do the job, I thought of a well-known psychologist in Taipei. None of our faculty knew him personally, but a graduate assistant allowed as how the gentleman was a personal friend. I immediately asked whether he might write a personal note requesting assistance in this admittedly minor matter — as looking over the list and checking the translations might conceivably occupy twenty or thirty minutes of precious time. I was ill-prepared for the graduate assistant's response, "Oh, but it would be inappropriate for someone of my (lowly) position to write such a letter, even though we are personal friends."

As I have suggested, one learns of such proprieties frequently through hard experience. After a previous furlough I returned to Tunghai University and in my role as department chairman convened the first semestral meeting of the sociology faculty. I asked one of my former students to take the minutes. He had served almost four years as graduate assistant before being promoted (in my absence). I was amazed that he immediately handed to one of the new graduate assistants the book in which the minutes were entered — this without a single word of explanation! I suddenly realized that I had committed a faux pas. Graduate assistants — not lecturers — were appropriate minutes-takers!

5. *Face*

Of all core values, "face" is, perhaps, the most central. He who causes a Chinese seriously to lose "face" can be considered guilty of an absolutely reprehensible act. What is this concept of "face"? It is something which all societies have, yet which differs from society to society in terms of scope, content, and importance in social life. My Chinese friends have often had great difficulty in communicating the meaning of the term in words, particularly since it is so implicit in their thinking. There are numerous definitions, but for my present purpose I have selected or formulated one of my own. "Face" refers to the public presentation of the self. I say "public" because this is the arena in which, because the attention of others is focused, however briefly, upon your words and actions, "face" is preserved, gained, or lost. If an embarrassing situation occurs privately, as among peers who are close friends, no particular harm is done. When the public dimension is added, however, the results of a faux pas for the individual committing it may be utterly devastating. "Face" is also a function of one's status, rank, or class: the higher the position, the more sensitive the "face" of the person occupying it — and, hence, the greater the care taken to maintain it or to avoid losing it.

Most Westerners, in considering non-Western societies, tend to focus on the negative aspects of "face," and much of the literature available in English has this emphasis. At the same time, we should not be blinded to recognition of its positive aspects: A student may work especially hard to gain "face" as the best in his class or to pass the joint university entrance examination at a level where he may be assigned his first choice of university and department in the preference list he filled out when he registered. In other words, the positive motivational dimensions of "face" should not be overlooked. Nevertheless, it is the negative side which, too frequently, is salient.

Once "face" has been established or gained, care must be taken to preserve it, i.e., to avoid losing it. Thus, in ambiguous or "sticky" administrative matters, officials may fall all over themselves to avoid showing "face." (The Chinese term here is *ch'u mien.*) In this context, while the rewards of success may be great, the risks of failure and, therefore, loss of "face," may be estimated as too high. "The boss can never be wrong" syndrome is, admittedly, an oversimplification, but it contains some element of substantive truth, as research by Robert Silin has indicated.[6]

"Face" can also be "given," as when the convener of a conference asks a lesser-known participant to "perform" by raising a question or making a penetrating comment. The "other side" of this behavioral pattern is that the person asked to perform should perform well or not at all. (He may politely pass up the opportunity to speak.) This is where the distinction between "those who have something to say" and "those who have to say something" is very much to the point. The high-ranking ones are expected to perform passably well, and their "performance," in a sense, gives "face" to the assemblage of those who observe it. Ideally, the best possible situation is where all speakers "have something to say." Otherwise it is preferable if the two types of performers are combined in a single person, i.e., a high-status individual who can take nothing and make it sound as though it's really something.

One of the most poignant examples of loss of "face" in my experience occurred after a meeting attended by students and faculty at Tunghai University. On the way back to the dormitory a student openly criticized the procedure of discussion. I unwittingly gave qualified support for that procedure, at which the student retorted, "That's wrong!" For a moment nothing happened, and then the student realized the import of what he had said, namely, that he had disagreed publicly with a professor — and a Ph.D., at that. The balmy night air was suddenly filled with sounds of highly emotional apologies. I repeatedly asserted that I did not care about the student's disagreeing with me in public, but all to no avail. The next day one of his English lecturers, a man who had been present at our exchange but had not "shown his face," came to me in an inter-cessory role. I repeated what I had said the previous evening, and, presumably, the message was relayed — again, to no avail. Fortunately, the student in question was enrolled in another department and was not taking any of my courses. Still, whenever we chanced to meet on campus, he inevitably looked the other way.

6. *Male-centeredness*

In an era when "women's lib" has made remarkable headway in this country, it is, perhaps, appropriate to comment, at least in passing, on this aspect of Chinese values.

In traditional China women were brought up to accept at a very early age their subordinate status to men — as epitomized in the "three obediences" regarded as their cardinal virtues: (1) obedience to one's father before marriage, (2) obedience to one's husband after marriage, and (3) obedience to one's son in widowhood. There were, of course, some noted exceptions such as the Emperor (she refused the title of Empress) Wu Tze-t'ien, General Hua Mu-lan, and Tz'u-hsi, the empress dowager who pulled the strings behind her son's throne and, presum-ably hastened the end of the Ch'ing Dynasty. It can be argued that these were exceptions which proved the rule!

In any event, even up to fairly recent times, it was customary on being introduced for a person to ask the number of children a man had, the term "children" meaning implicitly the number of sons, for girls did not count. Even gentry women (in today's jargon, upper- or upper-middle-class women) typically internalized, while growing up, self-concepts of themselves as "nonchildren", i.e., second-class children. More than sixty years ago, when mission schools for girls were founded in some of the larger cities of China, virtually all students were enrolled on scholarship. In other words, their families had to be paid to send them to school. Why? Because everyone knew that women were basically uneducable, being fit essentially for running the household and for bearing sons.

Much has changed in the past six decades — on both sides of the Strait of Taiwan — and yet numerous inequities remain. In the academic

community of Taiwan the number of full professors who are women
(except in departments such as child welfare) is almost miniscule. This is
not to claim that women have not come to the fore in many lines of
academic and professional endeavor — only that "tokenism" is widespread and even where it exists "equal pay for equal work" tends to be
the exception rather than the rule.

Discriminatory treatment of women can be very subtle, indeed, as
when a history professor remarks at a committee meeting, called to
evaluate teaching methods, that the male students in his department
are actually better than the female students, though their grades are
generally lower. Not so subtle is the "quota system" adopted by various
academic departments, whereby no more than half of the entering
freshman can be women. This means taking more men whose entrance
examination grades are lower and, therefore, extending to them decidedly
preferential treatment in the highly competitive university entrance
process. Why? Because having too many girl students is a blot on the
department's public image.

IV. CONCLUDING COMMENTS

In conclusion, let me comment briefly on prospects for these core
values, as I have called them, in the future. It is a truism, to assert that
Taiwan has been undergoing extremely rapid social and economic
change since 1950. Does this mean that these values will become less
salient as time goes on and, eventually, be replaced by other, new core
values? The question is not to be answered easily. Commercialism and
materialism seem rampant on Taiwan and, with a burgeoning population (the island has one of the highest densities in the world), the
freedom for individual assertiveness which anonymity gives in an urban
setting is receiving more and more emphasis.

Let us take familism as a case in point. Family size, defined as the
number of persons registered in a dwelling unit, is diminishing, in large
measure due to successfully implemented family planning. What was a
rural society thirty years ago has now become urban. The nuclear family
is increasingly the typical family unit. This poses problems as to what to
do with the living arrangements for the grandparents. In the traditional
village the existence of the extended family of at least three generations
was strongly supported by residential proximity: all grown sons and
their families lived next door to each other and grandparents, usually as

parts of a large house. The pattern today features one son back on the farm and the others, including daughters, living separately with their families in town — frequently half the length of the island away. The grandparents may remain with the farming son more or less permanently, though the new pattern of rotational residence (for several weeks at a time with each family of a son or daughter) is becoming fairly widespread. Neither arrangement is particularly ideal and there is speculation as to how the family itself is being affected.

A social historian-cum-sociologist, Cho-yun Hsü, of the University of Pittsburgh, has argued that the Chinese family is indestructible.[7] He recognizes the ongoing changes in family size, residential arrangements, and the technology of daily family life, yet he maintains that the quality of relationships continues without substantive modification. This view is not without its critics, though very few would deny the continuing predominance of the family in the individual's set of relationships. The question is one of time in the context of ongoing urbanization.

It seems reasonable to predict that thirty years from now the core values just discussed will still be around and occupying a strategic place in the constellation of Chinese values. Quite likely, however, their scope will have narrowed and their content somewhat changed.

Tunghai University

NOTES

* This paper was presented to a seminar held at Radcliffe College in Cambridge, Massachusetts, on March 3, 1983.
[1] A comment made by Canon P. T. Chandi during the opening address at a conference which he convened in Hong Kong during August 1970.
[2] The brief discussion which follows is largely drawn from Robin M. Williams, Jr., "The Concept of Values," in *International Encyclopedia of the Social Sciences*, 16:283–87.
[3] In social science, *values* generally refer to conceptions of the desirable and standards of desirability.
[4] Hans L. Zetterberg, *On Theory and Verification in Sociology* (Totowa, N.J.: Bedminster Press, 1963), p. 34.
[5] Lin Yu-tang, *My Country and My People* (New York: John Day, 1939).
[6] Robert H. Silin, *Leadership and Values: The Organization of Large-Scale Taiwanese Enterprises* (Cambridge: Harvard University Press, 1976).
[7] This statement was made during discussion which followed a presentation by Professor Hsü at Tunghai University in spring 1980.

HANS MARTIN SASS

THE MORAL A PRIORI AND THE DIVERSITY
OF CULTURES

I. MORALITY – NATURAL LAW OR SOCIAL CONTRACT?

In 1851, twenty-one white pioneers settled in the fertile area called
Puget Sound near the Pacific Ocean, about 120 miles south of where
the Canadian border now runs. After peaceful negotiations with the
native Duwamish Indians, they named the settlement Seattle after the
chief of the Duwamish tribe. Three years of more or less harmonious
coexistence followed; then Franklin Pierce, fourteenth president of the
United States, urged the Duwamish to make room for more settlers by
withdrawing to an island reservation.

The tribe acquiesced, but only after Chief Seattle had questioned the
president's moral position and the immorality of contracts for the sale
or ownership of land in general.

How can you buy or sell the skies – or the warmth of the earth? If we don't own the
freshness of the air and the glittering of the water, then how can you buy it from us. We
are a part of the earth, and it is part of us. The fragrant flowers are our sisters, the deers,
the horse, the great eagle – they are our brothers.[1]

Seattle's moral arguments assume a given order to which we humans
belong and must act in accordance with. Western intellectual thought
has termed such a position the *lex naturale* or natural law position.
Human morality is understood to be part of a given cosmic order; the
task of being moral is to fit as harmoniously as possible into this eternal
system.

Chief Seattle's address contains at least three guidelines which are
basic to moral behavior and to a further structuring of rewarding social
relationships. First, everything is interrelated and must show solidarity,
just as the same blood unites and dignifies a family. Therefore, "hurting
the earth means to despise the creator." Second, each individual human
being has the right to a life according to his personal values, without
having to "take into account how different he is in regard to his
brothers." And, third, *pacta sunt servanda*: covenants and contracts
must be faithfully observed, and must be as reliable as the turning of the
seasons. "My words are like stars, they never will pass away."

407

A-T. Tymieniecka (ed.), Analecta Husserliana, Vol. XX, 407–422.
© *1986 by D. Reidel Publishing Company.*

Seattle's moral attitude toward man and environment does not — in Seattle's understanding — presuppose an elaborated epistemology or ethics. Chief Seattle understands his position to be a result of deduction from observed phenomena, of sharing and communing with nature as a whole, and of solidarity with all living things within nature. His theses on the rights and dignity of the individual, as well as of social and political contracts, are based on this general understanding of natural law. The rules of this natural law are so basic and essential to Seattle that, had he been familiar with Western intellectual terminology, he might very well have used the Kantian term of a priori for these values.

Having always lived within the Western tradition, the settlers must have disagreed with the essence of Seattle's views on man's harmony with mother nature. The Judaic-Christian tradition of Adam's fall and Paradise Lost, the Age of Reason tradition of basing social relations on a social contract, the Baconian analyzing, transforming, and engineering approach to Nature, all put their general stance toward a given situation into a totally different situation. Their approach to nature is a culti-vating, an engineering one, which seeks to change raw nature into a cultivated environment for the educated and cultivated person. They strive to free man from the dependencies and insufficiencies of nature, to protect him from natural disaster and unfriendly seasons, and to increase the standard of living and enrich all areas of the arts and culture.

Many Western thinkers in the past assumed that people who did not share a specific cultural heritage could not share the same values and therefore were not equal. By denying equality to those of different cultural background, it is also possible to restrict their human rights, as well as to ignore any obligation to observe contracts. A sense of responsibility is felt only toward those individuals and groups who share the same specific human values and morality.

It is not without reason that recent Western thought has shifted from natural law arguments to the notion of a social contract as the basis of morality. Nature, including "natural" values and laws, is open to interpretation and hermeneutics, and is often used in both political rhetoric and ideological indoctrination. It is a concept easily exploited. Western intellectual life and politics provide numerous examples of how "interpreting" the creator's or nature's laws, or the best interests of the "workers" or "the people," serve the purpose of political domination and ideological alienation. This is the principal reason why, in modern

Western political thought, the theory of the social contract has increasingly replaced natural law argumentation in establishing grounds for moral argumentation. The main idea was to free nonformal moral and other values in what has been called an Open Society, which itself is safeguarded by a state exercising power and law, and protecting the free flow of ideas and the embodiment of nonformal values in formal law.

The fruitful elaboration of this tension and disunity between state and society may be found in Locke's discussion, in his *Letter Concerning Toleration*, of political and religious values. It is next impressively presented in Hegel's concept of the open society in his *Philosophy of Right*. The complicated nature of even the nonformal understanding of "formal" is demonstrated in Locke's model. Here, the acceptance of a personal God and other so-called essentials of Christian belief is a precondition for granting a person the dignity of being fully human and of enjoying and exercising his/her freedom and citizenship. Locke's model remains based on a reduced form of natural law. In Hegel's model, formal law is not any law, but is the result of a dialectical emergence in history of culture, politics, and morality. The claimed opposition between formal law and morality is not strictly observed in either Locke's or Hegel's models.

But in insisting on a legal and political framework, which is to be as formal and as minimal as possible, to serve as the basis for morality and culture and for diversity in culture, these models try to protect the individual's dignity against the *lex naturale* hermeneutics of others — individuals, groups, or the entire society, state, or church. As a means of achieving, whether intentionally or unintentionally, a dominating competence, it is possible to interpret the moral a priori to suit a given social and political context. The weakness of the latter has been summed up by, among others, Machiavelli. But the strategy for emancipation embodied of formal law was intended to create, to liberate, and to preserve an open space for the individual, within which the moral a priori could be autonomously interpreted and embodied into a personal life-style.

Kant's formal and categorical imperative and his establishment of legal — and moral — obligation as the foundation of an open society is the purest form of distrust toward any interpretation of the moral a priori. There may be a moral a priori in all human beings, as Kant believed, or, as Hobbes probably felt, there may be none. Either way, the social contract theory is usable. According to the theory, formal law is understood either as a minimal outgrowth of natural laws which must be obeyed in an open society, or as a tool created to protect each person from the more uncivilized, and therefore ineffective methods of egotism and aggression. This lays the groundwork for a better regulated or more civilized, in any case more effective, form of selfishness or egotism.

However, it is important that both positions of the social contract theory reach the same conclusions Chief Seattle drew from his specific natural laws position. These conclusions are (1) that covenants and contracts must be observed, and (2) that there exists a definite number of rights for each human being.

The conclusion that the same or similar practical principles can be established on different metaphysical or theoretical grounds raises the question of whether a flexible set of such general and practical principles can be established. These principles would serve as a basis for more detailed and specialized mid-level diversification in different cultures. This would immensely increase the importance of the general practical principles, and the demand to be able to understand and describe them phenomenologically or dialectically. It would decrease the role of metaphysical or theoretical means, which only give subsequent evidence to values otherwise established. But before we address that question, we must analyze in greater detail the nature of these general moral principles, and what it would mean to call them a priori values.

II. THE MORAL A PRIORI

Kant may have been correct in assuming that a logical and orientational a priori must exist in all human beings which serves as a precondition to logical argumentation and the understanding of each other, to an everyday orientation in space and time, and to activities in the natural sciences. But he is negligent in denying similar a priori preconditions for human beings in areas other than logic and the natural sciences. The thesis of a religious a priori, caused by the numinorum, was first elaborated by the Kantian Rudolf Otto in his book *The Holy* (1917). His successor in Marburg, my mentor Friedrich Heiler, provided further evidence in support of Otto's thesis of a religious a priori with his extensive studies on the phenomenology of *The Prayer* (1918).

In *Formalism in Ethics and Non-Formal Ethics of Values*,[2] Max Scheler, however, draws Kant's approach to anthropology into question. On the basis of phenomenology, he condemns it as too intellectual, too rational, and too mechanical, and instead presents a broad survey of the rich presence of a nonformal a priori in moral, cultural, and religious life, which is evident in every human being. Scheler blames the mechanical, rational psychology of the eighteenth century for not having enabled Kant to recognize pure or even formal relationships among values and between values and their bearers. But, at the same time, he adheres to the Kantian concept of a priori in his presentation of a "pure theory of values" and a "pure theory of valuations" which corresponds to Kant's "logical theory of thinking." A "basic moral tenor" is present in all humans, just as we all possess a basic capacity for logic.[3] Scheler elaborates on the existence of tables of values (*Wertetafeln*), which are analogous to the table of categories (*Kategorientafel*) presented by Kant in his *Critique of Pure Reason*. Most important among these tables is one showing a sequence of a priori preferences between higher and lower values. Scheler believes that there exists an a priori intuition of preferences in regard to sets or systems of qualities of values, which he styles value modalities.

The interconnections of these value modalities prove to be independent of actual value feelings or of factual objects and will not change locally or historically. The first two a priori value modalities are the a priori of agreeable-disagreeable and the a priori of vital feelings. The values of agreeable-disagreeable represent the particular responsiveness of living entities to approving or disapproving changes in their environment in regard to survival or maintenance of a better life-style as a self-regulating system. Vital feelings a priori, carried only by living entities, represent the genuine, and not just empirical, quality of life. The self-values of life are both noble and vulgar; its consecutive values would be feelings of health, strength, aging, death, and their opposites.

Of greater importance are the two other value modalities of spiritual values and of the holy. Spiritual values such as beautiful, right, and truth, its consecutive values, and its opposites cannot be simply reduced to biological values. They are a law within themselves, and are apparent in acts of spiritual feeling, love, preferences, and sympathy. Friendship, or admiration for the performance and embodiment of spiritual values are forms of spiritual sympathy. There seems to be an a priori tendency to sacrifice vital values for the realization of spiritual values. This is especially true of the a priori self-value of the holy, to which adoration, faith, hope, and despair stand linked as values in reaction. According to Scheler, we apprehend the divine a priori by means of a specific kind of love different from all other forms of love and sympathy. Vital values must have adequate objectives, but in regard to the holy any object can become an absolute one, while all other objects must then be understood as derivative or consecutive.

In Scheler's view, it is also a priori self-evident that the self-value of the holy is to be seen as the value of a person, not a thing.[4] The a priori existence of a personal God, as experienced by the individual, therefore represents the highest value among the analyzed value modalities. It simultaneously safeguards the specific ranking of the few levels of value modalities. These few levels serve as a guiding framework, within which value bearers retain a certain flexibility on the basis of individuality, history, or location. But, both this flexibility and the precise setting of the few ranks of value modalities are restricted or immunized by (1) a theory of perversion, and (2) the idea of a personal god.

Scheler's concept of perversion, i.e., of vices instead of virtues, serves the purpose of stabilizing the presented patterns of values and their relationships. It also acts as a necessary link in his phenomenology of persons in action and of the influence of value-bearing persons on actions. A person cannot be understood as a substance vested with good or bad attributes; rather a person is represented by means of his actions — "living in each of his acts" and "permeating every act with his peculiar character."[5] The acting person does not linger behind or float above his or her actions. Such misleading images are taken from the time-space framework of substantialization. It is the entire person which "is contained in every full and concrete act, and the whole person 'varies' in and through every act, without having exhausted itself by having performed these acts, and without 'changing' like a thing in time."[6]

This means that the identity of the person is inherent in his or her reaching out by means of actions, living into time, and becoming different through the passage of time. In all these acts, the person is the bearer of the above-mentioned values. It is impossible to determine directly the essence of the human being. Only indirectly does it become apparent that it is the value-bearing acts he or she performs which lay bare the human dignity and identity of the individual. At the same time, these value-bearing acts reveal the fact that, for different reasons, the individual may morally, logically, or aesthetically act in a perverse fashion — failing to achieve the a priori rank among the values.

The true relationship among values, and between values and their bearers a priori, is established and safeguarded by the personal God, who in his external acts is related to the entire world of values. Similarly, the individual value-bearing person is related to his or her individual world. According to Scheler, the a priori of values and the a priori of their ranking, as well as their objectivity, are ultimately protected by the personal God.[7]

Scheler's argument in favor of the a priori of nonformal values and their a priori ranking contains both strengths and weaknesses. By using the Kantian concept of a priori as something which can also be blended into a *tableau a valeur, Wertetafel*, or table of values, Scheler succeeds in creating a form of objectivity within an a priori value system. But this does not make the most advantageous use of his underlying concept of act-phenomenology. On the other hand, he avoids the problems that confronted Husserl in his attempts to identify the ground, foundation, and subject of phenomenological reduction. For Scheler, as originally for Kant, there exists an a priori precondition for acting persons. A formal as well as nonformal a priori forms the precondition for the development of cultivated social, moral, political, and aesthetic environments.

The Kantian notion of an a priori, which serves as a precondition for, as well as being realized in, human action, may be construed as a preliminary answer to what the real *ground* is of the *self-consciousness*. But the specific interpretation given it by Scheler makes it somewhat inflexible as a means toward understanding the global diversity of cultures, and toward establishing a global dialogue and cooperation between different cultures. The few ranks of value modalities presented by Scheler, which in fact truly represent the mainstream of the Western metaphysical tradition, are perceived and experienced in a totally different way by a Hindu or Taoist. In experiencing essential forms of harmony, the Taoist presents his moral and political argumentation differently. Scheler's demonstration of the a priori of a personal God would be meaningless. Instead the Taoist would maintain that the divine cannot be of a personal character, and that it does not labor under the restrictions binding the individual. The same view is presented by Scheler himself in his later *Man's Place in Nature*.[8]

In the Eastern tradition, the value of the Divine seems to be essentially nonpersonal. It is true that the sequence of honor, longevity, and harmony in Eastern culture has its opposite in similar, but not identical values in Western thought. Neither of these systems, however, should be understood as a priori, for both are results of cultural activities. Max Scheler's importance as the most eminent moral thinker of our century is not entirely based on his specific arrangement of a hierarchy of nonformal values, for which he draws on Western tradition. Rather, it lies in his overcoming of Kant's narrow rationalistic anthropology through his inclusions of morality, aesthetics, and religion into Kant's concept of the a priori. Scheler's expansion of the concept of acting and interacting persons from the field of logic to all other fields of human activities is extremely valuable.

We experience ourselves and others only through actions, to which we and other human beings are motivated as bearers of values — sometimes of different values. It is only consistent with the general concept of act-phenomenology to understand the rich diversity of factually existing cultures as a result of the acting and interacting of persons and groups motivated by human capacities given to all of us. These capacities do not necessarily manifest themselves within a specific prearranged ranking, which is established as a result of specific traditions and forms of interaction.

Both a logical and a moral a priori exist. But the specific form of the logical a priori, as formulated by Kant, is preconditioned by Euclidean geometry and Newtonian mechanics, while Scheler's specific form of the moral a priori is preconditioned by the Judaic-Christian tradition in metaphysics and ethics. Non-Euclidean geometry and the various contemporary forms of logic are handled by that same human logical capacity which Kant, in the late eighteenth century, considered identical to traditional logic. Non-Western forms of morality, and the rich variety of historically and locally realized individual and social moral values, must also be related to one a priori human capacity for morality, i.e., for arguing and acting according to moral values. The Western Christian form is only one specific elaboration among many others.

Kant's transcendental deduction of "*reine Verstandsbegriffe*" in his *Critique of Pure Reason* therefore does no more than serve the self-deceptive purpose of reconfirming an already-established form of traditional philosophical logic. Scheler, too, in his theory of value modalities only reaffirms the specific Western setting and ranking of nonformal values. We must be aware, therefore, that these aspects of both Kant's and Scheler's theories on acting persons merely reestablish epistemologically conclusions drawn from metaphysical argumentation. Nonetheless, whether argued on metaphysical or epistemological grounds, it is a fact that logical, as well as moral, capacities exist. The richness of their diversification may be grasped by reaching a phenomenological under-standing of the solidarity of reasoning, calculating, sympathizing, caring, and loving.

The logical, as well as the moral a priori — in which should be included the religious a priori, though it demands more elaboration — should not be seen as identical with a specific set or table of formal and nonformal values. Rather, the a priori should be understood as the precondition, the basic capacity, or tenor for human beings, which enables the development of such sets or tables. The logical a priori would be a precondition to both Aristotelian and other forms of contemporary logic, in order to argue reasonably, and to condition and engineer the environment. The moral a priori, on the other hand, would be a precondition to the establishment of ranks or sets of value systems specific to each and every culture. This would enable humans to argue morally and cooperate with others in a cultivated and educated fashion, to maintain a dialogue relationship, and to develop institutions as forms of embodied and settled value argumentation. The religious a priori would be the necessary foundation of different sets of achieving a dignified understanding of the world as a whole and acting accordingly. We would mediate without existence in general, just as we and our fellow human beings seek mediation on the basis of a common moral tenor of relatedness.

Such an understanding of the a priori would preclude its assimilation into, or identification with, any of the logical, moral, or religious systems based on it. The "in itself", the separate being of the a priori, would allow it to be adressed and experienced only in forms specific to the acts of persons, cultures, and institutions. As a consequence, the moral a priori can only be recognized within the different forms in which it has been embodied and developed locally and historically. Its main role is to elaborate on and acknowledge the individual's dignity and his solidarity with other human beings.[9]

III. SOLIDARITY AND DIVERSITY

Scheler has noted that unity a priori is the highest value for the holy, while diversity a priori is essential for the development of culture.[10] The diversity of moral culture can be understood as being related to the moral a priori. This means a focus on the most basic forms of interpersonal relationships in all cultures, such as the I-thou relationship, and the neighborhood or immediate community. In his or her specific being-acting-becoming relationship to others in society — Karel Wojtyła in *The Acting Person*[11] describes this as "genuine dynamism" — the person is not necessarily operating according to a specific table of values. These values are instead established in the interpersonal relationship of giving and accepting, of participation, and of authentic forms of solidarity.[12] Within these authentic forms of participation, the individual dynamically transcends former positions of its own self, and of involvement.

Specific attitudes toward community and the common good are created in a responsible fashion by such an actively participating individual. Two forms of nonparticipation are possible, however. Non-participation in this process can signify the unwillingness of the individual to become involved. But it can also be a result of enforced conformism. Within such a system, the individual is forced to submit to a preestablished set of values without being able to exert a personal dynamism or consciousness. Both forms of nonparticipation not only result in nonauthentic forms of community, but also lead to a miscon-ception of the common good. Most importantly, they cause an aliena-tion of individuals from each other and from their own identities, since the individual lacks "his authentic fulfillment in the community of being and acting together with others." As essentially social beings, human beings rely on forms of community as the essential element in which and through which they act, transcend, and fulfill themselves, embodying the common good as well as specific forms of social, cultural, and political institutions.

The most fundamental question in moral argumentation then remains how to make moral judgments on the basis of which certain acts can be considered to be authentically moral, and certain others can be perceived to not express the basic moral tenor. Taking into account the diversity of cultures, the answer must be given in the broadest and most liberal sense. Only extreme cases, which are unacceptable because of their violation of basic human rights, should be singled out. Concerning diversity within a single culture, it is possible to give a more precise answer, using specific reference systems developed within that particular culture. General metaphysical principles, however, cannot be used as references, since cultures and developments within any one culture disagree on exactly these generalia and their interpretation and application.

Instead of referring to general principles, it would be preferable to establish a system of mid-level principles. These principles would be founded on metaphysical principles, but would exist factually and culturally as social phenomena, regardless of a conscious reliance on orientational frameworks. These mid-level principles might be used phenomenologically or dialectically to develop a middle-of-the-road understanding of authenticity. They represent essential forms of togetherness and relatedness.

A look at the diversity of historically and culturally realized forms of morality shows at least half a dozen reference systems, within which it is possible to check the authenticity of moral actions, of mutual understanding, and of cooperation. Those mid-level principles which might serve as references are (1) the thou, (2) the family, (3) the neighbor, (4) the compatriot, (5) the friend, and (6) humanity. Within these systems of reference, both the thou and humanity as a whole refer to the equality of all human beings in a very essential sense. There is only one mankind, not a black or white one, or a contemporary and medieval one. This is absolutely self-evident, a priori, and a precondition to each and any act which wants to be understood as morally authentic.

The thou,[13] on the other hand, accepts the individual and his needs and relationships as a unique case of reference. This is especially true in regard to what might be called basic human rights. These basic human rights address not only the rights of others, but should focus on the personal responsibility to protect other human beings from starvation, pain, torture, or enslavement. This holds true even if the other is not a friend or a family member, bearing perhaps no other relationship than that of being a fellow human being. These human needs or rights are most fundamental. They are by no means the result of a culturally specific moral tradition. Such basic human rights are primarily related to the free functioning of body and mind, and to situations which pose extreme threats to a person's dignity in any culture.

More elaborated and culturally specific so-called human rights, for example, in the form of the *Universal Declaration of Human Rights*, cannot be used easily as a reference system because of the diversity of cultures. The *Universal Declaration of Human Rights*, first declared by the United Nations in 1947, is influenced above all by the Age of Reason concept of man drawn from European tradition. It therefore stresses the role of the subjective individual, while other traditions might focus on natural or cultural forms of solidarity, such as the family or neighborhood.

The family is the closest, most natural system of reference when raising the question of what specific act would be an authentic moral one in a specific situation. The complexity of interpersonal relations, the relatedness and common needs of the persons involved, make it easy to separate negative from acceptable actions. Insofar as the family serves as a nucleus within wide social bodies in culture, and, in its specifically realized form, always is a part of the surrounding culture, its use as a reference system is applicable only within a given culture. But it may serve a general use by drawing on a given culture as a reference for the formulation of basic human needs which in that specific culture might be violated.[14] The family can also be used as a reference system in order to determine the authenticity of our contemporary Western life-style, and to determine how authentic our actual forms of "being together with others" really are.

A neighborhood lacks all closer ties of blood, kinship, and other family properties, but it is the most eminent system of reference, enabling solidarity in a normally already culturally enriched and diversified human community. The basic relevance of mutual aid becomes a cultural asset while, on the other hand, the dignity of the neighbor as an individual requires respect for his or her values of privacy. This neighborhood, though a natural arrangement of people living close together, represents the transcendence of simple natural forms of living together. As the preeminent field for achieving participation and privacy, the neighborhood is the nucleus of all larger forms of community and provides a basic reference system in any cross-cultural dialogue.[15]

Naturally, a neighborhood shows common values only to a certain extent. Its members share common standards of conviviality but also seek to protect the diverse private values of each neighbor. The opposite, however, holds true for compatriots in a nation, educated people of the same level, and members of a club or church, who all share common values to the highest possible degree. Sharing the same high goals, paying reference to the same close cultural tradition, bearing the same values and thus being essentially a part of a value-related community, establishes a reference system which is most sensitive to spiritual and cultural values. Yet it can also make it difficult sometimes to realize the authenticity and originality of moral acting in reference systems other than one's own. The use of the common reference system of neighborhood between cultures makes it possible, therefore, to address questions of mutual aid, privacy and participation, and educated forms of living together.

Finally, the friend as a reference to one's own authentic acting is the most valuable reference man possesses. True friendship does not demand reaffirmation in regard to its authenticity. It can give all other forms of coexistence a high goal of mutual sympathy and of enriched forms of understanding, acting, transcending. The friend may also provide information on the authenticity of the person's actions in regard to others. The mutual experience of being together with a friend enables a fuller participation. A mutual exchange of friendship makes it possible to transcend former acts and to realize other forms of community on the highest and most intense level. Friendship serves to illuminate an individual's life and actions. It is neither a simple form of natural sympathy nor a culturally introduced form of being together; rather it transcends the difference between nature and culture, between natural law and social contract. It also may serve as an essential reference in regard to those communities which violate or disregard the conditions or challenges of either partner in the friendship.

The six above-mentioned mid-level principles or reference systems, by which an authentic human coexistence can be achieved, do not present any hierarchy between themselves or between forms of participation. Nor do they represent new forms of metaphysics or pseudo-metaphysics. They start with the point at which the individual realizes himself or herself in acts of being together with others, of being and becoming through participation or transcendence.[16] They are inter-related and compensate for each other.

IV. THE DIALOGUE AND THE A PRIORI

The above-described mid-level principles, which represent phenomeno-logically different forms of coexistence, are not static guidelines. They are highly flexible values centered around one or two key issues. In moral argumentation, they may be of aid in creating a dialogue approach for intensifying (1) participation within the community, (2) a cross-cultural dialogue, and (3) a dialogue relationship toward tradition.

The following will briefly outline what this threefold dialogue approach would mean in regard to the moral a priori and the diversity of cultures. Only such a dialogue approach could be indirect enough and would not confuse the a priori with any one of the realized sets of values.

(1) It is vital for Western intellectuals that community participation should above all not be a question of intellectual participation or of organizational structuring. Dialogue within a community involves the

entire person and brings about the opening up of the person through acts of dialogue, understanding, and cooperation. Such an "open" attitude will certainly eliminate pseudorational generalizations, which demand conformity or reject an active involvement in authentic acts of participation. It also will demonstrate that scientific or metaphysical frameworks using value tables may prove to be fair instruments for comprehending intellectually what has occurred or what is occurring within acts of authentic community in the widest possible sense.

(2) This would be even more valid for the as yet overdue dialogue between the main global cultural traditions. It is not merely a question of comparing systems of metaphysical orientation or of battling specific forms of social, political, or cultural institutions. It will require those same acts of solidarity as are necessary in neighborhoods of individuals or families, as well as toleration and mutual aid, privacy and participation, an "open" attitude toward the richness and needs of the other, and a willingness to transcend oneself in the new acts of becoming and participating. This not yet initiated intercultural dialogue cannot be a mere intellectual comparison of orientation systems. The a priori of acting morally, i.e., the diversity of authentically acting in solidarity, can only be grasped by acting together in various forms and fields of the community. The dynamics inherent in the variety of cultural and moral traditions can only be understood as the result of acting persons, and will serve as a driving force for further dynamics in acting and transcending. One of the results of such an intercultural dialogue will be the understanding that moral goals and genuine aspects of participation are the same or similar, regardless of the culture in which these issues have arisen or of how they are treated intellectually.

(3) The dialogue approach to one's own community, as well as to other cultures, might not be authentic, if one does not take the same approach toward one's own tradition. The dialogue approach toward tradition understands tradition as, for the main part, happily, and sometimes unhappily established standards for coexistence. Tradition, including moral tradition, is not an inflexible static system of commandments which must be obeyed. Such an understanding would violate the genuine dynamism and, within it, the personal responsibility of the acting person. Authentic interaction with others includes contact with other cultural traditions, as well as one's own tradition.

Dialogue, as Wojtyła points out, often uses the attitude of opposition. It definitely does not intend uniformity; instead, it is the authentic expression of dynamically, not statically, "strengthening interhuman solidarity." It "favors the development of the person and enriches the community."[17] No happy and constructive communal life is possible without a dialogue in the broadest sense. Successful and happy dialogue is proof of a basic social and moral tenor in the human being, just as the diversity of cultures illuminates the existence of the a priori and the dynamism of the acting person.

Such a dialogue approach could have been available to Chief Seattle of the Duwamish Indians and to the white settlers. Both positions could have been open to such a being together with others. The a priori for mutually shared moral goals and for acting morally toward each other would not, however, be exclusively the natural law position or the social contract position. There is a more fundamental, a more essential a priori. This a priori is a precondition to any form of solidarity. It also makes clear that we cannot avoid acting, and transcending ourselves, as well as seeking to establish a common good by being together with others. Human beings are bound to fellow humans, just as the centaur is bound to his horse. There is no means of disengagement.

The moral a priori, which encompasses a confrontation with morality and other areas of culture, and for the individual signifies an acting out and becoming part of it, results in a diversity of educated persons, rich neighborhoods, and the global variety of cultures. If there is a need for a demonstration of the a priori of acting morally, then it lies in this diversity of cultures, in which the fundamental needs of fellow humans are perceived and described within differently constructed theoretical frameworks, but are approached in the same authentic act of existing in solidarity with others. Jesus, in the parable of the good Samaritan, Luke 10, 29ff, displays such an authentic concept of moral a priori in presenting a case study in which the a priori moral requirements are essentially unquestionable, no matter how, why and whether they would be argued for by different ideological or metaphysical positions. He also stresses the a priori priority of practise over theory in morality.

When Chief Seattle expressed his concept of solidarity, he did not seek to establish new theses. Rather, he was expressing precisely that kind of "open" attitude which is necessary for the development of solidarity and individuality. "Maybe we are all brothers?" In asking this, Seattle was aware that being a person and being moral is not solely an intellectual issue. Such a question cannot be answered by means of metaphysical or other theses. The answer lies in active participation or acting, in a true solidarity with the rest of mankind. The moral a priori, and also the logical and the religious a priori, are not epistemological, theoretical, or metaphysical problems. These a priori pose practical questions, and are of practical, immediate importance. They therefore must be dealt with respectively in the phenomenology of the person, of professional ethics, of culture and cross-cultural activities, and in public policy, dialectically, i.e. in dialogue.

University of Bochum, Federal Republic of Germany
and Georgetown U., Washington DC

NOTES

[1] *Environmental Action*, November 11, 1972.
[2] 1913; English translation by Manfred S. Frings and Roger L. Funk, Northwestern University Press, Evanston, 1973.
[3] Scheler, loc. cit., p. 117f.
[4] loc. cit., pp. 109 and 396.
[5] loc. cit., p. 386.
[6] loc. cit., p. 385.
[7] loc. cit., p. 386f.
[8] 1928, English translation by Hans Meyerhoff (New York: The Noonday Press, 1961).
[9] For further elaboration of this argument, cf. H.-M. Sass "Kant's a priori, Scheler's Theories of Value, and Wojtyla's *Acting Person*" in *Kant and Phenomenology*, ed. by Th. M. Seebohm and J. J. Kockelmans (The University Press of America, 1984), 115–127.
[10] Scheler, loc. cit., pp. 545 and 555.
[11] *The Acting Person*, definitive edition by A-T. Tymieniecka of *Osoba i czyn*, Volume 10 of *Analecta Husserliana* (Dordrecht: D. Reidel, 1979).
[12] Wojtyla, loc. cit., pp. 163, 149f.
[13] One of the most neglected aspects of interhuman relationships in western culture is the non-intellectual, non-logical one. Most recently Ludwig Feuerbach and Martin Buber recommended a Copernican Turn in establishing grounds for an anti-subjectivistic I-Thou philosophy which has to be developed further; cf. H.-M. Sass, *Ludwig Feuerbach* (Hamburg: Rowohlt, 1978), especially pp. 80–90.
[14] Family and friendship in Aristotle's 'phenomenology' of the social world are such mid-level principles which are instrumental in achieving the rich benefits of a good life.
[15] I am aware that the term neighbor, used in Wojtyla, loc. cit., pp. 292ff is rooted in the Christian concept of the good Samaritan, Luke 10, 29, while I am using the term neighborhood for describing the interrelatedness of respecting the other's privacy while at the same time being available for help if called or needed, thus developing a very special form of solidarity in pluralistic societies of high mobility.
[16] See especially Wojtyla's concept of "intersubjectivity by participation", loc. cit., pp. 261–299. Participation will also be a prime principle in the new understanding of professional ethics replacing traditional forms of paternalism by informed consent.
[17] loc. cit., p. 287.

INDEX OF NAMES

423

ANALECTA HUSSERLIANA
The Yearbook of Phenomenological Research

Editor:
ANNA-TERESA TYMIENIECKA
The World Institute for Advanced Phenomenological Research and Learning
Belmont, Massachusetts